AF692880

Mineral Nutrition on Human Health and Disease

Mineral Nutrition on Human Health and Disease

Guest Editors

James H. Swain
Salvatore Minisola

MDPI

Basel • Beijing • Wuhan • Barcelona • Belgrade • Novi Sad • Cluj • Manchester

Guest Editors

James H. Swain
Department of Nutrition
Case Western Reserve
University
Cleveland
USA

Salvatore Minisola
Department of Medical and
Cardiovascular Sciences
"Sapienza" Rome University
Rome
Italy

Editorial Office
MDPI AG
Grosspeteranlage 5
4052 Basel, Switzerland

This is a reprint of the Special Issue, published open access by the journal *Nutrients* (ISSN 2072-6643), freely accessible at: https://www.mdpi.com/journal/nutrients/special_issues/6J8BXD3IZ0.

For citation purposes, cite each article independently as indicated on the article page online and as indicated below:

Lastname, A.A.; Lastname, B.B. Article Title. *Journal Name* **Year**, *Volume Number*, Page Range.

ISBN 978-3-7258-7046-2 (Hbk)
ISBN 978-3-7258-7047-9 (PDF)
https://doi.org/10.3390/books978-3-7258-7047-9

© 2026 by the authors. Articles in this reprint are Open Access and distributed under the Creative Commons Attribution (CC BY) license. The reprint as a whole is distributed by MDPI under the terms and conditions of the Creative Commons Attribution-NonCommercial-NoDerivs (CC BY-NC-ND) license (https://creativecommons.org/licenses/by-nc-nd/4.0/).

Contents

About the Editors

James H. Swain

James H. Swain is Professor and Director of the Dietetics Program in the Department of Nutrition, School of Medicine, at Case Western Reserve University, Cleveland, OH, USA. His professional activities in the U.S. include serving as Past-Chair of the Research Dietetics Practice Group (R-DPG) of the Academy of Nutrition and Dietetics (AND), Past-Chair of the Vitamin and Mineral Research Interest Section (VM-RIS) of the American Society for Nutrition (ASN), a past-member the US Food and Drug Administration's Center for Food Safety and Applied Nutrition (CFSAN) Advisory Committee, and Past-Chair of the national Accreditation Council for Education in Nutrition and Dietetics (ACEND) Board. He currently also serves as a global council member of the International Society for Trace Element Research in Humans (ISTERH).

Salvatore Minisola

Salvatore Minisola is an Honorary Professor of Internal Medicine at "Sapienza" Rome University, with a distinguished career marked by significant clinical, academic, and leadership roles. He served as Chief of the Division of Internal Medicine at Policlinico Umberto I from 2011 to 2022 and as Director of Emergency Medicine Residency from 2019 to 2022. His academic contributions include teaching Internal Medicine in the bachelor's degree program in Medicine and serving as a faculty member for the PhD program "Tecnologie Innovative nelle Malattie dello Scheletro, della Cute e del Distretto Oro-Cranio-Facciale". His specialized expertise lies in the clinical investigation of metabolic bone diseases, particularly osteoporosis and primary hyperparathyroidism, and disorders of phosphate metabolism, with a keen focus on tumor-induced osteomalacia. Professionally, he has held prominent positions such as President of the Italian Society of Osteoporosis, Mineral Metabolism and Skeletal Diseases (SIOMMMS) (2009-2011), member of the Professional Practice Committee of the American Society of Bone and Mineral Research (2009-2011), member of the Executive Committee of the Italian Society of Internal Medicine (2014-2018), and member of the Board of Directors of the European Calcified Tissue Society (2021-2026). He also holds key editorial roles, including Associated Editor of the *Journal of Bone and Mineral Research and Hormones*, and serves on the Editorial Board of *Calcified Tissue International*, *Bone Research*, the *Journal of Clinical Endocrinology and Metabolism*, and *Osteoporosis International*. With an extensive publication record, he has authored and co-authored 400 articles in international peer-reviewed journals, along with numerous reviews and book chapters, and boasts an H-index of 78.

nutrients

MDPI

Editorial

Mineral Nutrition and Human Health and Disease

Salvatore Minisola [1,*] and James H. Swain [2]

[1] Department of Medical and Cardiovascular Sciences, "Sapienza" Rome University, 00161 Rome, Italy
[2] Department of Nutrition, School of Medicine, Case Western Reserve University, 10900 Euclid Avenue, SOM WG-48, Cleveland, OH 44106, USA; jhs31@case.edu
* Correspondence: salvatore.minisola@uniroma1.it

Healthy nutritional intake is fundamental for the physiological development of organs and tissues and, later in life, for their optimal function. In this context, mineral nutrition, including both major elements (i.e., calcium, phosphorus, magnesium, potassium and sodium) as well as trace elements (i.e., iron iodine, zinc), is a major component of the preservation of a number of functions. These include those related to the skeleton, endocrine, and the haematopoietic system.

To provide couple of examples, inadequate vitamin D levels have detrimental effects on the skeleton through the insufficient absorption of both calcium and phosphate in the intestine [1]. The decrease in serum calcium levels is compensated by a secondary increase in parathyroid hormone secretion. Long-term consequences on the skeleton are well-known in terms of reduced bone mineral density, increased risk for fractures, and, depending on the severity and duration of vitamin D deficiency, osteomalacia.

Iron is not only important in the production and function of haemoglobin and myoglobin but is also an essential co-factor in many other processes, such as mitochondrial energy production and neurotransmitters synthesis. Individuals with iron deficiency are sometimes asymptomatic but may have a number of signs and symptoms as a consequence of reduced oxygen levels. Among these, fatigue and reduced exercise tolerance are prominent. Recent findings also highlight the negative effect of hypoxia on bone. Indeed, both skeletal and muscle tissue implement specific responses to hypoxemia [2]. It has also been hypothesised that oxygen-sensing can control osteocyte function, with possible implications for the ageing-mediated dysregulation of oxygen bioavailability. This can have a role in the pathogenesis of reduced bone mass with advancing age [3].

In the Special Issue of this journal, targeting "Mineral Nutrition and Human Health and Disease", we provide ten contributions, both experimental and review studies, aiming to better understand current issues and offer balanced solutions. Various topics are discussed; however, in the modern era, a couple of observations need to be emphasised.

There is now a large body of evidence that the quality of our foods is deteriorating. Ultra-processed foods seem to have an increasing role in the pathogenetic mechanisms of chronic non-transmissible diseases [4]. The manuscript by Souza et al. [5] addresses these problems, showing how processed foods modify the Na/K ratios in manufacturing workers. In this context, it is important to refer to a recent paper published by Greatorex Brooks et al. Analysing the NHANES data, they showed an association between osteoporosis and osteopenia and the intake of ultra-processed foods as a proportion of total daily energy [6].

Finally, some papers in this series refer to the importance of the microbiota for adequate mineral nutrition. There is now substantial evidence that changes in gut microbiome with ageing may be instrumental in osteoporosis, a typical disease of ageing [7]. Future therapies should also consider this aspect to guarantee a healthier transition in the last decades of life.

Looking ahead, the future of research in the area of mineral nutrition should provide exciting new insights, especially with the introduction of new technologies and an increased understanding of the interplay between genetics and dietary intake. This Special Issue contributes to this effort, and the increased knowledge it provides promises to assist in the development of more precise recommendations, better prevention strategies, and innovative solutions to global nutrition challenges.

Author Contributions: Conceptualization, S.M. and J.H.S.; S.M. and J.H.S.; writing—original draft preparation, S.M. and J.H.S.; writing—review and editing, S.M. and J.H.S. All authors have read and agreed to the published version of the manuscript.

Funding: This paper received no external funding.

Institutional Review Board Statement: Not applicable.

Informed Consent Statement: Not applicable.

Conflicts of Interest: The authors declare no conflicts of interest.

References

1. Souberbielle, J.C.; Brazier, F.; Piketty, M.L.; Cormier, C.; Minisola, S.; Cavalier, E. How the reference values for serum parathyroid hormone concentration are (or should be) established? *J. Endocrinol. Investig.* **2017**, *40*, 241–256. [CrossRef] [PubMed]
2. Mendoza, S.V.; Genetos, D.C.; Yellowley, C.E. Hypoxia-Inducible Factor-2alpha Signaling in the Skeletal System. *JBMR Plus* **2023**, *7*, e10733. [CrossRef] [PubMed]
3. Janeczko, K.; Agoro, R. Oxygen Sensing in Osteocytes: From Physiology to Age-related Osteoporosis. *Curr. Osteoporos. Rep.* **2025**, *23*, 28. [CrossRef] [PubMed]
4. Monteiro, C.A.; Louzada, M.L.; Steele-Martinez, E.; Cannon, G.; Andrade, G.C.; Baker, P.; Bes-Rastrollo, M.; Bonaccio, M.; Gearhardt, A.N.; Khandpur, N.; et al. Ultra-processed foods and human health: The main thesis and the evidence. *Lancet* **2025**, *406*, 2667–2684. [CrossRef] [PubMed]
5. Souza, A.M.; Bezerra, I.W.L.; Torres, K.G.; Pereira, G.S.; Costa, R.M.; Oliveira, A.G. Association of Dietary Sodium-to-Potassium Ratio with Nutritional Composition, Micronutrient Intake, and Diet Quality in Brazilian Industrial Workers. *Nutrients* **2025**, *17*, 2483. [CrossRef] [PubMed]
6. Brooks, E.L.G.; Tangney, C.C.; Ritz, E.M. Ultra-processed food intake and prevalence of osteoporosis in US adults aged 50 years and older: A cross-sectional analysis. *Osteoporos. Int.* **2025**, *36*, 455–464. [CrossRef] [PubMed]
7. Ghosh, T.S.; Shanahan, F.; O'Toole, P.W. The gut microbiome as a modulator of healthy ageing. *Nat. Rev. Gastroenterol. Hepatol.* **2022**, *19*, 565–584. [CrossRef] [PubMed]

Disclaimer/Publisher's Note: The statements, opinions and data contained in all publications are solely those of the individual author(s) and contributor(s) and not of MDPI and/or the editor(s). MDPI and/or the editor(s) disclaim responsibility for any injury to people or property resulting from any ideas, methods, instructions or products referred to in the content.

https://doi.org/10.3390/nu18030370

nutrients

MDPI

Article

Time to Consider Potassium Intake in Saudi: A Cross-Sectional Assessment Using 24 h Urinary Excretion

Salwa Ali Abdullah Albar * and Merfat Abdulrahman Almaghrabi *

Food and Nutrition Department, King Abdulaziz University, P.O. Box 80200, Jeddah 21589, Saudi Arabia
* Correspondence: salbar1@kau.edu.sa or albarsalwa@hotmail.com (S.A.A.A.); mamaghrabi@kau.edu.sa (M.A.A.); Tel.: +966-50-531-7160 (S.A.A.A.)

Abstract: Background: Evaluating potassium intake can be a powerful tool in epidemiologic studies to reduce the burden of noncommunicable diseases (NCDs). In Saudi Arabia, NCDs are responsible for 35% of deaths in 2023. Monitoring people's potassium intake can be a powerful tool to reduce the burden of NCDs. There is a significant lack of information on potassium intake. The aim is to assess potassium intake using 24 h urinary excretion; to investigate the urinary sodium-to-potassium (Na/K) excretion ratio among Saudi adults; and to explore other lifestyle factors that influence potassium intake. **Methods:** A cross-sectional survey was conducted among young adults (19–29 years old) residing in Jeddah, Saudi Arabia. Data collection included a self-reported questionnaire regarding participants' general attitudes and practices related to potassium consumption (n = 600) of whom 173 participated in 24 h urine collection. Descriptive analyses and regression models were used to evaluate the associations between urinary potassium excretion (mmol/24 h), daily potassium intake (g/day), and the Na/K ratio (dependent variables), and descriptive variables such as age and gender (predictor variables). A p value < 0.05 indicated statistical significance for all tests. **Results:** The mean urinary potassium excretion was 48.6 $\pm$ 23 mmol/24 h, equivalent to a mean daily potassium intake of 1.9 $\pm$ 0.89 g/day, and only 4.1% of participants met the World Health Organization-recommended potassium intake of $\geq$90 mmol/day ($\geq$3.90 g/day). The average potassium intake was significantly lower in females compared with males by 0.52 g (95% CI: -0.78 to -0.25; $p < 0.001$). Physical activity was also a significant factor, associated with both urinary potassium excretion ($p = 0.039$) and intake ($p = 0.006$). Besides the low potassium intake, the mean Na/K ratio was 3.2 $\pm$ 1.4, and the ratio differed significantly by physical activity habits ($p = 0.050$). Only 13% of participants consumed fruit 5–7 days per week (mean portion intake 1.4/day; 95% CI: 1.3–1.5), and 34.7% consumed vegetables 3–4 days per week (mean portion intake 1.5/day; 95% CI: 1.3–1.5). These findings reflect low adherence to recommended fruit and vegetable intake in the study population. **Conclusions:** The findings of this study can be used to create evidence-based nutritional strategies to help people achieve the recommended potassium intake. The study underscores the need for more research on potassium intake across Saudi Arabia.

Keywords: dietary intake; fruit; vegetables; dietary potassium; sodium; potassium ratio; urinary excretion; Saudi Arabia

1. Introduction

Worldwide, noncommunicable diseases (NCDs) are the main contributors to mortality and morbidity [1]. NCDs are the leading cause of death in Saudi Arabia; according to a

2023 report, around 22,000 Saudis die every year from NCDs, which account for 35% of deaths, many of which occur between 30 and 70 years of age [2]. In 2024, the prevalence of leading NCDs in Saudi Arabia is higher than previously reported in national estimates [3]. Heart disease, stroke, kidney disease, and diabetes mellitus are the main contributors to the NCD burden. However, high blood pressure (hypertension) is a major risk factor for cardiovascular diseases, especially coronary heart disease, stroke, and renal failure. The prevalence of hypertension is 34% among Saudi adults, in addition to a high prevalence of obesity and diabetes (40.6% and 39.5%, respectively) [4,5]. As in many parts of the world, NCDs cost Saudi Arabia around 91.6 billion Saudi Riyal (USD24.4 billion) every year to provide healthcare, social care, and welfare support [2]. According to the World Health Organization (WHO), 80% of premature heart disease and diabetes, as well as 40% of cancers, can be prevented by eliminating their risk factors [1,4].

Evaluating potassium intake can be a powerful tool in epidemiologic studies to reduce the burden of NCDs [1,6,7]. Potassium (K+) is an essential nutrient naturally present in many foods and obtained primarily from fruits and vegetables; it helps maintain intracellular fluid volume and electrolyte balance, regulates muscle and nerve function, and supports cardiovascular health [1,6]. Epidemiologic and clinical studies have shown that adequate potassium intake is associated with lower risk factors for several chronic diseases—including reduced blood pressure and lower risk of stroke and type 2 diabetes—and with reduced cardiovascular disease (CVD) mortality [6,8–11]. Thus, potassium intake has been inversely related to NCD risk. Historically, potassium intake was very high—exceeding 200 mmol/day—in preagricultural (hunter gatherer) and early agricultural diets [12]. However, modern diets—often high in processed foods and low in fresh fruits and vegetables—have led to a significant decrease in potassium intake across many populations, including Saudi Arabia [7,12]. This decline is largely attributed to urbanization and lifestyle transitions, which have altered dietary patterns and been accompanied by increasingly sedentary behaviors [13].

A systematic review and Bayesian meta-analysis of 104 studies from 52 countries reported a global mean potassium intake of 2.25 g/day (57 mmol/day), with the highest intakes in Eastern and Western Europe and the lowest in East Asia (mean intake 1.89 g/day); only 31% of the global population had an estimated potassium intake of >2.5 g/day [14].

Urinary potassium excretion reflects broader lifestyle patterns, including dietary intake, physical activity, and demographic factors. Among these, dietary intake is considered the primary determinant, as approximately 85% of ingested potassium is excreted in urine. Higher consumption of potassium-rich foods such as fruits, vegetables, fish, milk, nuts, and legumes, and lower consumption of low-potassium diets dominated by ultra-processed foods, are strongly associated with greater urinary potassium excretion [15]. Physical activity also influences potassium excretion through sweat losses, altered renal handling, and hormonal regulation [16,17]. In addition, other lifestyle factors, including socioeconomic status, smoking, and alcohol consumption, have been linked to variations in potassium intake and excretion [17–19].

A recent report by the WHO and the Food and Agriculture Organization found that a majority of countries fall short of the recommended daily potassium intake of 70–80 mmol/day [20]. Even fewer countries, such as the United Kingdom, Spain, Mexico, and Belgium, report average potassium consumption of 90 mmol/day. Data from the National Health and Nutrition Examination Survey indicate that the average daily potassium intake was 3016 mg for men and 2320 mg for women in the United States [21,22]. Lower potassium intake has been associated with a number of NCDs and risk factors, such as blood pressure, hypertension, and stroke, whereas higher levels of consumption may be protective against these diseases [1,23].

WHO guidelines and recent studies have highlighted the importance of evaluating the sodium-to-potassium (Na/K) intake ratio, as this index is considered more reliable than assessing either sodium or potassium intake alone. A Na/K ratio ≤ 1.0 is reported to be optimal for preserving cardiovascular health [24–26]. This ratio results from consuming at least 3510 mg/day of potassium and no more than 2000 mg/day of sodium [27]. Thus, understanding the connections between nutrients and their relationships to one another within the human body is important for preventing such diseases.

Objective estimates of potassium intake from 24 h urine collections are considered accurate and reliable compared with subjective estimates based on 24 h dietary recalls, which may lead to over- or underestimation of actual intake [28,29]. There is limited evidence using 24 h urine collections to assess potassium intake in Saudi Arabia. To our knowledge, there are only two studies that evaluated potassium intake among the Saudi population: one conducted among 25 young soccer players in Riyadh [30] and another among adults in northern Saudi Arabia [31]. Therefore, this study aims to assess 24 h urinary potassium excretion and other lifestyle factors that influence potassium intake, and to investigate the urinary Na/K excretion ratio among Saudi adults.

2. Materials and Methods

2.1. Study Design and Participants

This cross-sectional study was conducted among residents of Jeddah, Saudi Arabia, between October and December 2024. The sample size was calculated using Cochrane's formula: $n = z^2 \times p \times (1 - p)/e^2$, with z = 95% confidence level (z = 1.96); p = estimated proportion of the population that presents the characteristic $p = 0.5$; and e = margin of error (e = 0.05 or 5%). The required sample size was 384 participants; adding a 25% attrition rate, a total of 480 participants were required for estimating a proportion of the general dietary behaviors questionnaire.

Based on the WHO recommendation, a minimum of 120 individuals per specific age and gender group is required (this number is based on the expected mean and standard deviation of potassium excretion from the previous literature which is necessary for quantitative analysis). To compensate for the anticipated high rate of participant dropout and incomplete urine sample collection, which are common challenges in this type of study [32], a total of 240 participants were invited from the initial sample pool to participate in the 24 h urine collection.

A convenience sample of 600 young adults (19–29 years old) was recruited via email and social applications such as WhatsApp and Telegram to assess participants' eligibility, general characteristics, and practices related to general dietary behavior that related to potassium intake. Internet penetration in Saudi Arabia reached 99% of the population in 2024, reflecting an exponential increase. Simultaneously, social media use is nearly universal, with 94.3% of the population identified as active users [33]. Participants who fit the study inclusion criteria were then asked to take part in the urine execration collection, and we stopped inviting them after reaching 240 participants (this was achieved after inviting 500 participants). Thus, the response rate for the urine execration portion was 48%.

The exclusion criteria were females who were pregnant or lactating and participants who reported chronic diseases or long-term prescription medication use. Moreover, participants undergoing treatment with diuretics were also excluded because loop diuretics alter potassium excretion by inhibiting reabsorption in the thick ascending limb, which can lead to an inaccurate assessment of participants' true potassium status [34]. The study details were explained so that participants understood why the research was being conducted and what it would involve. The participants were informed that participation was voluntary and that they could withdraw at any time without giving a reason.

A total of 240 participants agreed to provide a 24 h urine sample; however, 24 did not return their samples. An additional 38 participants were excluded because of improper collection procedures, such as failing to complete a time sheet, omitting the first-morning specimen, or providing a urine volume of less than 500 mL. Five further samples were also excluded based on low urinary creatinine levels: three women with creatinine excretion less than 4 mmol/day and two men with creatinine excretion less than 6 mmol/day. Thus, urinary excretion data from 173 participants were included.

The study was conducted in accordance with the Declaration of Helsinki, and the Ethical Committee of King Abdulaziz University, Unit of Biomedical Ethics, approved the protocol (Ethics reference: 167–22). Written informed consent was obtained from all participants.

2.2. Data Collection and Study Procedures

An electronic semi-structured questionnaire was developed by adapting items from validated international surveillance instruments, including the WHO STEPS diet module [35,36]. It was translated into Arabic using forward–back translation procedures. Content and cultural validation were performed by a panel of experts. Following this review, a pilot study was conducted with 27 participants to refine the instrument prior to its use with the main sample; based on the pilot, some items were refined, and the pilot data were excluded from the final analysis. The questionnaire demonstrated acceptable internal consistency, as indicated by Cronbach's alpha coefficient of 0.70. The final instrument comprising information about sociodemographic characteristics (age, gender, education, and income) and health characteristics (physical activity, smoking, and eating habits) was collected via a self-reported survey. The participants were asked about the frequency of consuming fruits, vegetables, fast food, ultra-processed foods, and snacks as meals. All items were multiple-choice questions.

The participants who agreed to provide a 24 h urine sample were asked to attend an interview, which was conducted by trained researchers in the university research center. Systolic and diastolic blood pressure (BP) measurements using a random-zero sphygmomanometer were taken four times for each participant (twice at each of two visits), and the mean was used for analysis. Measurements were taken after the participant had rested in a seated position for 15 min, with 3 min of rest between measurements.

2.3. Collection of 24 h Urine

For 24 h urine data collection, this study followed the WHO protocol for estimating potassium via urinary excretion [32]. Trained researchers provided participants with verbal and written instructions and sterile urine-collection bottles and answered any questions. The participants were instructed to start the 24 h urine collection from the second morning specimen and continue until the first specimen of the following day. The female participants were instructed to collect urine on non-menstruating days.

The participants completed a time sheet, recording the date of the urine sample and the start and end times, and reported any urine volume missed during the collection period. A 24 h urine sample was considered complete if the total urine volume was >500 mL, the collection period was >20 h, and (for female participants) no menstruation occurred during the collection. Analysis of the 24 h urine samples was conducted at certified laboratories.

The urinary potassium samples were transported in portable thermoelectric coolers and analyzed in certified laboratories using the ion-selective electrode (ISE) method with an Abbott ARCHITECT c8000 Chemistry Analyzer (Abbott Diagnostics, Abbott Park, IL, USA), following the manufacturer's protocols. This method is a standard and validated laboratory technique for electrolytes, specifically used to quantify urinary potassium (K) [37].

2.4. Statistical Analysis

Descriptive statistics were used to summarize participants' general characteristics. For continuous variables (e.g., age and weight), means and standard deviations (SDs) were used, and the Shapiro–Wilk test was used to assess normality. For categorical variables (e.g., education level and smoking status), the number (n) and percentage (%) in each category were reported. To test the proportion of differences between males and females in categorical variables (education, smoking, and body mass index [BMI] classification), Pearson's chi-square test was used. For continuous variables, independent two-sample t-tests were used to compare genders.

BMI was calculated as weight (kg) divided by height (m^2). The participants were then classified according to WHO adult guidelines as underweight (BMI < 18.5 kg/m^2), normal weight (18.5 to <25 kg/m^2), overweight (25 to <30 kg/m^2), and obese (BMI $\geq$ 30 kg/m^2). Guidelines from the American Heart Association were used to categorize blood pressure (BP): normal BP defined as systolic BP <120 mm Hg and diastolic BP <80 mm Hg; elevated as systolic 120–129 mm Hg and diastolic <80 mm Hg; and high BP when systolic $\geq$130 mm Hg or diastolic >80 mm Hg [38].

To convert urinary output to dietary potassium intake, urinary K excretion (mmol/day) was multiplied by 39/1000 g (1 mmol K = 39 mg), based on WHO conversion [38]. Associations between urinary K excretion (mmol/24 h), daily potassium intake (g/day), and the Na/K ratio (dependent variables) and descriptive predictors such as age and gender were investigated using regression (model 1).

In addition, associations of these dependent variables with BMI, education, income, physical activity, BP, and fruit and vegetable consumption were investigated using regression (model 2), adjusted for age and gender. One-sample t-tests were performed to test whether the mean potassium intake for the total sample, and after stratification by gender, differed from the WHO recommendation of $\geq$3.51 g/day. Statistical analyses were carried out using Stata Statistical Software, Release 12 (StataCorp, College Station, TX, USA) [39]. A p value < 0.05 indicated statistical significance for all tests.

3. Results

3.1. Sample Characteristics

Table 1 presents the study participants' general characteristics overall and stratified by gender.

A total of 600 adults participated in the study; 60.1% were female. The mean age for the total sample was 23 $\pm$ 3 years. Significant differences were found between males and females in mean age (Coef. = $-$0.74; 95% CI: $-$1.2 to $-$0.2; and p = 0.004). The majority of participants were of normal weight (62%), and there were significant differences in the proportion of males and females across BMI categories (p < 0.001). Most of the sample was single (85%), and 61.5% held bachelor's degrees. A total of 30.7% reported a family income ranging from 10,999 to 20,000 SAR, followed by 26% with an income range from 5000 to 10,000 SAR. Furthermore, significant differences were observed between males and females in social status, education level, and family economic status.

Most of the sample (84.7%) had no diagnosed health issues, and 25% were smokers. There were significant differences between genders in smoking status and health issues. Significant differences between males and females were also found in physical activity (p < 0.001). However, no significant differences were found between genders in BP readings or in following a special diet. Regarding the general characteristics of the participants who provided 24 h urine samples, the study included 173 individuals, 58.4% of whom were males. Their average age was 23.8 $\pm$ 3.3 years, with a significant difference observed between males and females in their mean age (mean difference 1.2 years; p = 0.018).

Table 1. General characteristics of study participants by total sample (*n*= 600), subsample (*n* = 173), and gender.

General Characteristics †	All Participants				Subsample of 24 h Urine Execration			
	Total (*n* = 600)	Females (*n* = 364)	Males (*n* = 236)	*p* Value †	Total (*n* = 173)	Females (*n* = 72)	Males (*n* = 101)	*p* Value †
	n (%)	*n* (%)	*n* (%)		*n* (%)	*n* (%)	*n* (%)	
Age (years) (mean ± SD) †	23.2 (3.1)	22.9 (2.7)	23.6 (3.5)	0.004	23.8 (3.3)	23.1 (2.6)	24.3 (3.7)	0.018
Height (cm) (mean ± SD) †	163.3 (8.9)	163.3 (8.9)	168.7 (9.2)	0.000	166.3 (10.4)	165.1 (9.8)	167.1(10.9)	0.193
Weight (kg) (mean ± SD) †	65.7 (15.4)	62.3 (14.7)	70.8 (15.1)	0.000	69.3 (15.1)	68.1 (20.7)	70.1 (16.1)	0.484
BMI (kg/m^2) (mean ± SD) †	24.5 (4.7)	24.3 (4.9)	24.7 (4.3)	0.265	24.8 (5.3)	24.8 (6.2)	24.9 (4.6)	0.839
BMI (kg/m^2) ‡								
Normal weight	372 (62)	251 (68.9)	121 (51.3)	0.000	95 (54.9)	42 (58.3)	53 (52.5)	0.455
Overweight	159 (26.5)	66 (18.1)	93 (39.4)		57 (32.9)	20 (27.8)	37 (5.9)	
Obese	69 (11.5)	47 (12.9)	22 (9.3)		21 (12.1)	10 (13.9)	11 (10.9)	
Social statues ‡								
Single	510 (85)	295 (81.1)	215 (91.1)	0.002	153 (88.4)	61 (84.7)	53 (91.1)	0.416
Mired	76 (12.7)	60 (16.5)	16 (6.8)		14 (8.1)	8 (11.1)	37 (5.9)	
Divorce	14 (2.3)	9 (2.5)	5 (2.1)		6 (3.5)	3 (4.2)	11 (2.9)	
Education ‡								
Diploma	30 (5)	9 (2.5)	21 (8.9)	0.002	10 (5.8)	3 (4.2)	7 (6.9)	0.613
High school	181 (30.2)	108 (29.7)	73 (30.9)		57 (32.9)	26 (36.1)	31 (30.7)	
Bachelor degree	369 (61.5)	237 (65.1)	132 (55.9)		106 (61.3)	43 (59.7)	63 (62.4)	
Postgraduate degree	20 (3.3)	10 (2.8)	10 (4.2)		-	-	-	
Economic statues (for family) ‡								
Less than 5000 SAR	117 (19.5)	62 (17.0)	55 (23.3)	0.000	40 (23.1)	15 (20.8)	25 (24.8)	0.606
5000–10,000 SAR	157 (26.2)	106 (29.1)	51 (21.6)		47 (27.2)	23 (31.9)	24 (23.76)	
10,999–20,000 SAR	184 (30.7)	128 (35.2)	56 (23.7)		42 (24.3)	18 (25.0)	24 (23.76)	
More than 20,999 SAR	142 (23.7)	68 (18.7)	74 (31.4)		44 (25.4)	16 (22.2)	28 (27.7)	
Smoking statues ‡								
Smoker	150 (25)	49 (13.5)	101 (42.8)	0.000	57 (32.9)	12 (16.7)	45 (44.6)	0.000
Former smoker	18 (3)	9 (2.5)	9 (3.9)		9 (5.2)	2 (2.8)	7 (6.9)	
Special diet (yes)	100 (16.7)	59 (15.9)	42 (17.8)	0.550	21 (12.1)	6 (8.3)	15 (14.9)	0.196
Health Statues ‡								
No health issue	508 (84.7)	297 (81.6)	211 (89.4)	0.018	143 (82.7)	54 (75)	89 (88.1)	0.025
Other diseases	92 (15.3)	67 (18.4)	25 (10.6)		30 (17.3)	25.0 (12)	12 (11.9)	
Blood pressure ‡ (*n* = 122)								
Normal	55 (45.1)	28 (54.9)	27 (38.1)	0.159	55 (45.1)	28 (54.9)	27 (38.1)	0.159
Pre-hypertensions	56 (45.9)	20 (39.2)	36 (50.7)		56 (45.9)	20 (39.2)	36 (50.7	
Stage I	11 (9.1)	3 (5.9)	8 (11.3)		11 (9.1)	3 (5.9)	8 (11.3)	
Physical activity ‡								
Never	217 (36.2)	159 (43.4)	59 (25)	0.000	57 (32.9)	34 (47.2)	23 (22.8)	0.003
Once a week	21 (3.5)	10 (2.8)	11 (4.7)		73 (42.2)	28 (38.9)	45 (44.6)	
Twice a week	243 (40.5)	138 (37.9)	105 (44.5)		34 (19.7)	8 (11.1)	26 (25.7)	
>four times week	119 (19.8)	58 (15.9)	61 (25.8)		9 (5.2)	2 (2.8)	7 (6.9)	

† Differences between gender were assessed by using independent sample t-test. ‡ Significant differences between the proportion of males and females in other variables were assessed using Chi2 test. SAR—Saudi Arabian Riyal.

There were no significant differences found between males and females in education level, economic status, or blood pressure. However, a significant difference was noted in the proportion of males to females regarding physical activity levels ($p = 0.003$).

3.2. General Dietary Behavior

Table 2 illustrates participants' general dietary behavior.

Table 2. General dietary behavior across the total sample (n = 600), subsamples (n = 173), and gender.

General Dietary Behavior	All Participants				Subsample of 24 h Urine Execration			
	Total (n = 600)	Females (n = 364)	Males (n = 236)	p Value (Mean ± 95%CI)	Total (173)	Females (n = 72)	Males (n = 1 01)	p Value (Mean ± 95%CI)
	n (%)	n (%)	n (%)		n (%)	n (%)	n (%)	
Frequency of eating fruits per week †								
Once a week	62 (10.3)	55 (15.1)	7 (2.9)		20 (11.6)	13 (18.1)	7 (6.9)	
Twice a week	292 (48.7)	171 (46.9)	121 (51.3)		88 (50.9)	35 (48.6)	53 (52.5)	
3–4 days a week	170 (28.3)	96 (26.4)	74 (31.4)		42 (24.3)	10 (13.9)	32 (31.7)	
5–7 days a week	76 (12.7)	42 (11.5)	34 (14.4)	0.000	23 (13.3)	14 (19.4)	9 (8.9)	0.003
Portion intake of fruit/day *	1.4 (1.3, 1.5)	1.3 (1.2, 1.5)	1.5 (1.3, 1.7)	−0.10 (−0.35, 0.14) p = 0.401	1.3 (1.2)	1.1 (0.95)	1.4 (1.2)	0.32 (−0.03, 0.67) p = 0.073
Frequency of eating vegetable †								
Once a week	50 (8.3)	28 (7.7)	22 (9.3)		22 (12.7)	8 (11.1)	14 (13.9)	
Twice a week	174 (29.0)	101 (27.8)	73 (30.9)		56 (32.4)	24 (33.3)	32 (31.7)	
3–4 days a week	208 (34.7)	119 (32.7)	89 (37.7)		59 (34.1)	17 (23.6)	42 (41.6)	
5–7 days a week	168 (28)	116 (31.9)	52 (22.0)	0.005	36 (20.8)	23 (31.9)	13 (12.9)	0.009
Portion intake of vegetable/day *	1.5 (1.4, 1.6)	1.4 (1.3, 1.5)	1.6 (1.4, 1.7)	−0.15 (−0.42, 0.13) p = 0.301	1.4 (1.3)	1.4 (1.3)	1.4 (1.3)	−0.06 (−0.47, 0.35) p = 0.773
Frequency of fast food †								
Once a week	158 (26.3)	117 (32.1)	41 (17.4)		33 (19.1)	20 (27.8)	13 (12.9)	
Twice a week	209 (34.8)	141 (38.7)	68 (28.8)		58 (33.5)	28 (38.8)	30 (29.7)	
3–4 days a week	130 (21.7)	23 (6.3)	26 (11.1)		82 (47.4)	24 (33.3)	58 (47.4)	
5–7 days a week	97 (16.2)	65 (17.9)	32 (13.6)	0.008	-	–	-	0.001
Eating processed food †								
Always	86 (14.3)	64 (17.6)	22 (9.3)		23 (13.3)	15 (20.8)	8 (7.9)	
Often	217 (36.2)	121 (33.2)	96 (40.7)		68 (39.3)	27 (37.5)	41 (40.6)	
Sometimes	200 (33.3)	114 (31.3)	86 (36.4)		59 (34.1)	18 (25)	41 (40.6)	
Rarely	86 (14.3)	59 (16.2)	27 (11.4)		18 (10.4)	10 (13.9)	8 (7.9)	
Never	11 (1.8)	6 (1.6)	5 (2.1)	0.014	5 (2.9)	2 (2.3)	3 (2.9)	0.044
Number of snacks/day †								
1–2 times/day	404 (67.3)	257 (40.6)	147 (62.3)		103 (59.5)	44 (61.1)	59 (58.4)	
2–3 times/day	118 (19.7)	71 (19.5)	47 (19.9)		42 (24.3)	17 (23.6)	25 (24.8)	
3–4 times/ day	63 (10.5)	28 (7.7)	35 (14.8)		28 (16.2)	11 (15.3)	17 (16.8)	
4–5 times/day	15 (2.5)	8 (2.2)	7 (2.9)	0.033	-	-	-	0.093
Eating snacks as a meal †								
Yes	138 (23)	87 (23.9)	51 (21.6)		51 (29.5)	23 (31.9)	28 (27.7)	
No	232 (38.7)	127 (34.9)	105 (44.5)		71 (41.0)	25 (34.7)	46 (45.5)	
Sometimes	230 (38.3)	150 (41.2)	82 (33.9)	0.050	51 (29.5)	24 (33.3)	27 (26.7)	0.354

* Differences between genders were assessed by using independent sample t-test; $p = 0.001$. † Significant differences between the proportion of males and females in other variables were assessed using Chi2 test; $p < 0.001$.

Only 13% of participants consumed fruits 5–7 days per week, and nearly half of the sample (48.9%) consumed fruits twice per week, with an average portion size of 1.4 (95% CI: 1.3, 1.5) portions. While 34.7% of participants consumed vegetables 3–4 days per week, 29% consumed vegetables twice per week. The average portion of vegetable consumption was 1.5 (95% CI: 1.3, 1.5) portions. There was a significant difference between males and females in the frequency of fruit and vegetable consumption ($p < 0.001$ and $p = 0.005$, respectively). The frequency of fast-food consumption among participants showed that 34.8% consumed fast food "rarely," followed by 26% who consumed it "once a week." Processed food consumption data indicated that 36% of participants consumed it "often," followed by 33% who consumed it "sometimes." Significant differences were found between genders in the frequency of fast-food and processed food consumption ($p = 0.008$ and $p = 0.014$, respectively). The general dietary behavior observed in the subsample (n = 173) followed a similar trend to that of the main sample.

3.3. Twenty-Four-Hour Urinary Volume, Potassium, and Equivalent Dietary Potassium (Supsample)

The average urinary volume excretion was 1642.8 mL/day, and there were significant differences between males and females in urinary volume excretion (mL/24 h), with females having significantly lower urine volume than males (coef. −13.3; 95% CI: −20.1, −6.5; and $p = 0.014$). The mean potassium excretion was 48.6 ± 23 mmol/24 h, which is equivalent to a mean daily potassium intake of 1.9 ± 0.89 g (median: 2.02 g; IQR: 1.2, 2.2). Potassium intake in females was significantly lower than in males by −0.52 g (95% CI: −0.78, −0.25; $p < 0.001$). Only 4.1% of the study participants met the recommended potassium intake levels of >90 mmol/day (>3.90 g/day), as advised by the WHO.

Furthermore, physical activity habits were significantly associated with urinary potassium excretion (mmol/24 h) and daily potassium intake (g) ($p = 0.039$ and $p = 0.006$, respectively), after adjustment for important confounders (age and gender). Participants who consumed more than five portions of fruits and vegetables had significantly higher urinary potassium excretion by 12.9 mmol/24 h (95% CI: 6.6, 19.2; $p < 0.001$) and significantly higher daily potassium intake by 0.50 g (95% CI: 0.26, 0.75; $p < 0.001$).

The average sodium-to-potassium (Na/K) ratio was high at 3.2 ± 1.4, (median: 2.95; IQR: 2.2, 3.9), and the ratio was significantly different according to physical activity habits ($p = 0.050$). However, no significant associations were observed between age, BMI, BMI category, education level, income status, or any other dietary behavior and average urinary potassium excretion (mmol/24 h), daily potassium intake (g), or the Na/K ratio (Table 3).

Table 3. Urine potassium excretion, daily potassium intake, sodium/potassium ratio, and their associations with descriptive variables (n = 173).

	n	Urine K Excretion (mmol/24 h)			Daily Potassium Intake (g)			Ratio Between Na/K		*p*-Value
		Mean (±SD)	Coef. (95% CI)	*p*-Value	Mean (±SD)	Coef. (95% CI)	*p*-Value	Mean (±SD)	Coef. (95% CI)	
Age †	173	48.6 (23.1)	0.86 (−0.14, 1.89)	0.093	1.9 (0.89)	0.034 (−0.01, 0.1)	0.093	3.16 (1.41)	−0.10 (−0.15, 0.02)	0.149
Gender †**: Male**	101	54.5 (21.3)	−13.3 (−20.1, −6.5)	0.000	2.1 (0.83)	−0.52 (−0.78, −0.25)	0.000	3.18 (1.29)	0.02 (−0.55, 0.58)	0.950
Female	72	40.3 (23.1)			1.6 (0.89)			3.14 (1.56)		
BMI	173	48.6 (23.1)	0.18 (−0.45, 0.81)	0.577	1.9 (0.89)	0.01 (−0.02, 0.03)	0.577	3.16 (1.41)	−0.02 (−0.06, 0.19)	0.314
BMI Category ‡										
Normal weight	95	4606 (21.7)	1		1.8 (0.84)			3.3 (1.45)		
Overweight	57	51.9 (23.9)	3.4 (−3.9, 10.7)	0.659	2.1 (0.93)	0.13 (−0.15, 0.41)	0.659	3.2 (1.38)	−0.11 (−0.57, 0.35)	0.138
Obese	21	48.6 (26.7)	1.8 (−8.7, 12.3)		1.8 (1.03)	0.07 (−0.34, 0.48)		2.6 (1.57)	−0.67 (−1.34, −0.01)	
Education ‡										
Diploma	10	45.6 (16.9)			1.8 (0.66)			2.4 (0.83)		
High school	57	42.5 (22.8)	0.11 (−14.9, 15.2)	0.114	1.7 (0.89)	0.004 (−0.58, 0.59)	0.114	3.29 (1.48)	0.75 (−0.22,1.71)	0.185
Bachelor degree	106	52.2 (23.1)	7.8 (−6.4, 22.1)		2.03 (0.9)	0.31 (−0.25, 0.86)		3.17 (1.39)	0.85 (−0.10, 1.76)	
Economic statues ‡										
Less than 5000 SAR	40	46.6 (20.6)			1.8 (0.80)			3.13 (1.49)		
5000–10,000 SAR	47	49.2 (30.5)	4.6 (−4.8, 13.9)		1.9 (1.19)	0.18 (−0.19, 0.54)		3.32 (1.62)	0.169 (−0.42, 0.76)	
10,000–20,000 SAR	42	50.9 (20.9)	4.5 (−5.2, 14.1)	0.758	1.9 (0.81)	0.17 (−0.20, 0.55)	0.758	3.31 (1.18)	0.27 (−0.34, 0.87)	0.101
>20,000 SAR	44	47.5 (17.9)	2.5 (−7.3, 12.3)		1.9 (0.69)	0.10 (−0.28, 0.48)		2.88 (1.29)	−0.46 (−1.07, 0.16)	
Physical activity habits ‡										
Do not workout	57	42.3 (20.2)			1.41 (0.78)			3.26 (1.42)		
Workout once a week	9	58.9 (44.1)	15.2 (−0.6, 30.9)		1.68 (0.62)	0.15 (−0.27, 0.58)		3.39 (1.99)	−0.10 (−0.56, −0.44)	
Workout 2–4 times a week	73	47.2 (17.9)	3.7 (−4.1, 11.5)	0.039	2.28 (0.99)	0.66 (0.16, 1.15)	0.006	3.31 (1.32)	−0.74 (−1.35, −0.12)	0.050
Workout daily	34	58.8 (27.3)	12.7 (2.8, 22.6)		2.69(1.00)	1.11 (0.33,1.88)		2.61 (1.32)	−0.12 (−1.12, −0.89)	
Blood pressure ‡										
Normal	55	47.1 (30.5)			1.8 (1.19)			3.2 (1.89)		
Pre-hypertensions	56	48.5 (18.3)	−2.2 (−11.2, 6.7)	0.832	1.9 (0.71)	−2.2 (−11.2, 6.7)	0.832	3.16 (1.15)	0.06 (−0.52,0.65)	
Stage 1	11	47.6 (24.3)	−3.7 (−19.1, 11.7)		1.9 (0.95)	−3.7 (−19.1, 11.7)		2.9 (1.36)	−0.17 (−1.18, 0.85)	0.895
Total consumption of fruit and vegetables ‡										
Consumed ≤ 5 portion /day	90	42.01 (16.1)	12.9 (6.6, 19.2)	0.000	1.6 (0.63)	0.50 (0.26, 0.75)	0.000	3.4 (1.7)	−0.49 (−1.04, 0.05)	0.076
Consumed > 5 portion/ day	83	55.7 (27.1)			2.2 (1.1)			2.9 (1.3)		
Frequency of fast food ‡										
Once a week	33	45.54 (19.5)			1.8 (0.76)			3.06 (1.17)		
Twice a week	58	45.99 (25.14)	−1.51 (−11.11, 8.05)		1.79 (0.98)	−0.06 (−0.43, 0.31)		3.18 (1.76)	0.13 (−0.47, 0.74)	
3- 7 days a week	82	51.67 (22.73)	1.24 (−8.04, 10.52)	0.775	0.02 (0.87)	0.05 (−0.31, 0.41)	0.775	3.19 (1.22)	0.16 (−0.45, 0.76)	0.856

Table 3. *Cont.*

	n	Urine K Excretion (mmol/24 h)			Daily Potassium Intake (g)			Ratio Between Na/K		p-Value
		Mean (±SD)	Coef. (95% CI)	p-Value	Mean (±SD)	Coef. (95% CI)	p-Value	Mean (±SD)	Coef. (95% CI)	
Eating processed food ‡										
Always/Often	91	46.2(22.9)			1.8 (0.9)			3.3 (1.6)		
Sometimes	59	52.6 (22.4)	3.3 (−4.2, 10.7)		2.1 (0.9)	0.123 (−0.16, 0.42)		2.9 (1.1)	−028 (−0.76, 0.19)	
Rarely/Never	23	47.8 (25.2)	1.3 (−9.1, 11.6)	0.684	1.9 (0.9)	0.05 (−0.35, 0.45)	0.685	3.2 (1.4)	0.03 (−0.63, 0.68)	0.454
Number of snacks per day ‡										
1–2 times/day	103	49.1 (21.67)			1.9 (0.85)			3.02 (1.17)		
2–3 times/day	42	46.6 (23.96)	−2.53 (−10.5, 5.51)		1.8 (0.93)	−0.09 (−0.41, 0.21)		3.63 (1.84)	0.59 (0.09, 1.1)	
3–4 times/day	28	49.6 (27.09)	−1.04 (−10.4, 8.31)	0.821	1.9 (1.05)	−0.04 (−0.41, 0.32)	0.820	2.97 (1.36)	0.27 (−0.56, 0.61)	0.064
Eating snacks as a meal ‡										
Yes	51	46.8 (31.26)			1.8 (1.22)			3.3(1.77)		
No	71	50.5 (19.2)	2.6 (−5.44, 10.66)		1.9 (0.74)	0.10 (−0.21, 0.42)		3.3 (1.44)	0.04 (−0.47, 0.55)	
Sometimes	51	47.7 (18.09)	1.3 (−7.32, 9.92)	0.814	1.9 (0.71)	0.05 (−0.29, 0.39)	0.814	2.8 (1.03)	−0.43 (−0.97, 0.11)	0.145

† Age- and gender-adjusted regression (model 1). ‡ Age- and gender-adjusted regression (model 2).

4. Discussion

Assessing current dietary intake is fundamental to crafting evidence-based nutritional strategies that enable individuals to meet recommended nutrient intakes. To our knowledge, this is the first study conducted in the Western Province and one of only two across the Kingdom to assess potassium status using the gold-standard 24 h urinary excretion method. The data reveals a critical nutritional shortfall in the population. Only 4.1% of participants met the WHO-recommended potassium intake level (≥90 mmol/day) ≥3.51 g/day, with an average potassium intake of 1.9 ± 0.89 g (median intake of 2.02 g (IQR: 1.2, 2.2)), a prevalence that is substantially low. The Na/K ratio was high at 3.2 ± 1.4, which strongly indicates a high risk for hypertension and cardiovascular disease within this young population. Furthermore, findings reveal a significant and positive association between physical activity and fruit and vegetable consumption with both urinary potassium excretion and estimated daily potassium intake. This precisely quantifies the linkage between these healthy lifestyle components and overall potassium status.

In Saudi Arabia, only two studies to date have used 24 h urinary potassium excretion as a tool to assess potassium intake, which were conducted in the Riyadh and Al-Jouf regions. Findings from the Riyadh region reported an optimal urinary potassium concentration of 64.3 ± 23.7 mmol/L [30], likely due to the specific nature of the study population, as participants were professional soccer players who were more likely to adhere to optimal dietary intake. Additionally, the study included only 25 players, a small sample size that limits generalizability. On the other hand, in the Al-Jouf region (n = 392; aged 25–64 years), the average potassium intake was 2.3 g/day 4 [31].

Similar to our findings, a similar trend was observed in the Gulf region. A nationally representative survey of 569 Omani participants found that the average population potassium intake was 2.36 ± 1.46 g/day, with fewer than 10% meeting the adequate intake level. Significant gender differences were also observed, with females having a lower potassium intake (2.27 ± 1.46 g/day) than males (2.53 ± 1.45 g/day) [15]. A similar finding was reported in the UAE (n = 190; aged 20–60 years), where the mean urinary potassium intake was 2.5 ± 0.615 g/day, with significant differences by gender (p = 0.007) [40]. This difference may be attributable to the greater energy intake among males [15] as well as the higher physical activity observed in this study.

Most of the world's population consumes less than the recommended level of potassium [1]. A systematic review and meta-analysis that included 104 studies from 52 countries examined potassium intake assessed by 24 h urinary excretion, dietary methods, or spot urine samples. The mean global potassium intake was found to be 2.25 g/day (57 mmol/day) [14]. These global data indicate significant regional variation, with the highest mean intakes observed in Eastern and Western Europe at 3.53 g/day (95% CI: 3.05–4.01 g/day) and 3.29 g/day (95% CI: 3.13–3.47 g/day), respectively, while East Asia reported the lowest mean intake at 1.89 g/day (95% CI: 1.55–2.25 g/day). Globally, only 14% (95% CI: 11–17%) of the population reached the higher threshold of 3.5 g/day [14].

The general dietary behavior of this study's participants shows that only 13% consumed fruits 5–7 days per week, with an average portion intake of 1.4 portions/day (95% CI: 1.3, 1.5), and 34.7% consumed vegetables 3–4 days per week, with an average portion intake of 1.5 portions/day (95% CI: 1.3, 1.5). This is significantly lower than the WHO recommendation of at least 400 g (five portions) of fruits and vegetables per day [23] and the USDA recommendation of approximately 4.5 cups/day (2 cups of fruit + 2.5 cups of vegetables) for an average 2000 kcal/day diet. At the same time, a considerable proportion of participants—35%—consumed fast food twice per week, and 36% consumed processed food more often. Changes in lifestyle and dietary patterns have led to a notable rise in the intake of various sodium-rich foods [41,42]. This is consistent with current dietary trends in Western populations, where there is a steady decrease in fruit and vegetable intake and an increase in the consumption of processed (sodium-rich) foods [22]. This results in reduced potassium intake and increased sodium intake; the higher the sodium intake, the greater the urinary potassium loss. Thus, the ratio of sodium to potassium intake may be more important than either nutrient alone [7].

Although potassium intake has received less attention than sodium intake as a public health strategy to reduce the risk of CVDs, results from the SSaSS trial showed that increasing potassium intake may be of greater importance for public health strategies. Increasing potassium by 57% (803 mg) and reducing sodium by 8.1% (350 mg), without modifying participants' dietary quality, resulted in a 3.34 mmHg reduction in systolic blood pressure [43,44]. A cluster-randomized controlled trial of potassium-enriched salt conducted in Taiwan reported a significant reduction in cardiovascular mortality in the experimental group. Therefore, the Na/K ratio has been reported to be a superior metric compared to either sodium or potassium alone in relation to blood pressure prevention [20,25,45–47].

In the current study, the Na/K intake ratio was as high as 3.16 ± 1.41, which is consistent with findings from epidemiological and clinical studies across various regions and populations worldwide. In an Omani study, the average urinary Na/K ratio was 3.3, with no difference by gender [15]. China reported a high average Na/K ratio of 3.93 (95% CI: 3.31–4.64), while the Netherlands and the United Kingdom reported lower levels at 1.60 (95% CI: 1.27–1.90) and 1.60 (95% CI: 1.33–1.87), respectively. Globally, the average Na/K ratio was 2.88 (95% CI: 2.56–3.23), with significant regional differences: the highest ratios were reported in East Asia (3.85; 95% CI: 3.04–4.81) [48], Eastern Sub-Saharan Africa (3.42; 95% CI: 2.12–5.52), and South Asia (3.39; 95% CI: 2.54–4.45) [14]. In contrast, the lowest Na/K ratios were found in Western Europe (1.68; 95% CI: 1.46–1.92) and Eastern Europe (1.82; 95% CI: 1.23–2.48) [14]. There are no accepted guidelines for the optimal Na/K ratio; however, the WHO suggests that meeting individual intake guidelines for both sodium and potassium would result in a ratio of approximately 1.0 [1]. Another study has suggested that a ratio less than two may be an appropriate goal for reducing blood pressure and CVD risk [49,50].

The observed high Na/K ratio in our study population is highly concerning from a public health perspective. Evidence demonstrates that a high Na/K ratio is a better predictor of elevated blood pressure and subsequent CVD mortality than sodium intake alone. This indicates that the population is facing a profound dual dietary imbalance (excessive sodium intake coupled with insufficient potassium intake). This combined imbalance places the individuals in this study at a substantially increased, quantifiable risk for developing hypertension and other CVD complications, necessitating urgent targeted nutritional interventions to reduce CVD burden in Saudi Arabia.

Similar to the findings of this study, engagement in physical activity significantly influences urinary potassium excretion by altering electrolyte balance, leading to reduced urinary sodium and increased urinary potassium through hormonal regulation [30]. One study shows that there are high potassium losses during heavy physical activities in hot climates [51]. Although socioeconomic status did not influence potassium excretion in this study, many studies found that lower urinary potassium excretion was reported among those with the greatest socioeconomic disadvantage compared with higher socioeconomic groups [19]. Another study found that urinary potassium excretion was lower among individuals with lower income and education levels [52]. Evidence from Italy also indicated that individuals in low-skilled employment had lower potassium excretion than those in senior managerial roles [53]. In addition, lower household expenditure has been associated with reduced potassium intake and higher Na/K ratios. Taken together, these studies suggest that socioeconomic status may play an important role in shaping dietary patterns, potassium intake, and overall diet quality [18,54].

Our study has several strengths. It is the first study conducted in the Western Province of Saudi Arabia and the second conducted among the general population that included both genders. It used the gold standard for measuring urinary excretion, applying the WHO's current preferred technique for estimating potassium consumption [32]. Furthermore, this study collected information on participants' general lifestyle and dietary behaviors related to potassium intake. On the other hand, some limitations exist in this study. First, the cross-sectional nature of the study prevented the determination of direct association. The study was conducted on a relatively small sample size and covered a narrow age range (19–29 years) in Jeddah, which restricts the generalizability of the results to the nationwide scale. In addition, the use of a convenience sampling technique meant that urine samples were collected from volunteers, which may have introduced self-selection bias. Although we administered a self-reported questionnaire on participants' general attitudes and practices to explore lifestyle factors influencing potassium intake, the dietary behavior data were reported only in general terms and therefore do not provide a reliable estimate of actual potassium intake. Finally, relying on a single 24 h urine collection may not adequately reflect typical potassium excretion, since multiple samples are generally needed to account for daily variations in intake. However, applying such an approach would have significantly increased the participants' burden and workload.

5. Conclusions

This study assessed 24 h urinary potassium excretion among young Saudi adults aged 19–29 years, addressing a critical gap in regional nutrition research. The findings establish baseline values for this age group and reveal insufficient potassium intake and low adherence to international recommendations, accompanied by elevated sodium-to-potassium ratios. This imbalance may have significant and immediate implications for the prevalence of CVDs and other non-communicable diseases within this young population.

The study provides direct evidence that can aid the health professional in crafting targeted, evidence-based nutritional strategies to mitigate this cardiovascular risk. It is imperative that future public health programs focus on increasing fruit and vegetable consumption to boost population potassium intake and prioritizing the high Na/K ratio—rather than treating sodium or potassium in isolation—when developing interventions to combat CVDs in Saudi Arabia. Further comprehensive studies utilizing 24 h urinary excretion are warranted to clarify the situation across other regions of the Kingdom.

Author Contributions: Conceptualization, M.A.A. and S.A.A.A.; methodology, S.A.A.A. and M.A.A.; formal analysis, investigation, and data curation, S.A.A.A.; writing—review and editing, M.A.A. and S.A.A.A. All authors have read and agreed to the published version of the manuscript.

Funding: The project was funded by KAU Endowment (WAQF) at King Abdulaziz University, Jeddah, Saudi Arabia. The authors, therefore, acknowledge with thanks WAQF and the Deanship of Scientific Research (DSR) for technical and financial support.

Institutional Review Board Statement: The study was conducted in accordance with the Declaration of Helsinki and approved by King Abdulaziz University, Unit of Biomedical Ethics (Ethics reference: 167–22, date of approval 7 April 2022).

Informed Consent Statement: Informed consent was obtained from all the subjects involved in the study.

Data Availability Statement: The data presented in this study are available at the research center, King Abdulaziz University, or upon request from the corresponding author.

Conflicts of Interest: The authors declare no conflicts of interest.

References

1. WHO. *Potassium Intake for Adults and Children*; World Health Organization: Geneva, Switzerland, 2012.
2. Grafton, D.; Elmushara, K.; Jung, J. *Prevention and Control of NonCommunicable Diseases in Saudi Arabia: The Case for Investment*; UNDP, WHO, UNIATF, GHC: Geneva, Switzerland, 2023.
3. Alghnam, S.; Bosaeed, M.; Aljouie, A.; Alshahrani, S.M.; Alshenqeety, O.; Atun, R.; Alqahtani, S. Estimating the prevalence of select non-communicable diseases in Saudi Arabia using a population-based sample: Econometric analysis with natural language processing. *Ann. Saudi Med.* **2024**, *44*, 329–338. [CrossRef]
4. World Health Organization. Data. Available online: https://data.who.int/countries/682 (accessed on 21 July 2024).
5. Aljulifi, M.Z. Prevalence and reasons of increased type 2 diabetes in Gulf Cooperation Council Countries. *Saudi Med. J.* **2021**, *42*, 481. [CrossRef]
6. Aburto, N.J.; Hanson, S.; Gutierrez, H.; Hooper, L.; Elliott, P.; Cappuccio, F.P. Effect of increased potassium intake on cardiovascular risk factors and disease: Systematic review and meta-analyses. *BMJ* **2013**, *346*, f1378. [CrossRef] [PubMed]
7. Frassetto, L.A.; Goas, A.; Gannon, R.; Lanham-New, S.A.; Lambert, H. Potassium. *Adv. Nutr.* **2023**, *14*, 1237–1240. [CrossRef] [PubMed]
8. McCarron, D.A.; Reusser, M.E. Are low intakes of calcium and potassium important causes of cardiovascular disease? *Am. J. Hypertens.* **2001**, *14*, 206S–212S. [CrossRef]
9. Zhang, Z.; Cogswell, M.E.; Gillespie, C.; Fang, J.; Loustalot, F.; Dai, S.; Carriquiry, A.L.; Kuklina, E.V.; Hong, Y.; Merritt, R. Association between usual sodium and potassium intake and blood pressure and hypertension among US adults: NHANES 2005–2010. *PLoS ONE* **2013**, *8*, e75289. [CrossRef]
10. O'Donnell, M.J.; Yusuf, S.; Mente, A.; Gao, P.; Mann, J.F.; Teo, K.; McQueen, M.; Sleight, P.; Sharma, A.M.; Dans, A. Urinary sodium and potassium excretion and risk of cardiovascular events. *JAMA* **2011**, *306*, 2229–2238. [CrossRef]
11. Huang, L.; Neal, B.; Wu, J.H.; Huang, Y.; Marklund, M.; Campbell, N.R.; He, F.J.; Yoshimura, S.; Chalmers, J.; Trieu, K. The impact of baseline potassium intake on the dose–response relation between sodium reduction and blood pressure change: Systematic review and meta-analysis of randomized trials. *J. Hum. Hypertens.* **2021**, *35*, 946–957. [CrossRef]
12. Cordain, L.; Eaton, S.B.; Sebastian, A.; Mann, N.; Lindeberg, S.; Watkins, B.A.; O'Keefe, J.H.; Brand-Miller, J. Origins and evolution of the Western diet: Health implications for the 21st century. *Am. J. Clin. Nutr.* **2005**, *81*, 341–354. [CrossRef]
13. Alzeer, H.M.; Alhumaidan, O.; Alkhunein, S.; Alkhaldy, A. Development of the Saudi healthy eating index (SHEI): A first step toward national diet quality monitoring. *BMC Nutr.* **2025**, *11*, 153. [CrossRef]
14. Reddin, C.; Ferguson, J.; Murphy, R.; Clarke, A.; Judge, C.; Griffith, V.; Alvarez, A.; Smyth, A.; Mente, A.; Yusuf, S.; et al. Global mean potassium intake: A systematic review and Bayesian meta-analysis. *Eur. J. Nutr.* **2023**, *62*, 2027–2037. [CrossRef]
15. Al-Mawali, A.; D'Elia, L.; Jayapal, S.K.; Morsi, M.; Al-Shekaili, W.N.; Pinto, A.D.; Al-Kharusi, H.; Al-Balushi, Z.; Idikula, J.; Al-Harrasi, A. National survey to estimate sodium and potassium intake and knowledge attitudes and behaviours towards salt consumption of adults in the Sultanate of Oman. *BMJ Open* **2020**, *10*, e037012. [CrossRef] [PubMed]

16. Okuda, M.; Asakura, K.; Sasaki, S.; Shinozaki, K. Twenty-four-hour urinary sodium and potassium excretion and associated factors in Japanese secondary school students. *Hypertens. Res.* **2016**, *39*, 524–529. [CrossRef] [PubMed]
17. Chlíbková, D.; Knechtle, B.; Rosemann, T.; Žákovská, A.; Tomášková, I. The prevalence of exercise-associated hyponatremia in 24-hour ultra-mountain bikers, 24-hour ultra-runners and multi-stage ultra-mountain bikers in the Czech Republic. *J. Int. Soc. Sports Nutr.* **2014**, *11*, 3. [CrossRef] [PubMed]
18. Livingstone, K.M.; Olstad, D.L.; Leech, R.M.; Ball, K.; Meertens, B.; Potter, J.; Cleanthous, X.; Reynolds, R.; McNaughton, S.A. Socioeconomic Inequities in Diet Quality and Nutrient Intakes among Australian Adults: Findings from a Nationally Representative Cross-Sectional Study. *Nutrients* **2017**, *9*, 1092. [CrossRef]
19. Bolton, K.A.; Trieu, K.; Woodward, M.; Nowson, C.; Webster, J.; Dunford, E.K.; Bolam, B.; Grimes, C. Dietary intake and sources of potassium in a cross-sectional study of Australian adults. *Nutrients* **2019**, *11*, 2996. [CrossRef]
20. Intersalt Cooperative Research Group. Intersalt: An international study of electrolyte excretion and blood pressure. Results for 24 hour urinary sodium and potassium excretion. *BMJ* **1988**, *297*, 319–328. [CrossRef] [PubMed] [PubMed Central]
21. U.S. Department of Agriculture, A.R.S. What We Eat in America. 2013–2014. External Link Disclaimer. 2017. Available online: https://pubmed.ncbi.nlm.nih.gov/36501082/ (accessed on 4 September 2025).
22. Neil, C.E.; Keast, D.R.; Fulgoni, V.L.; Nicklas, T.A. Food Sources of Energy and Nutrients among Adults in the US: NHANES 2003–2006. *Nutrients* **2012**, *4*, 2097–2120. [CrossRef]
23. World Health Organization. Healthy Diet. Available online: https://www.who.int/news-room/fact-sheets/detail/healthy-diet (accessed on 10 June 2025).
24. Vulin, M.; Magušić, L.; Metzger, A.-M.; Muller, A.; Drenjančević, I.; Jukić, I.; Šijanović, S.; Lukić, M.; Stanojević, L.; Davidović Cvetko, E.; et al. Sodium-to-potassium ratio as an indicator of diet quality in healthy pregnant women. *Nutrients* **2022**, *14*, 5052. [CrossRef]
25. Iwahori, T.; Miura, K.; Ueshima, H. Time to consider use of the sodium-to-potassium ratio for practical sodium reduction and potassium increase. *Nutrients* **2017**, *9*, 700. [CrossRef]
26. Bailey, R.L.; Parker, E.A.; Rhodes, D.G.; Goldman, J.D.; Clemens, J.C.; Moshfegh, A.J.; Thuppal, S.V.; Weaver, C.M. Estimating sodium and potassium intakes and their ratio in the American diet: Data from the 2011–2012 NHANES. *J. Nutr.* **2016**, *146*, 745–750. [CrossRef] [PubMed]
27. World Health Organization. *Guideline: Sodium Intake for Adults and Children*; World Health Organization: Geneva, Switzerland, 2012.
28. Leiba, A.; Vald, A.; Peleg, E.; Shamiss, A.; Grossman, E. Does dietary recall adequately assess sodium, potassium, and calcium intake in hypertensive patients? *Nutrition* **2005**, *21*, 462–466. [CrossRef] [PubMed]
29. Huang, Y.; Van Horn, L.; Tinker, L.F.; Neuhouser, M.L.; Carbone, L.; Mossavar-Rahmani, Y.; Thomas, F.; Prentice, R.L. Measurement error corrected sodium and potassium intake estimation using 24-hour urinary excretion. *Hypertension* **2014**, *63*, 238–244. [CrossRef] [PubMed]
30. Alshuwaier, G.O. Assessment of Urinary Potassium Concentration Among Saudi Young Soccer Players. *J. Sport Sci. Phys. Educ.* **2021**, *7*, 1–9.
31. Saeedi, M.; AlMadani, A.; Alsafi, Y.; Arif, S.; Mustafa, S.; Jamo, A. Estimation of sodium and potassium intake in 24-hours urine, Aljouf Region, Northern Saudi Arabia. *Chronic. Dis. Int* **2017**, *4*, 1026.
32. WHO. *How to Obtain Measures of Population-Level Sodium Intake in 24-Hour Urine Samples*; World Health Organization: Geneva, Switzerland, 2018.
33. STATISTICS. K.S.M. Global Media Insight.
34. Wilcox, C.S. Diuretics and Potassium. In *Current Topics in Membranes and Transport*; Giebisch, G., Ed.; Academic Press: Cambridge, MA, USA, 1987; Volume 28, pp. 331–350.
35. World Health Organization. *WHO STEPS Instrument Question-by-Question Guide (Core and Expanded)*; World Health Organization: Geneva, Switzerland, 2008.
36. Birukov, A.; Rakova, N.; Lerchl, K.; Engberink, R.H.G.O.; Johannes, B.; Wabel, P.; Moissl, U.; Rauh, M.; Luft, F.C.; Titze, J. Ultra-long–term human salt balance studies reveal interrelations between sodium, potassium, and chloride intake and excretion. *Am. J. Clin. Nutr.* **2016**, *104*, 49–57. [CrossRef]
37. Oesch, U.; Ammann, D.; Simon, W. Ion-selective membrane electrodes for clinical use. *Clin. Chem.* **1986**, *32*, 1448–1459. [CrossRef]
38. Yano, Y.; Reis, J.P.; Colangelo, L.A.; Shimbo, D.; Viera, A.J.; Allen, N.B.; Gidding, S.S.; Bress, A.P.; Greenland, P.; Muntner, P. Association of blood pressure classification in young adults using the 2017 American College of Cardiology/American Heart Association blood pressure guideline with cardiovascular events later in life. *JAMA* **2018**, *320*, 1774–1782. [CrossRef]
39. *S.S.R. Statistical Software Release 11*; StataCorp: College Station, TX, USA, 2012.

40. Jarrar, A.H.; Thondre, P.S.; Cheikh Ismail, L.; Lightowler, H.; Bataineh, M.a.F.; Al Baloushi, A.K.; Al Braiki, A.Y.; Al Halabi, S.; Hajouz, J.; Souka, U.; et al. Salt-related knowledge, attitudes and practices and their relationship with 24-h urinary sodium and potassium excretions among a group of healthy residents in the UAE: A cross-sectional study. *Public Health Nutr.* **2024**, *27*, e235. [CrossRef]
41. Adam, A.; Osama, S.; Muhammad, K.I. Nutrition and food consumption patterns in the Kingdom of Saudi Arabia. *Pak. J. Nutr.* **2014**, *13*, 181. [CrossRef]
42. Tash, A.A.; Al-Bawardy, R.F. Cardiovascular disease in Saudi Arabia: Facts and the way forward. *J. Saudi Heart Assoc.* **2023**, *35*, 148. [CrossRef]
43. Neal, B.; Wu, Y.; Feng, X.; Zhang, R.; Zhang, Y.; Shi, J.; Zhang, J.; Tian, M.; Huang, L.; Li, Z. Effect of salt substitution on cardiovascular events and death. *N. Engl. J. Med.* **2021**, *385*, 1067–1077. [CrossRef]
44. Chang, H.-Y.; Hu, Y.-W.; Yue, C.-S.J.; Wen, Y.-W.; Yeh, W.-T.; Hsu, L.-S.; Tsai, S.-Y.; Pan, W.-H. Effect of potassium-enriched salt on cardiovascular mortality and medical expenses of elderly men. *Am. J. Clin. Nutr.* **2006**, *83*, 1289–1296. [CrossRef]
45. Stamler, J.; Rose, G.; Stamler, R.; Elliott, P.; Dyer, A.; Marmot, M. INTERSALT study findings. Public health and medical care implications. *Hypertension* **1989**, *14*, 570–577. [CrossRef]
46. Tzoulaki, I.; Patel, C.J.; Okamura, T.; Chan, Q.; Brown, I.J.; Miura, K.; Ueshima, H.; Zhao, L.; Van Horn, L.; Daviglus, M.L. A nutrient-wide association study on blood pressure. *Circulation* **2012**, *126*, 2456–2464. [CrossRef]
47. Perez, V.; Chang, E.T. Sodium-to-potassium ratio and blood pressure, hypertension, and related factors. *Adv. Nutr.* **2014**, *5*, 712–741. [CrossRef] [PubMed]
48. Powles, J.; Fahimi, S.; Micha, R.; Khatibzadeh, S.; Shi, P.; Ezzati, M.; Engell, R.E.; Lim, S.S.; Danaei, G.; Mozaffarian, D. Global, regional and national sodium intakes in 1990 and 2010: A systematic analysis of 24 h urinary sodium excretion and dietary surveys worldwide. *BMJ Open* **2013**, *3*, e003733. [CrossRef] [PubMed]
49. Cook, N.R.; Obarzanek, E.; Cutler, J.A.; Buring, J.E.; Rexrode, K.M.; Kumanyika, S.K.; Appel, L.J.; Whelton, P.K.; Trials of Hypertension Prevention Collaborative Research Group. Joint effects of sodium and potassium intake on subsequent cardiovascular disease: The Trials of Hypertension Prevention follow-up study. *Arch. Intern. Med.* **2009**, *169*, 32–40. [CrossRef] [PubMed]
50. Weaver, C.M. Potassium and Health. *Adv. Nutr.* **2013**, *4*, 368S–377S. [CrossRef]
51. Lane, H.W.; Roessler, G.S.; Nelson, E.W.; Cerda, J.J. Effect of physical activity on human potassium metabolism in a hot and humid environment. *Am. J. Clin. Nutr.* **1978**, *31*, 838–843. [CrossRef]
52. Loftfield, E.; Yi, S.; Curtis, C.J.; Bartley, K.; Kansagra, S.M. Potassium and fruit and vegetable intakes in relation to social determinants and access to produce in New York City. *Am. J. Clin. Nutr.* **2013**, *98*, 1282–1288. [CrossRef]
53. Cappuccio, F.P.; Ji, C.; Donfrancesco, C.; Palmieri, L.; Ippolito, R.; Vanuzzo, D.; Giampaoli, S.; Strazzullo, P. Geographic and socioeconomic variation of sodium and potassium intake in Italy: Results from the MINISAL-GIRCSI programme. *BMJ Open* **2015**, *5*, e007467. [CrossRef]
54. Miyagawa, N.; Okuda, N.; Nakagawa, H.; Takezaki, T.; Nishi, N.; Takashima, N.; Fujiyoshi, A.; Ohkubo, T.; Kadota, A.; Okamura, T.; et al. Socioeconomic Status Associated with Urinary Sodium and Potassium Excretion in Japan: NIPPON DATA2010. *J. Epidemiol.* **2018**, *28*, S29–S34. [CrossRef]

Disclaimer/Publisher's Note: The statements, opinions and data contained in all publications are solely those of the individual author(s) and contributor(s) and not of MDPI and/or the editor(s). MDPI and/or the editor(s) disclaim responsibility for any injury to people or property resulting from any ideas, methods, instructions or products referred to in the content.

nutrients

MDPI

Article

Dietary Iron Intake Impacts the Microbial Composition of the Murine Intestinal and Lung Microbiome

Ama-Tawiah Essilfie [1,2,†], Alison Smith [1,†], Rebecca Watts [1], Pramila Maniam [1], Iain L. Lamont [3], David M. Frazer [1,4,5], Gregory J. Anderson [1] and David W. Reid [1,2,6,*]

1 QIMR Berghofer Medical Research Institute, Brisbane, QLD 4006, Australia; ama-tawiah.essilfie@qimrb.edu.au (A.-T.E.)
2 The School of Medicine, Herston Campus, The University of Queensland, Brisbane, QLD 4006, Australia
3 The Department of Biochemistry, University of Otago, Dunedin 9012, New Zealand
4 The School of Biomedical Sciences, The University of Queensland, St Lucia, Brisbane, QLD 4067, Australia
5 The School of Biomedical Sciences, Queensland University of Technology, Gardens Point, Brisbane, QLD 4000, Australia
6 The Prince Charles Hospital, Brisbane, QLD 4032, Australia
* Correspondence: david.reid@qimrb.edu.au
† These authors contributed equally to the work.

Abstract: Background: Iron is an essential nutrient for many bacterial pathogens and normal cellular function and homeostasis of their hosts. Studies suggest that iron deficiency or overload may contribute to the pathogenesis of several chronic conditions and modify host–microbial interactions. In this study, we assessed the impact of varying dietary iron intakes on the microbiota of the intestinal tract and lungs of wild-type mice. Methods: Male C57BL/6J mice were fed either a standard pellet chow (high iron diet), a ferrous ammonium sulfate (FeAS)-supplemented diet or an iron-deficient diet for four weeks. Tissue from the lung, duodenum and colon was collected, and 16S rRNA gene fragments were pyrosequenced. Results: Total serum iron levels were negatively associated with richness of the lung microbiome ($p = 0.035$). In the murine lungs, there was no association between the iron diet and the overall lung microbiota community composition, but *Bacteroides* spp. were significantly enriched in the lungs of mice fed the FeAS diet (LDA score > 4, $p < 0.05$). The community composition of the intestinal microbiota changed significantly depending on the iron diet, with increased richness in the low-iron compared to the iron-supplemented groups ($p = 0.053$). In the duodenum, *Prevotella* spp. were reduced (Mean = 7.869, SEM = 3.464, $p < 0.05$), and *Desulfovibrio* species increased (Mean = 5.343, SEM = 1.362, $p < 0.001$) in iron-supplemented groups compared to the low-iron-diet group. In the colon, *Bifidobacterium* and *Bacteroides* species were reduced (Mean = 7.175, SEM = 2.246, $p < 0.01$ and Mean = 6.967, SEM = 1.834, $p < 0.01$ respectively), and *Pseudomonas* increased (Mean = 24.03, SEM = 8.919, $p < 0.05$) in mice on higher-iron diets compared to the low-iron diet. Discussion: This study demonstrates that dietary iron intake significantly impacts the intestinal microbiota and has a small, yet significant, effect on the lung microbiome in C57BL/6J mice. Whilst dietary iron content per se did not significantly modulate the composition of the lung microbiota, serum iron levels had subtle impacts on the community composition of the lung microbiota.

Keywords: dietary iron; microbiome; colon; duodenum; lung; mouse

1. Introduction

Several human and animal studies have demonstrated that diet, specifically the intake of different proteins, fats and carbohydrates, can have a significant effect on intestinal microbiome composition [1]. Alterations in diet can change the gut microbial composition in as little as 24 h, but microbiome composition is also able to revert to baseline within 48 h of diet discontinuation [2].

Changes to dietary iron intake in human and animal studies can influence the population structure of bacterial communities that make up the gut microbiome, and they can be correlated with downstream impacts on immune regulation. This microbial dysbiosis is associated with diseases such as hypertension, Parkinson's disease, asthma, type 2 diabetes, cancer and inflammatory bowel disease [3,4].

Iron is an essential biometal for the host in both health and disease, and also for almost all gut bacteria. As a result, several studies have investigated the effect of dietary iron on the gut microbiome, comparing the consequences of low versus high iron levels in the gut. Studies in East- and West-African children receiving either iron-supplemented biscuits or iron-fortified porridge showed increases in *Enterobacteria* and decreases in *Lactobacilli*, suggesting a shift towards a more pathogenic profile [5,6]. These children also showed signs of increased intestinal inflammation and a trend towards an increase in diarrhoea. Animal studies support these findings; iron-deprived mice and rats develop an increased abundance of *Lactobacillus* and a decrease in *Bacteroides* in the gut microbiota [7,8]. Recent studies have compared changes in the gut microbiome when iron-deficient animals received different iron supplements in the form of either conventional ferrous sulfate or Sucrosomial® iron, which is a relatively new supplement [9]. The form of iron ingested significantly changed the microbial profile in the gut, with Sucrosomial® iron increasing the abundance of potentially beneficial short-chain fatty acid–producing bacteria while decreasing more potentially pathogenic species.

The recognition of the inter-connectedness of the lung and gut microbiomes as part of the "gut–lung axis" suggests that diet could also potentially influence the lung microbiome and respiratory health outcomes [10–12]. Whether dietary iron impacts the lung microbiome is an important question, as there is strong evidence that lung iron homeostasis is abnormal in a number of chronic lung diseases, including the genetic disorder cystic fibrosis (CF), asthma, pulmonary fibrosis and chronic obstructive pulmonary disease (COPD) [13–16], and these diseases also have significant alterations in their microbiome. As a result, changes in diet which cause dysbiosis may influence pathology and clinical outcomes.

Although several studies have investigated the effect of either dietary iron or iron supplementation on gut microbiota [5,9,17–26], few have linked iron-related changes in the gut to changes in the lung microbiome [27–29]. The gut–lung axis (GLA) is proposed to work in a bi-directional manner, with the lung potentially being the source of some gut microbes and vice versa [30,31]. Apart from microbes, there are several factors that influence and regulate the GLA, including immune cells and their mediators, microbe-derived metabolites and microbe crosstalk between the lung and gut. Gut dysbiosis, caused by diet, antibiotics, infection or smoking, can cause an inflammatory response that damages the epithelial membrane barrier, allowing metabolites, commensal microbes and inflammatory cytokines to leak into the lung via systemic circulation [32]. This cross-communication is important in both health and disease, so we hypothesise that higher levels of dietary iron intake will adversely affect the gut microbiome and that these changes will induce a change in the lung microbiome via the gut–lung axis. Despite the large number of studies investigating the human gut microbiome, these are still far from exhaustive, and there are always many cofounders in disease settings, such as the impact of drug treatments.

We have, therefore, chosen to use an animal model to investigate our hypothesis, as this is a well-established, proof-of-concept starting point. The precise effect of different dietary iron intake levels on the gut microbiome is a field of research still in its infancy, and although there are no obvious adverse effects from making small changes to iron intake on the human body as a whole, the changes to the microbiome of the gut and the downstream effects of those changes are largely unknown, but likely to be significant. The mouse model gives us a controlled environment within which this can be investigated. We, therefore, investigated how dietary iron intake in wild-type C57BL/6J mice would influence the composition of the microbiota of the gastrointestinal tract (GIT) and whether those changes would alter the lung microbiome.

2. Materials and Methods

2.1. Experimental Approach

Upon weaning, specific-pathogen-free, three-week-old C57BL/6J mice were randomly assigned into three groups of six mice and fed one of three diets. The mice were housed in groups of three mice/cage, with the same cage components, such as bedding material, on the same animal housing rack for the entirety of the study. Animals in group 1 were fed an iron-deficient diet ad libitum based on AIN93G (Specialty Feeds, Glen Forrest, Australia; about 1 mg/kg iron, n = 6); those in group 2 were fed a ferrous ammonium sulfate-supplemented diet ad libitum (AIN93G supplemented with FeAS; 50 mg iron/kg, n = 6); and those in group 3 were fed a standard mouse pellet chow ad libitum (Norco Stockfeed, Lismore, Australia), which is intrinsically high in iron content (120 mg iron/kg as ferric citrate, n = 6). After four weeks, the mice were weighed and euthanised, and faeces and tissue samples from the duodenum, colon and lung were collected, snap frozen and stored at −80 °C (Scheme 1 below). All mice were housed in the animal facility at the QIMR Berghofer Institute under identical conditions, and all processes and procedures were approved by The QIMR Berghofer Animal Ethics Committee.

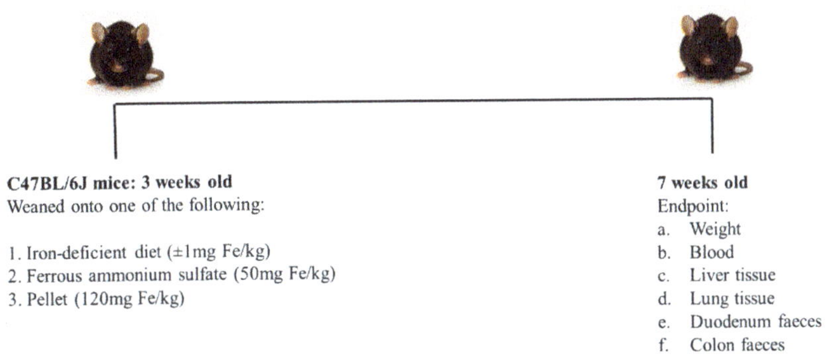

Scheme 1. Timeline of experimental interventions. Upon weaning, 3-week-old mice were placed on different iron diets. At the endpoint, 4 weeks later, weight was measured and blood, liver and lung tissue and faeces collected for further 16S analysis.

2.2. Iron Concentration Measurements

The iron status of the mice at the end of the dietary intervention was assessed by measurement of their liver iron concentration and total serum iron levels [33,34]. Approximately 20 mg of dried liver was used to measure hepatic iron levels, as previously described by Frazer and colleagues [33]. Briefly, 1 mL of 3 M hydrochloric acid and 0.6 M trichloroacetic acid solution were added to each sample and to the relevant amounts of the iron stock to make up a standard curve (0–90 µg/mL). The samples and iron standards were incubated overnight at 65 °C, then 4 μL of the sample (standard or liver sample) was added to freshly prepared chromogen reagent. After a 30 min incubation at room temperature, the absorbance was measured at 535 nm with a plate reader, and, using the standard curve, the amount of iron was determined. Serum iron was measured using a colorimetric assay according to the manufacturer's instructions (Pointe Scientific, Canton, OH, USA).

2.3. Microbiome Analysis

Genomic DNA was extracted from the tissue samples using the PureLink® Genomic DNA extraction kit according to the manufacturer's instructions (Invitrogen, Waltham, MA, USA). 16S rRNA was amplified using the universal primer set 939F (TTGACGGGGGCC-CGCAC) and 1492R (TACCTTGTTACGACTT) that binds to the conserved region flanking the variable region V5-V9 and sequenced using bacterial tag-encoded FLX amplicon pyrosequencing, as previously described [35]. A sample comprising only the extraction kit buffer was co-sequenced to assess reagent contamination (signal was subtracted from experimental samples). The data were processed using the QIIME2 (v2020.11) software [36] and analysed using the Calypso platform [37].

2.4. Data Handling

The sample size was determined by the primary outcome measure 16S sequencing. All six mice/group were included in each experiment; however, the final results contained n = 4–6 data points. Data points were removed after an outlier test (GraphPad Prism, v9.3.1) revealed outliers in the data. The authors conducting the experiments were aware of the different group allocations while conducting the experiments and preparing the final results' graphs and figures. Microbiome analysis was conducted in a blinded fashion by an external service provider.

2.5. Statistical Analysis

Iron indices and weight data were analysed using a one-way analysis of variance (ANOVA) with a Bonferroni post-test or Pearson correlation. The data are displayed as mean ± SEM.

The microbiome data were imported into Calypso using total sum normalisation combined with square root transformation. Differences in taxa between different diet groups were compared using ANOVA with a *p*-value correction using the false discovery rate (FDR). Pearson correlations were calculated to identify the relationships between the abundance of genera in the duodenum, colon and lung. The microbial diversity at each anatomical site, depending on the different iron diets, was assessed with Shannon diversity and richness and evaluated using a one-way ANOVA test.

3. Results

3.1. Iron Status and Weight

Three-week-old mice were weaned onto either an iron-deficient diet (Fe-Def), an FeAS-containing diet (50 mg/kg iron), or a relatively high iron standard pellet diet (120 mg/kg) and maintained on these diets for four weeks before samples were collected. Liver iron levels were significantly increased in mice fed the pellet diet compared to those fed the iron-deficient diet (Figure 1A), while total serum iron levels showed no significant differences but trended towards an increase in the pellet diet ($p = 0.09$; Figure 1B). Mice on the iron-deficient diet had significantly lower weights than the mice on the FeAS and pellet diets (Figure 1C). This trend may be attributed to the lower amount of iron in the diet of these mice, though there is also a slight difference in the make-up of the iron-deficient diet compared to the FeAS diet, so the variation in weight may also be affected by a variation in protein and carbohydrates. There was no correlation between mouse weight, liver iron content and serum iron levels (Figure 1D–F).

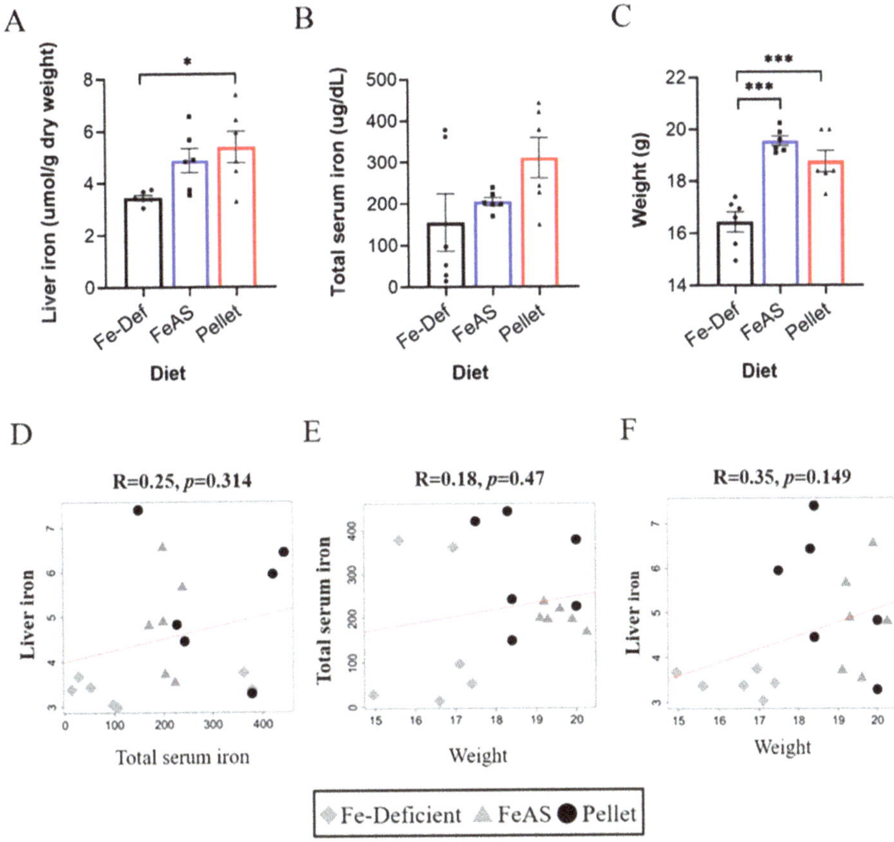

Figure 1. Iron indices and weight of mice fed different dietary iron intakes. (**A**) Liver iron (η_p^2 = 0.60), (**B**) total serum iron (η_p^2 = 0.40) and (**C**) weight (η_p^2 = 0.71) of mice fed an iron-deficient (Fe-Def), pellet (high-iron) or FeAS (ferrous ammonium sulfate)-supplemented diet. Relationship between liver iron and total serum iron (**D**), serum iron and weight (**E**) and liver iron and weight (**F**) were assessed. Pairwise group comparisons were performed using t-tests and are annotated using asterisks (p-value: * < 0.05, *** < 0.001). Correlation between liver iron, total serum iron and weight was calculated using Pearson correlation (**D–F**); iron concentration (µM/g dry weight), total serum iron (µg/dL) and weight (g). For each group in every analysis, n = 6. η_p^2 = effect size (partial eta squared).

3.2. Iron Supplementation Has a Small Effect on the Microbiome of the Lung

At the phylum level, the mouse lung microbiome was dominated by *Firmicutes* and *Bacteroidetes*, which has been shown by others, and this remained largely unchanged after feeding mice diets of differing iron contents (Figure 2A). A redundancy analysis (RDA) plot demonstrated that while the different diets led to some clustering, this did not change significantly when the iron content of the diet was altered (Figure 2B). There was no difference in microbial diversity (Shannon index) or richness in the lung when diets of differing iron content were used (Figure 2C). Interestingly, total serum iron showed a significant negative correlation with richness (Figure 2D). Furthermore, a comparison of the lung microbiome in the mice fed the three different diets using linear discriminant analysis (LDA) effect size revealed an association between operational taxonomic units (OTUs) related to *Bacteroides* spp. and the FeAS diet (LEfSe, LDA score > 4; Supplementary Figure S1A–C). The family *Bacteroidaceae*, the genus *Bacteroides* and OTUs related to *Bacteroides* spp. were enriched in the lung microbiome of the mice fed the FeAS-supplemented diet compared to the iron-deficient and normal pellet diets (Supplementary Figure S1D,E).

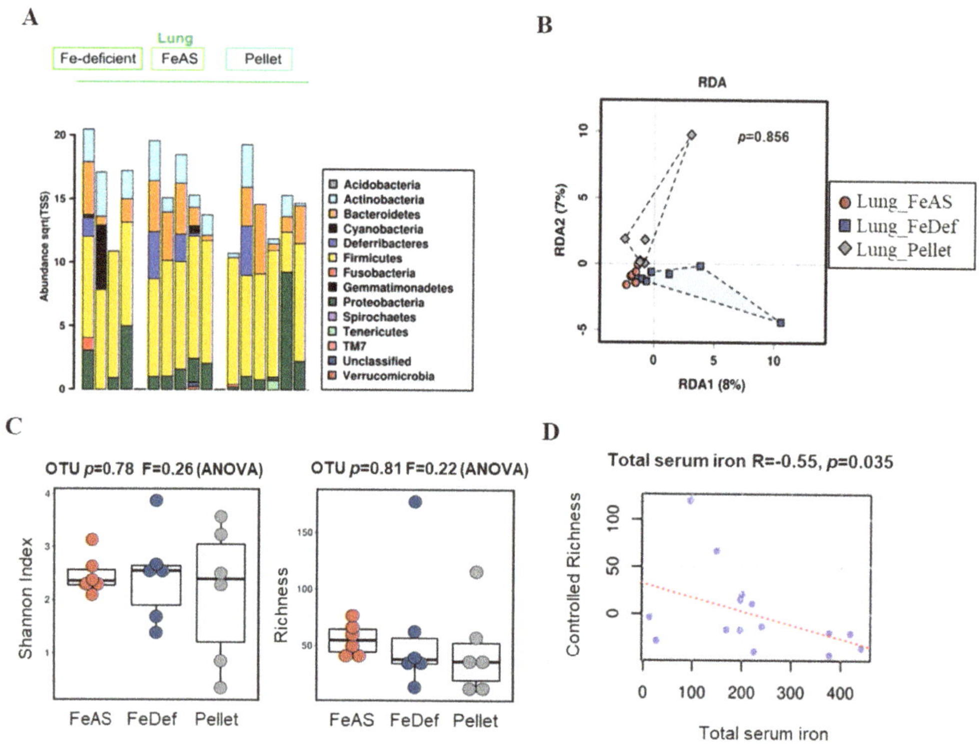

Figure 2. Taxonomic profile of the microbiome in the lung. (**A**) Bar chart of the microbiome composition at the phyla level in the lung after 4 weeks of different iron diets, (**B**) redundancy analysis (RDA) plot, (**C**) Shannon index and richness plots and (**D**) richness correlation with total serum iron after correcting for weight. Fe-deficient diet (FeDef, 0 mg/kg iron), ferrous ammonium sulfate diet (FeAS, 50 mg/kg iron) and pellet diet (120 mg/kg iron) and OTUs (operational taxonomic units). N = 4–6 for the phyla-level microbiome composition data, and n = 6 for all other groups in the remaining analyses.

3.3. Iron Supplementation Significantly Changes the Microbiome of the Duodenum

As in the lung, and as expected, the dominant bacterial phyla in the duodenum were *Firmicutes* and *Bacteroidetes*. In addition, *Proteobacteria* and *Actinobacteria* were increased in abundance in the duodenum compared to the lung. The iron-deficient and FeAS diets had a similar abundance of *Proteobacteria* and *Actinobacteria*; however, the pellet diet seemed to reduce the abundance of *Actinobacteria* (Figure 3A). Redundancy analysis showed significant differences between all diets (Figure 3B), while microbial diversity and richness were similar across the different diets (Figure 3C). Richness tended to be decreased in the pellet-fed mice group compared to the iron-deficient mice group ($p = 0.057$).

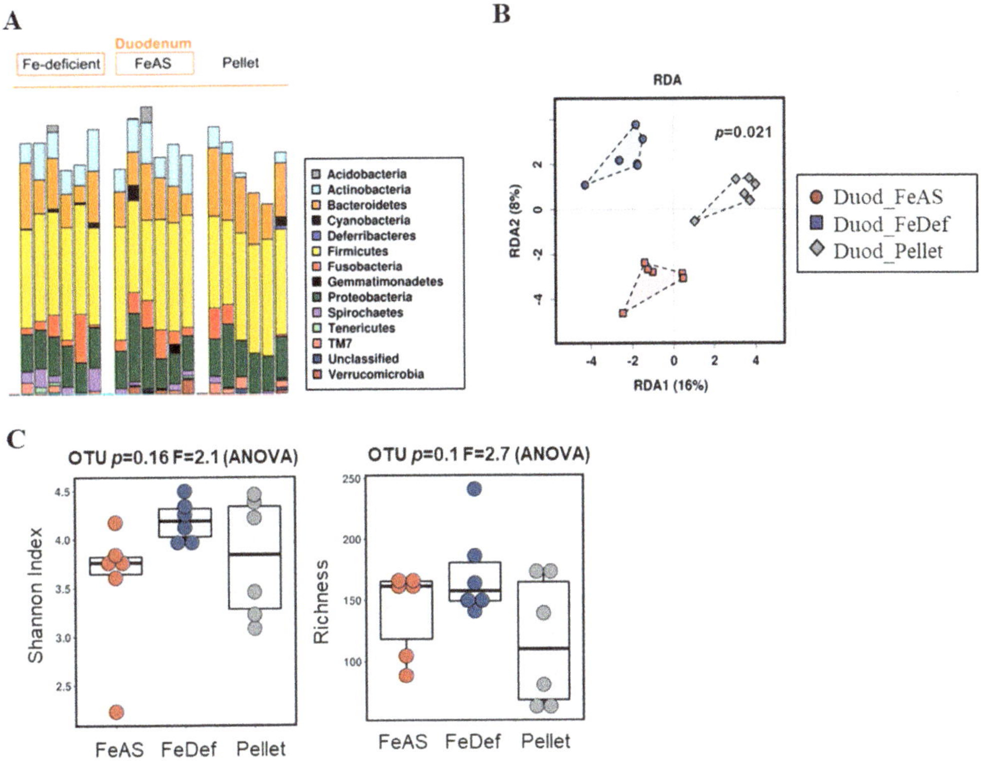

Figure 3. Microbiome profile in the duodenum. (**A**) Bar chart of the microbiome composition at the phyla level in the duodenum after 4 weeks of different iron diets, (**B**) redundancy analysis (RDA) plot and (**C**) Shannon index and richness plots for the duodenum. Fe-deficient diet (FeDef, 0 mg/kg iron), ferrous ammonium sulfate diet (FeAS, 50 mg/kg iron) and pellet diet (120 mg/kg iron), OTUs (operational taxonomic units). There was an n = 6 for each group in every analysis.

3.4. Iron Supplementation Significantly Changes the Microbiome of the Colon

In the colon, *Firmicutes*, *Proteobacteria*, *Bacteroidetes* and *Actinobacteria* were again the main bacterial phyla. *Proteobacteria* were increased and *Bacteroidetes* decreased in animals fed the higher-iron diets compared to those fed the FeDef diet (Figure 4A). Similar to the duodenum, redundancy analysis demonstrated distinct clustering and a significant difference between all three diets (Figure 4B). However, unlike the lung and duodenum, the diet significantly altered the microbial diversity and richness in the colon. Interestingly, the iron-deficient diet had the highest microbial diversity index and richness score, and the FeAS diet had the lowest diversity and richness score (Figure 4C).

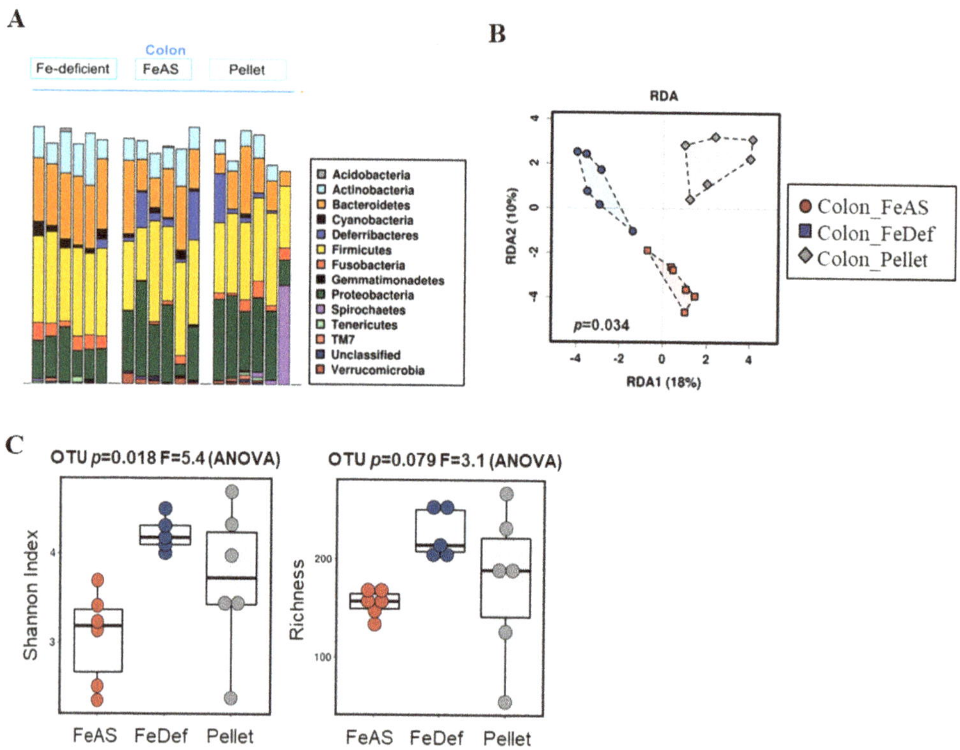

Figure 4. Microbiome profile in the colon. (**A**) Bar chart of the microbiome composition at the phyla level in the colon after 4 weeks of different iron diets, (**B**) redundancy analysis (RDA) plot and (**C**) Shannon index and richness plots for the colon. Fe-deficient diet (FeDef, 0 mg/kg iron), ferrous ammonium sulfate diet (FeAS, 50 mg/kg iron) and pellet diet (120 mg/kg iron), OTUs (operational taxonomic units). N = 5–6 for microbiome diversity data, and n = 6 for all other groups in the remaining analyses.

3.5. Changes in Microbial Taxa Between Diets in the Lung, Duodenum and Colon

To narrow down the classification of microbes, several bacterial genera were assessed after the mice were fed a low-iron diet (FeDef), a medium-iron diet (FeAS) or a pellet diet. A small number of bacteria were changed in the lungs as a result of these diets; however, there were several significant changes in the gut (duodenum and colon). *Allobaculum* was significantly reduced in mice fed the pellet diet in both the duodenum and colon (Figure 5A). We showed that in the gut, the levels of bacteria in the class *Bacteroidia*, *Prevotella* and *Bacteroides* were significantly reduced in the mice fed the pellet diet compared to the iron-deficient animals (Figure 5B,C). When we further examined the *Bacilli* class, we found a variety of responses. There were no changes in *Streptococcus* (Figure 5D), *Staphylococcus* was lower in the lungs of mice on the FeAS-supplemented diet relative to those on the pellet diet (Figure 5E), there was an increase in *Lactobacillus* in the duodenum of mice fed the pellet diet (Figure 5F) and, finally, there was a decrease in *Veillonella* in the gut of mice fed the pellet diet compared to the iron-deficient animals (Figure 5G). When we examined the *Gamma-proteobacteria* class of pellet-fed animals, we found increases in *Pseudomonas* in the colon (Figure 5H) and *Desulfovibrio* in the duodenum (Figure 5I) and decreases in *Sutterella* in the duodenum (Figure 5J) but no changes in *Haemophilus* (Figure 5K). Finally, we showed that *Bifidobacterium* was decreased in the gut of mice fed the pellet diet compared to those fed the FeAS or FeDef diets (Figure 5L).

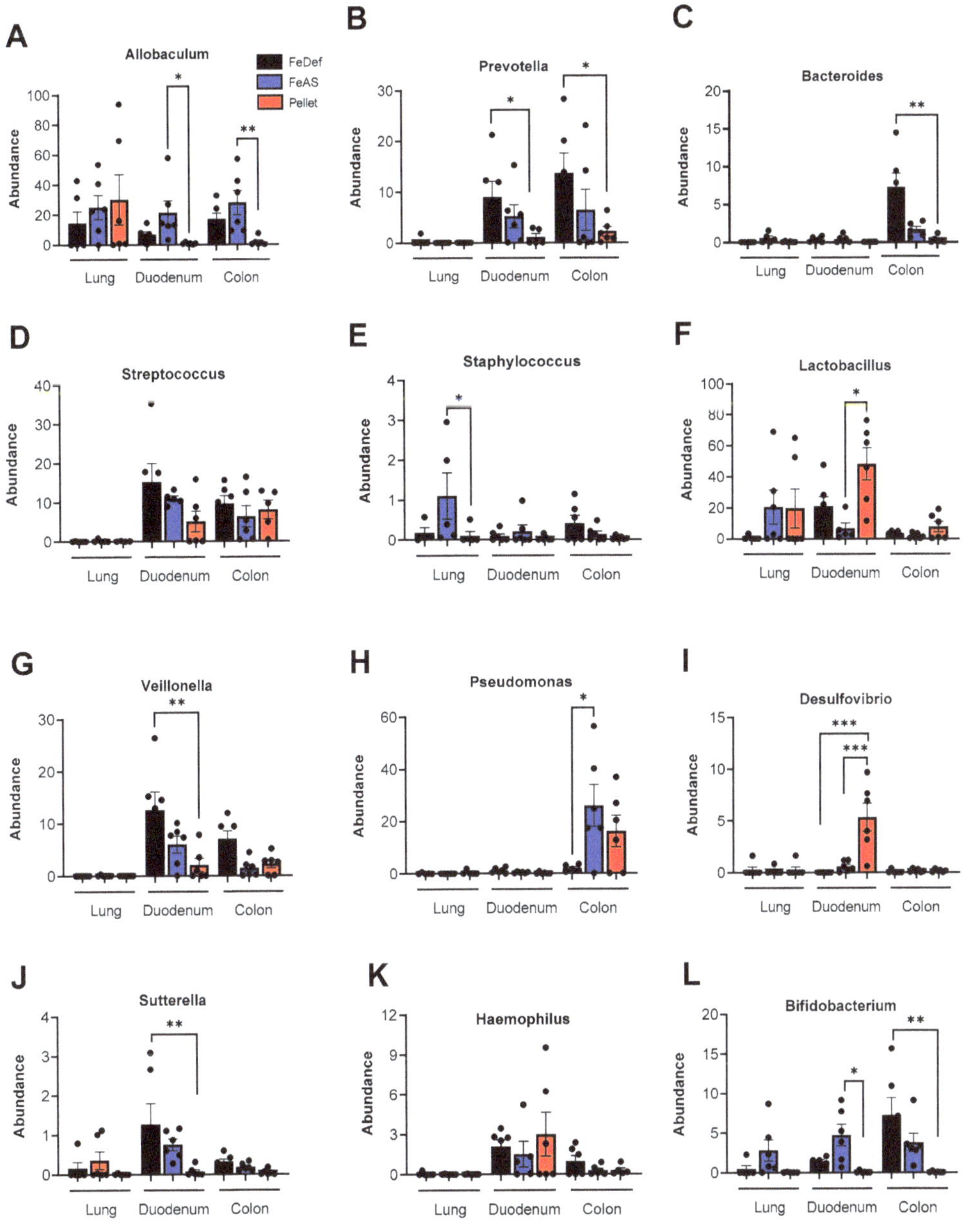

Figure 5. Changes in bacterial genera in lung, duodenum and colon. Mice fed either iron-deficient, FeAS or pellet diet were analysed for microbial genera in lung, duodenum and colon. Relative abundance of (**A**) *Allobaculum* (η_p^2 = 0.26), (**B**) *Prevotella* (η_p^2 = 0.76), (**C**) *Bacteroides* (η_p^2 = 0.71), (**D**) *Streptococcus* (η_p^2 = 0.48), (**E**) *Staphylococcus* (η_p^2 = 0.33), (**F**) *Lactobacillus* (η_p^2 = 0.44), (**G**) *Veillonella* (η_p^2 = 0.59), (**H**) *Pseudomonas* (η_p^2 = 0.58), (**I**) *Desulfovibrio* (η_p^2 = 0.70), (**J**) *Sutterella* (η_p^2 = 0.42), (**K**) *Haemophilus* (η_p^2 = 0.40) and (**L**) *Bifidobacterium* (η_p^2 = 0.50) were quantified. N = 6 for each group in every analysis. η_p^2 = effect size (partial eta squared), *p*-value * < 0.05, ** < 0.01, *** < 0.001.

4. Discussion

This study is one of the few to investigate the simultaneous effects of dietary iron intake on the microbiota of the lung, duodenum and colon in C57BL/6J mice. Our findings demonstrate that the community composition of the microbiota in the duodenum and colon changes significantly with dietary iron intake, which is consistent with previous studies [5,9,18,19,21,22,28]. The lung microbiota was not significantly affected by the different iron diets per se, but *Bacteroides* spp. showed a trend of increased abundance in the mice fed the FeAS-supplemented diet compared to the mice on a low-iron diet. Interestingly, as serum iron levels increased, microbial richness of the lung decreased, which suggests that systemic iron levels may potentially modulate the lung microbiome.

Reports on the most abundant bacteria in the murine lung microbiome are conflicting [38–40], which probably reflects different sample collection methods (bronchoalveolar lavage [32] vs. lung tissue), sequencing technology, PCR primer selection, mouse strain and the environment within which mice are housed, i.e., differing resident microbes in the pellet chow, cage bedding and drinking water. Sequencing of lung tissue in C57BL/6J mice has demonstrated a microbiome dominated by *Ralstonia, Lactobacillus, Enterobacteriaceae* and *Sphingomonas* species [38]. In another study, BAL and lung tissue samples demonstrated that the murine lung microbiome was mainly composed of the genera *Staphylococcus, Massilia, Corynebacterium, Pseudomonas* and *Streptococcus* [40]. A study by Scheiermann et al. categorised the murine lung microbiome in female Balb/cJ mice into three distinct pneumotypes [39]. The first pneumotype (a third of the lung samples) was considered sterile after correcting for contamination introduced by sequencing. A second pneumotype was characterised by a high abundance of the genera *Streptococcus, Haemophilus* and *Actinobacillus*, which overlapped with the microbiome of the murine oral cavity. The third pneumotype described was dominated by chow- and cage-bedding-associated bacteria. Thus, there is some variability in the murine lung microbiota depending on the study variables and environmental factors in experimental animals living under artificial conditions.

In our study, the genera *Allobaculum, Lactobacillus, Bifidobacterium* and *Streptococcus* were the most abundant taxa in the murine lung microbiome. *Lactobacillus* and *Streptococcus* have previously been reported in the murine lung microbiome [39], but to our knowledge, this is the first report of the presence of *Allobaculum* and *Bifidobacterium*. *Allobaculum* and *Bifidobacterium* are common members of the murine gut flora, and it is conceivable that their presence represents inhalation of faecal material by the mice from the environment in the cage. Future studies would need to investigate how bedding and faeces in the cage may contribute to the lung microbiome. The few studies that have investigated the role of *Allobaculum* in the gut demonstrated that *Allobaculum* induced an increase in T-helper 17 cells, disrupted the intestinal barrier and promoted a pro-inflammatory milieu. Although studies have shown correlations between *Allobaculum* abundance in the gut and changes in the lung microbiome composition, there are no data demonstrating its role specifically in the lung [41–43].

Bacteroides spp., which differed in relative abundance between low and high iron diets in the lung in this study, are also normal commensals of the gut, and the presence of *Bacteroides* spp. has been reported in other studies of the murine lung [38]. *Bacteroides* spp. have also been identified in BAL samples from human studies [44]. In our study, *Bacteroides* spp. were enriched in the lungs of animals fed the FeAS diet but not the iron-deficient or pellet mice dietary groups. *Bacteroides* spp. were also present in low abundance in the pellet-fed mice but were highly abundant in the colons of the mice receiving both the FeAS-supplemented and iron-deficient diets.

Within each dietary iron group, the lung microbiome was more heterogeneous between individual mice than in the duodenum or colon. This individual heterogeneity of bacterial communities in the lung microbiota is consistent with previous reports [39]. Interestingly, serum iron levels were negatively associated with overall diversity of the lung microbiome, whilst liver iron, which is generally considered a reliable measure of systemic iron stores, was not. A limitation of this lung microbiome proof-of-concept study is the small sample size and the variation of lung microbiota composition that we found within each diet group. However, although our study was not sufficiently powered to detect changes at an individual OTU level, the relationship between serum iron and lung microbiome diversity suggests that the lung microbiome is sensitive to systemic iron levels. As the lung microbiota represent a low-biomass sample, increasing the sample size will also decrease the within-group variation and strengthen the reliability of the data. We did not determine whether increased serum iron levels resulted in higher lung iron content, and the mechanism by which serum iron modulates the lung microbiome needs to be further investigated in a larger study. Our group and others have demonstrated an increase in iron levels and iron metabolism dysregulation in the airways and lungs of CF, asthma and COPD patients [13,14,16,45,46]. This dysfunction in iron homeostasis induces pro-fibrotic responses in fibroblasts and airway smooth muscle cells in asthma and fibrotic lung disease models [15,16], induces ferroptosis of airway epithelial cells in CF and COPD [47,48] and induces pro-inflammatory responses in all three airway diseases. These changes in iron are associated with the pathogenesis of severe disease and increased susceptibility to respiratory infections [16,49–51]. Interestingly, in humans, iron infusion resulted in a rapid rise in sputum iron levels in adult CF patients, suggesting that iron may transit from the vascular compartment into the airway lumen relatively easily, particularly in the setting of the mucosal inflammation that is present in CF [52]. The microbiome is also significantly altered in all three of these diseases, and this is also thought to play a critical role in disease pathogenesis [53–63]. Our findings in mice may, therefore, have implications for chronic airway diseases where abnormal lung iron homeostasis plays a pivotal role, and in these diseases, an abnormal lung microbiome likely contributes to airway inflammation and disease progression; in this context, iron moving from the circulation into the airway lumen may be an important modifying factor.

In contrast to the lung, the gut is very densely populated with diverse bacterial communities; in fact, the gut contains the largest microbiome in the human body [64]. Within the gut, the human host and resident microbiota compete for iron acquisition, and the microbiota are able to regulate host functions to increase access to iron. Dietary iron is, therefore, a strong regulator of the community composition and behaviour of the gut microbiome [65]. The phyla *Firmicutes* and *Bacteroidetes* dominate both the mouse and human gut from early life, with changes in the ratio between these two phyla occurring at different time points of life [66]. Our findings are consistent with these observations, and, interestingly, a low-iron diet in our study was associated with an increase in the relative abundance of both *Bacteroidetes* and *Firmicutes* in seven-week-old mice. This may be due to the fact that for many of these species, the heme requirement for survival is low due to their being able to produce energy by non-respiratory fermentative methods that do not rely on heme [67]. *Bifidobacterium* is also increased in the colon of mice fed a low-iron diet compared to those fed a pellet diet. There are several studies showing the health benefits of *Bifidobacterium*, with a high relative abundance in the gut reported during early infancy [68–71]. These early-life bacterial communities have been shown to decrease in abundance with age, and decreased *Bifidobacterium* is associated with asthma [72] and irritable bowel syndrome [73], and it may potentially contribute to dysbiosis and lung disease in CF [74]. Further research is needed to determine how constituents of the gut microbiota such as *Bifidobacterium* may interact with other species, such as *Prevotella* and *Bacteroides*, to influence disease pathogenesis and progression.

In the duodenal microbiota, iron supplementation in our study increased the relative abundance of the sulfate-reducing bacterium *Desulfovibrio*, consistent with a mouse model of Crohn's disease in which a low-iron diet was associated with a decreased *Desulfovibrio* abundance [75]. *Desulfovibrio* spp. are predominantly sulfate-reducing bacteria that produce the metabolite hydrogen sulfide, which has far-reaching effects on host health. *Desulfovibrio* spp. are one of the 20 most abundant genera in humans and mice; they are generally considered to be harmful in large amounts and are associated with a number of chronic diseases, such as ulcerative colitis, obesity and Parkinson's disease. However, they are also thought to provide some benefits to the host [76]. A handful of studies have shown in animal models that an increase in iron intake results in an increase in *Desulfovibrio;* however, it remains unclear why this occurs, and we cannot answer this question from our data [77,78]. Taken together with our results, these studies do, however, suggest that an increase in iron, irrespective of the source, may directly induce changes in the lung microbiome that contribute to disease pathogenesis but that also have several effects in distal sites due to the release of metabolites such as hydrogen sulfide. In contrast, we found that *Sutterella*, *Veillonella dispar* and *Prevotella melaninogenica* were increased in the duodenum in the low-iron-diet group. *Veillonella dispar* is normally isolated from the oral cavity and small intestine of humans [79], and it is known to possess a genome adapted to scavenging iron from its environment, while *P. melaninogenica* also has the capacity to scavenge iron through heme and hemoglobin uptake mechanisms. This allows it to survive in low-iron environments where only organic iron sources may be available [80].

Our observations that dietary iron changed the composition of the duodenal and colonic microbiota can be considered in light of the growing body of evidence that suggests the gut microbiome plays a role in the maintenance of human health [81–84]. The impact of dietary iron intake on the composition and behaviour of the gut microbiota may, therefore, have far-reaching effects on overall health. This needs further investigation, particularly given the widespread and sometimes indiscriminate use of iron supplements in developed countries and the high prevalence of iron deficiency as a component of malnutrition in many parts of the world. The putative "gut–lung axis" concept is also gaining substantial traction, and mouse studies have shown that changes in the composition of the gut microbiome modulate lung T cell population and activation profiles and, thus, may impact the host immunity and lung disease phenotype [10]. Although we did not assess downstream effects on host immunity, our demonstration of a significant perturbation of the dynamics of the gut microbiota depending on dietary iron may well have systemic implications, such as changes in the exacerbation of systemic and/or lung inflammation [85,86]. There are a number of pathways that may be important in how dietary iron may influence distal microbiota. As discussed above, studies have shown that diet (and other factors) can adversely affect gut microbe communities. This gut dysbiosis can directly disrupt epithelial tight junctions by altering junctional complexes to increase intestinal permeability [87–89] and also increase systemic exposure to PAMPs and bacterial metabolites that can induce a pro-inflammatory response, including increased T-helper cell (Th) type-1 and Th17 cells [90,91]. Alternatively, metabolites could also be transported from the intestinal lumen into the lamina propria and be transported to the lungs via the systemic circulation. Together, these potential mechanisms could explain how changes in nutrient intake could affect the lung microbiome.

To validate and expand on our data, future studies should include investigating the metabolite changes that occur as a result of dietary intervention as well as subsequent changes to inflammatory markers and immune cells. To further this work, future studies should also quantify iron levels and iron modulatory proteins in the lungs and gut across different timepoints. Our data capture a snapshot of the microbial communities after four weeks on the different diets; however, there may well be more significant changes that arise after several months on these diets. These data would inform how iron-induced changes to the microbiome affect the critical underlying processes in chronic airway diseases and, potentially, other distal organ diseases.

5. Conclusions

This proof-of-concept study demonstrates that dietary iron modulates the community composition of the gut microbiome and that systemic iron levels may influence the diversity of the lung microbiome in C57BL/6J mice. These findings may be of particular relevance in human diseases where an altered lung microbiome already exists and lung iron homeostasis may be intrinsically abnormal. The potential for inappropriate dietary iron supplementation to have adverse effects on the gut and lung microbiome in diseases such as CF and COPD needs to be considered and explored in further detail.

Supplementary Materials: The following supporting information can be downloaded at https://www.mdpi.com/article/10.3390/nu17162696/s1: Figure S1: Dietary iron intake changes the *Bacteroides* genus in the lung.

Author Contributions: Conceptualization, A.S., G.J.A. and D.W.R.; Data curation, A.-T.E., A.S. and R.W.; Formal analysis, A.-T.E., A.S. and R.W.; Funding acquisition, I.L.L. and D.W.R.; Investigation, R.W., P.M. and D.W.R.; Methodology, A.S., R.W., D.M.F. and G.J.A.; Resources, I.L.L., D.M.F. and D.W.R.; Supervision, D.W.R.; Validation, A.S. and R.W.; Writing—original draft, A.-T.E., A.S. and R.W.; Writing—review and editing, A.-T.E., A.S., P.M., I.L.L., D.M.F., G.J.A. and D.W.R. All authors have read and agreed to the published version of the manuscript.

Funding: This research was funded by an NHMRC Project grant (number APP1084130) and The Prince Charles Hospital Foundation Team Grant (TM2017-08).

Institutional Review Board Statement: The animal study protocol was approved by the Animal Ethics Committee of QIMR Berghofer (protocol code P1347, approved 13 October 2016).

Data Availability Statement: The original contributions presented in this study are included in the article/supplementary material. Further inquiries can be directed to the corresponding author.

Conflicts of Interest: The authors declare no conflicts of interest. The funders had no role in the design of the study; in the collection, analyses or interpretation of data; in the writing of the manuscript; or in the decision to publish the results.

References

1. Singh, R.K.; Chang, H.W.; Yan, D.; Lee, K.M.; Ucmak, D.; Wong, K.; Abrouk, M.; Farahnik, B.; Nakamura, M.; Zhu, T.H.; et al. Influence of diet on the gut microbiome and implications for human health. *J. Transl. Med.* **2017**, *15*, 73. [CrossRef] [PubMed]
2. David, L.A.; Maurice, C.F.; Carmody, R.N.; Gootenberg, D.B.; Button, J.E.; Wolfe, B.E.; Ling, A.V.; Devlin, A.S.; Varma, Y.; Fischbach, M.A.; et al. Diet rapidly and reproducibly alters the human gut microbiome. *Nature* **2014**, *505*, 559–563. [CrossRef] [PubMed]
3. Hou, K.; Wu, Z.-X.; Chen, X.-Y.; Wang, J.-Q.; Zhang, D.; Xiao, C.; Zhu, D.; Koya, J.B.; Wei, L.; Li, J.; et al. Microbiota in health and diseases. *Signal Transduct. Target. Ther.* **2022**, *7*, 135. [CrossRef] [PubMed]
4. Durack, J.; Lynch, S.V. The gut microbiome: Relationships with disease and opportunities for therapy. *J. Exp. Med.* **2019**, *216*, 20–40. [CrossRef]
5. Jaeggi, T.; Kortman, G.A.; Moretti, D.; Chassard, C.; Holding, P.; Dostal, A.; Boekhorst, J.; Timmerman, H.M.; Swinkels, D.W.; Tjalsma, H.; et al. Iron fortification adversely affects the gut microbiome, increases pathogen abundance and induces intestinal inflammation in Kenyan infants. *Gut* **2015**, *64*, 731–742. [CrossRef]
6. Zimmermann, M.B.; Chassard, C.; Rohner, F.; N'goran, E.K.; Nindjin, C.; Dostal, A.; Utzinger, J.; Ghattas, H.; Lacroix, C.; Hurrell, R.F. The effects of iron fortification on the gut microbiota in African children: A randomized controlled trial in Cote d'Ivoire. *Am. J. Clin. Nutr.* **2010**, *92*, 1406–1415. [CrossRef]
7. Tompkins, G.R.; O'dell, N.L.; Bryson, I.T.; Pennington, C.B. The effects of dietary ferric iron and iron deprivation on the bacterial composition of the mouse intestine. *Curr. Microbiol.* **2001**, *43*, 38–42. [CrossRef]

8. Dostal, A.; Chassard, C.; Hilty, F.M.; Zimmermann, M.B.; Jaeggi, T.; Rossi, S.; Lacroix, C. Iron depletion and repletion with ferrous sulfate or electrolytic iron modifies the composition and metabolic activity of the gut microbiota in rats. *J. Nutri.* **2012**, *142* 271–277. [CrossRef]
9. Zakrzewski, M.A.-O.; Wilkins, S.A.-O.; Helman, S.A.-O.; Brilli, E.A.-O.; Tarantino, G.A.-O.; Anderson, G.A.-O.; Frazer, D.A.-O Supplementation with Sucrosomial® iron leads to favourable changes in the intestinal microbiome when compared to ferrous sulfate in mice. *Biometals* **2022**, *35*, 27–38. [CrossRef]
10. Budden, K.F.; Gellatly, S.L.; Wood, D.L.A.; Cooper, M.A.; Morrison, M.; Hugenholtz, P.; Hansbro, P.M. Emerging pathogenic links between microbiota and the gut–lung axis. *Nat. Rev. Microbiol.* **2017**, *15*, 55–63. [CrossRef]
11. Marsland, B.J.; Trompette, A.; Gollwitzer, E.S. The Gut-Lung Axis in Respiratory Disease. *Ann. Am. Thorac. Soc.* **2015**, *12*, S150–S156. [CrossRef]
12. Trompette, A.; Gollwitzer, E.S.; Yadava, K.; Sichelstiel, A.K.; Sprenger, N.; Ngom-Bru, C.; Blanchard, C.; Junt, T.; Nicod, L.P.; Harris, N.L.; et al. Gut microbiota metabolism of dietary fiber influences allergic airway disease and hematopoiesis. *Nat. Med.* **2014**, *20*, 159–166. [CrossRef]
13. Reid, D.W.; Lam, Q.T.; Schneider, H.; Walters, E.H. Airway iron and iron-regulatory cytokines in cystic fibrosis. *Eur. Respir. J.* **2004**, *24*, 286–291. [CrossRef]
14. Ghio, A.J.; Roggli, V.L.; Soukup, J.M.; Richards, J.H.; Randell, S.H.; Muhlebach, M.S. Iron accumulates in the lavage and explanted lungs of cystic fibrosis patients. *J. Cyst. Fibros.* **2013**, *12*, 390–398. [CrossRef]
15. Ali, M.A.-O.; Kim, R.Y.; Brown, A.C.; Donovan, C.; Vanka, K.S.; Mayall, J.R.; Liu, G.A.-O.; Pillar, A.L.; Jones-Freeman, B.; Xenaki, D.; et al. Critical role for iron accumulation in the pathogenesis of fibrotic lung disease. *J. Pathol.* **2020**, *251*, 49–62. [CrossRef]
16. Ali, M.K.; Kim, R.Y.; Brown, A.C.; Mayall, J.R.; Karim, R.; Pinkerton, J.W.; Liu, G.A.-O.; Martin, K.L.; Starkey, M.R.; Pillar, A.L.; et al. Crucial role for lung iron level and regulation in the pathogenesis and severity of asthma. *Eur. Respir. J.* **2020**, *55*, 1901340. [CrossRef] [PubMed]
17. Baldi, A.; Braat, S.; Hasan, M.I.; Bennett, C.; Barrios, M.; Jones, N.; Abdul Azeez, I.; Wilcox, S.; Roy, P.K.; Bhuiyan, M.S.A.; et al. Effects of iron supplements and iron-containing micronutrient powders on the gut microbiome in Bangladeshi infants: A randomized controlled trial. *Nat. Commun.* **2024**, *15*, 8640. [CrossRef]
18. Celis, A.I.; Relman, D.A.; Huang, K.C. The impact of iron and heme availability on the healthy human gut microbiome in vivo and in vitro. *Cell Chem. Biol.* **2023**, *30*, 110–126. [CrossRef] [PubMed]
19. Coe, G.L.; Pinkham, N.V.; Celis, A.I.; Johnson, C.; DuBois, J.L.; Walk, S.A.-O. Dynamic Gut Microbiome Changes in Response to Low-Iron Challenge. *Appl. Environ. Microbiol.* **2021**, *87*, e02307-20. [CrossRef]
20. Elms, L.; Hand, B.; Skubisz, M.; Best, K.P.; Grzeskowiak, L.E.; Rogers, G.B.; Green, T.J.; Taylor, S.L. The Effect of Iron Supplements on the Gut Microbiome of Females of Reproductive Age: A Randomized Controlled Trial. *J. Nutr.* **2024**, *154*, 1582–1587. [CrossRef] [PubMed]
21. Kortman, G.A.M.; Reijnders, D.; Swinkels, D.A.-O. Oral iron supplementation: Potential implications for the gut microbiome and metabolome in patients with CKD. *Hemodial. Int.* **2017**, *21*, S28–S36. [CrossRef] [PubMed]
22. La Carpia, F.; Wojczyk, B.S.; Annavajhala, M.K.; Rebbaa, A.; Culp-Hill, R.; D'Alessandro, A.; Freedberg, D.E.; Uhlemann, A.C.; Hod, E.A. Transfusional iron overload and intravenous iron infusions modify the mouse gut microbiota similarly to dietary iron. *NPJ Biofilms Microbiomes* **2019**, *5*, 26. [CrossRef]
23. Liu, H.; Wu, W.; Luo, Y.A.-O. Oral and intravenous iron treatment alter the gut microbiome differentially in dialysis patients. *Int. Urol. Nephrol.* **2023**, *55*, 759–767. [CrossRef]
24. Paganini, D.; Zimmermann, M.B. The effects of iron fortification and supplementation on the gut microbiome and diarrhea in infants and children: A review. *Am. J. Clin. Nutr.* **2017**, *106*, 1688S–1693S. [CrossRef]
25. Parmanand, B.A.; Kellingray, L.; Le Gall, G.; Basit, A.W.; Fairweather-Tait, S.; Narbad, A. A decrease in iron availability to human gut microbiome reduces the growth of potentially pathogenic gut bacteria; an in vitro colonic fermentation study. *J. Nutr. Biochem.* **2019**, *67*, 20–27. [CrossRef]
26. Rusu, I.G.; Suharoschi, R.; Vodnar, D.C.; Pop, C.R.; Socaci, S.A.; Vulturar, R.; Istrati, M.; Moroșan, I.; Fărcaș, A.C.; Kerezsi, A.D.; et al. Iron Supplementation Influence on the Gut Microbiota and Probiotic Intake Effect in Iron Deficiency-A Literature-Based Review. *Nutrients* **2020**, *12*, 1993. [CrossRef]
27. Gifford, A.H.; Alexandru, D.M.; Li, Z.; Dorman, D.B.; Moulton, L.A.; Price, K.E.; Hampton, T.H.; Sogin, M.L.; Zuckerman, J.B.; Parker, H.W.; et al. Iron supplementation does not worsen respiratory health or alter the sputum microbiome in cystic fibrosis. *J. Cyst. Fibros.* **2014**, *13*, 311–318. [CrossRef]
28. Noordine, M.L.; Seyoum, Y.; Bruneau, A.; Baye, K.; Lefebvre, T.; Cherbuy, C.A.-O.; Canonne-Hergaux, F.; Nicolas, G.A.-O.; Humblot, C.A.-O.; Thomas, M.A.-O. The microbiota and the host organism switch between cooperation and competition based on dietary iron levels. *Gut Microbes* **2024**, *16*, 2361660. [CrossRef] [PubMed]

29. Healy, C.; Munoz-Wolf, N.; Strydom, J.; Faherty, L.; Williams, N.C.; Kenny, S.; Donnelly, S.C.; Cloonan, S.M. Nutritional immunity: The impact of metals on lung immune cells and the airway microbiome during chronic respiratory disease. *Respir. Res.* **2021**, *22*, 133. [CrossRef]
30. Bassis, C.M.; Erb-Downward, J.R.; Dickson, R.P.; Freeman, C.M.; Schmidt, T.M.; Young, V.B.; Beck, J.M.; Curtis, J.L.; Huffnagle, G.B. Analysis of the upper respiratory tract microbiotas as the source of the lung and gastric microbiotas in healthy individuals. *mBio* **2015**, *6*, e00037. [CrossRef] [PubMed]
31. Dickson, R.P.; Singer, B.H.; Newstead, M.W.; Falkowski, N.R.; Erb-Downward, J.R.; Standiford, T.J.; Huffnagle, G.B. Enrichment of the lung microbiome with gut bacteria in sepsis and the acute respiratory distress syndrome. *Nat. Microbiol.* **2016**, *1*, 16113. [CrossRef]
32. Eladham, M.W.; Selvakumar, B.; Saheb Sharif-Askari, N.; Saheb Sharif-Askari, F.; Ibrahim, S.M.; Halwani, R. Unraveling the gut-Lung axis: Exploring complex mechanisms in disease interplay. *Heliyon* **2024**, *10*, e24032. [CrossRef] [PubMed]
33. Frazer, D.M.; Wilkins, S.J.; Darshan, D.; Mirciov, C.S.G.; Dunn, L.A.; Anderson, G.J. Ferroportin Is Essential for Iron Absorption During Suckling, But Is Hyporesponsive to the Regulatory Hormone Hepcidin. *Cell. Mol. Gastroenterol. Hepatol.* **2016**, *3*, 410–421. [CrossRef]
34. Mirciov, C.S.G.; Wilkins, S.J.; Dunn, L.A.; Anderson, G.J.; Frazer, D.M. Characterization of Putative Erythroid Regulators of Hepcidin in Mouse Models of Anemia. *PLoS ONE* **2017**, *12*, e0171054. [CrossRef] [PubMed]
35. Dowd, S.E.; Sun, Y.; Wolcott, R.D.; Domingo, A.; Carroll, J.A. Bacterial tag-encoded FLX amplicon pyrosequencing (bTEFAP) for microbiome studies: Bacterial diversity in the ileum of newly weaned Salmonella-infected pigs. *Foodborne Pathog. Dis.* **2008**, *5*, 459–472. [CrossRef]
36. Caporaso, J.G.; Kuczynski, J.; Stombaugh, J.; Bittinger, K.; Bushman, F.D.; Costello, E.K.; Fierer, N.; Pena, A.G.; Goodrich, J.K.; Gordon, J.I.; et al. QIIME allows analysis of high-throughput community sequencing data. *Nat. Methods* **2010**, *7*, 335–336. [CrossRef]
37. Zakrzewski, M.; Proietti, C.; Ellis, J.J.; Hasan, S.; Brion, M.J.; Berger, B.; Krause, L. Calypso: A user-friendly web-server for mining and visualizing microbiome-environment interactions. *Bioinformatics* **2017**, *33*, 782–783. [CrossRef] [PubMed]
38. Yun, Y.; Srinivas, G.; Kuenzel, S.; Linnenbrink, M.; Alnahas, S.; Bruce, K.D.; Steinhoff, U.; Baines, J.F.; Schaible, U.E. Environmentally determined differences in the murine lung microbiota and their relation to alveolar architecture. *PLoS ONE* **2014**, *9*, e113466. [CrossRef]
39. Scheiermann, J.; Klinman, D.M. Three distinct pneumotypes characterize the microbiome of the lung in BALB/cJ mice. *PLoS ONE* **2017**, *12*, e0180561. [CrossRef]
40. Barfod, K.K.; Roggenbuck, M.; Hansen, L.H.; Schjorring, S.; Larsen, S.T.; Sorensen, S.J.; Krogfelt, K.A. The murine lung microbiome in relation to the intestinal and vaginal bacterial communities. *BMC Microbiol.* **2013**, *13*, 303. [CrossRef]
41. Wang, K.; Wu, W.; Jiang, X.; Xia, J.; Lv, L.; Li, S.; Zhuge, A.; Wu, Z.; Wang, Q.; Wang, S.; et al. Multi-Omics Analysis Reveals the Protection of Gasdermin D in Concanavalin A-Induced Autoimmune Hepatitis. *Microbiol. Spectr.* **2022**, *10*, e0171722. [CrossRef]
42. Luo, Y.; Zhou, S.; Zhang, X.; Lin, Y.; Liu, J.; Cheng, W.; Zeng, Y. The role of the microbiota and metabolites in the treatment of pulmonary fibrosis with UC-MSCs: Integrating fecal metabolomics and 16S rDNA analysis. *PLoS ONE* **2025**, *20*, e0313989. [CrossRef] [PubMed]
43. Hou, S.; Wang, X.; Guo, J.; Han, Y.; You, J.; Tian, Z.; Zheng, X.; Zheng, S.; Ling, Y.; Pei, L.; et al. Triangle correlations of lung microbiome, host physiology and gut microbiome in a rat model of idiopathic pulmonary fibrosis. *Sci. Rep.* **2024**, *14*, 28743. [CrossRef]
44. Sverrild, A.; Kiilerich, P.; Brejnrod, A.; Pedersen, R.; Porsbjerg, C.; Bergqvist, A.; Erjefalt, J.S.; Kristiansen, K.; Backer, V. Eosinophilic airway inflammation in asthmatic patients is associated with an altered airway microbiome. *J. Allergy Clin. Immunol.* **2017**, *140*, 407–417. [CrossRef] [PubMed]
45. Philippot, Q.; Deslée, G.; Adair-Kirk, T.L.; Woods, J.C.; Byers, D.; Conradi, S.; Dury, S.; Perotin, J.M.; Lebargy, F.; Cassan, C.; et al. Increased Iron Sequestration in Alveolar Macrophages in Chronic Obtructive Pulmonary Disease. *PLoS ONE* **2014**, *9*, e96285. [CrossRef] [PubMed]
46. Cloonan, S.M.; Mumby, S.; Adcock, I.M.; Choi, A.M.K.; Chung, K.F.; Quinlan, G.J. The "Iron"-y of Iron Overload and Iron Deficiency in Chronic Obstructive Pulmonary Disease. *Am. J. Respir. Crit. Care Med.* **2017**, *196*, 1103–1112. [CrossRef]
47. Maniam, P.A.-O.; Essilfie, A.T.; Kalimutho, M.; Ling, D.; Frazer, D.M.; Phipps, S.; Anderson, G.J.; Reid, D.W. Increased susceptibility of cystic fibrosis airway epithelial cells to ferroptosis. *Biol. Res.* **2021**, *54*, 38. [CrossRef]
48. Yoshida, M.; Minagawa, S.; Araya, J.; Sakamoto, T.; Hara, H.; Tsubouchi, K.; Hosaka, Y.; Ichikawa, A.; Saito, N.; Kadota, T.; et al. Involvement of cigarette smoke-induced epithelial cell ferroptosis in COPD pathogenesis. *Nat. Commun.* **2019**, *10*, 3145. [CrossRef]
49. Reid, D.W.; Carroll, V.; May, C.; Champion, A.; Kirov, S.M. Increased airway iron as a potential factor in the persistence of *Pseudomonas aeruginosa* infection in cystic fibrosis. *Eur. Respir. J.* **2007**, *30*, 286. [CrossRef]

50. Smith, D.J.; Lamont, I.L.; Anderson, G.J.; Reid, D.W. Targeting iron uptake to control *Pseudomonas aeruginosa* infections in cystic fibrosis. *Eur. Respir. J.* **2013**, *42*, 1723. [CrossRef]
51. Ghio, A.J.; Hilborn, E.D.; Stonehuerner, J.G.; Dailey, L.A.; Carter, J.D.; Richards, J.H.; Crissman, K.M.; Foronjy, R.F.; Uyeminami, D.L.; Pinkerton, K.E. Particulate matter in cigarette smoke alters iron homeostasis to produce a biological effect. *Am. J. Respir. Crit. Care Med.* **2008**, *178*, 1130–1138. [CrossRef]
52. Hoo, Z.H.; Wildman, M.J. Intravenous iron among cystic fibrosis patients. *J. Cyst. Fibros.* **2012**, *11*, 560–562. [CrossRef]
53. Norkina, O.; Burnett, T.G.; De Lisle, R.C. Bacterial overgrowth in the cystic fibrosis transmembrane conductance regulator null mouse small intestine. *Infect. Immun.* **2004**, *72*, 6040–6049. [CrossRef] [PubMed]
54. Burke, D.G.; Fouhy, F.; Harrison, M.J.; Rea, M.C.; Cotter, P.D.; O'Sullivan, O.; Stanton, C.; Hill, C.; Shanahan, F.; Plant, B.J.; et al. The altered gut microbiota in adults with cystic fibrosis. *BMC Microbiol.* **2017**, *17*, 58. [PubMed]
55. Wrigley-Carr, H.E.; van Dorst, J.M.; Ooi, C.Y. Intestinal dysbiosis and inflammation in cystic fibrosis impacts gut and multi-organ axes. *Med. Microecol.* **2022**, *13*, 100057. [CrossRef]
56. Hogan, D.A.; Willger, S.D.; Dolben, E.L.; Hampton, T.H.; Stanton, B.A.; Morrison, H.G.; Sogin, M.L.; Czum, J.; Ashare, A. Analysis of Lung Microbiota in Bronchoalveolar Lavage, Protected Brush and Sputum Samples from Subjects with Mild-To-Moderate Cystic Fibrosis Lung Disease. *PLoS ONE* **2016**, *11*, e0149998. [CrossRef]
57. Wang, Z.; Lai, Z.; Zhang, X.; Huang, P.; Xie, J.; Jiang, Q.; Zhang, Q.; Chung, K.F. Altered gut microbiome compositions are associated with the severity of asthma. *J. Thorac. Dis.* **2021**, *13*, 4322–4338. [CrossRef]
58. McCauley, K.; Durack, J.; Valladares, R.; Fadrosh, D.W.; Lin, D.L.; Calatroni, A.; LeBeau, P.K.; Tran, H.T.; Fujimura, K.E.; LaMere, B.; et al. Distinct nasal airway bacterial microbiotas differentially relate to exacerbation in pediatric patients with asthma. *J. Allergy Clin. Immunol.* **2019**, *144*, 1187–1197. [CrossRef] [PubMed]
59. Huang, Y.J.; Nelson, C.E.; Brodie, E.L.; Desantis, T.Z.; Baek, M.S.; Liu, J.; Woyke, T.; Allgaier, M.; Bristow, J.; Wiener-Kronish, J.P.; et al. Airway microbiota and bronchial hyperresponsiveness in patients with suboptimally controlled asthma. *J. Allergy Clin. Immunol.* **2011**, *127*, 372–381.e3. [CrossRef]
60. Bowerman, K.L.; Rehman, S.F.; Vaughan, A.; Lachner, N.; Budden, K.F.; Kim, R.Y.; Wood, D.L.A.; Gellatly, S.L.; Shukla, S.D.; Wood, L.G.; et al. Disease-associated gut microbiome and metabolome changes in patients with chronic obstructive pulmonary disease. *Nat. Commun.* **2020**, *11*, 5886. [CrossRef]
61. Einarsson, G.G.; Comer, D.M.; McIlreavey, L.; Parkhill, J.; Ennis, M.; Tunney, M.M.; Elborn, J.S. Community dynamics and the lower airway microbiota in stable chronic obstructive pulmonary disease, smokers and healthy non-smokers. *Thorax* **2016**, *71*, 795–803. [CrossRef]
62. Garcia-Nuñez, M.; Millares, L.; Pomares, X.; Ferrari, R.; Pérez-Brocal, V.; Gallego, M.; Espasa, M.; Moya, A.; Monsó, E. Severity-related changes of bronchial microbiome in chronic obstructive pulmonary disease. *J. Clin. Microbiol.* **2014**, *52*, 4217–4223. [CrossRef]
63. Millares, L.; Pascual, S.; Montón, C.; García-Núñez, M.; Lalmolda, C.; Faner, R.; Casadevall, C.; Setó, L.; Capilla, S.; Moreno, A.; et al. Relationship between the respiratory microbiome and the severity of airflow limitation, history of exacerbations and circulating eosinophils in COPD patients. *BMC Pulm. Med.* **2019**, *19*, 112. [CrossRef]
64. Turnbaugh, P.J.; Ley, R.E.; Hamady, M.; Fraser-Liggett, C.M.; Knight, R.; Gordon, J.I. The human microbiome project. *Nature* **2007**, *449*, 804–810. [CrossRef]
65. Ellermann, M.; Arthur, J.C. Siderophore-mediated iron acquisition and modulation of host-bacterial interactions. *Free Radic. Biol. Med.* **2017**, *105*, 68–78. [CrossRef]
66. Johansson, M.E.; Jakobsson, H.E.; Holmén-Larsson, J.; Schütte, A.; Ermund, A.; Rodríguez-Piñeiro, A.M.; Arike, L.; Wising, C.; Svensson, F.; Bäckhed, F.; et al. Normalization of Host Intestinal Mucus Layers Requires Long-Term Microbial Colonization. *Cell Host Microbe* **2015**, *18*, 582–592. [CrossRef] [PubMed]
67. Gruss, A.; Borezée-Durant, E.; Lechardeur, D. Environmental heme utilization by heme-auxotrophic bacteria. *Adv. Microb. Physiol.* **2012**, *61*, 69–124. [PubMed]
68. Arboleya, S.; Stanton, C.; Ryan, C.A.; Dempsey, E.; Ross, P.R. Bosom Buddies: The Symbiotic Relationship Between Infants and *Bifidobacterium longum* ssp. *longum* and ssp. *infantis*. Genetic and Probiotic Features. *Annu. Rev. Food Sci. Technol.* **2016**, *7*, 1–21. [CrossRef] [PubMed]
69. O'Callaghan, A.; van Sinderen, D. Bifidobacteria and Their Role as Members of the Human Gut Microbiota. *Front. Microbiol.* **2016**, *15*, 925. [CrossRef]
70. Tang, J.; Wei, Y.; Pi, C.; Zheng, W.; Zuo, Y.; Shi, P.; Chen, J.; Xiong, L.; Chen, T.; Liu, H.; et al. The therapeutic value of bifidobacteria in cardiovascular disease. *Npj Biofilms Microbiomes* **2023**, *9*, 82. [CrossRef]
71. Wong, C.B.; Odamaki, T.; Xiao, J.-z. Beneficial effects of *Bifidobacterium longum* subsp. *longum* BB536 on human health: Modulation of gut microbiome as the principal action. *J. Funct. Foods* **2019**, *54*, 506–519. [CrossRef]

72. Hevia, A.; Milani, C.; López, P.; Donado, C.D.; Cuervo, A.; González, S.; Suárez, A.; Turroni, F.; Gueimonde, M.; Ventura, M.; et al. Allergic Patients with Long-Term Asthma Display Low Levels of *Bifidobacterium adolescentis*. *PLoS ONE* **2016**, *11*, e0147809. [CrossRef] [PubMed]
73. Taverniti, V.; Guglielmetti, S. Methodological issues in the study of intestinal microbiota in irritable bowel syndrome. *World J. Gastroenterol. WJG* **2014**, *20*, 8821.
74. Duytschaever, G.; Huys, G.; Bekaert, M.; Boulanger, L.; De Boeck, K.; Vandamme, P. Dysbiosis of bifidobacteria and *Clostridium* cluster XIVa in the cystic fibrosis fecal microbiota. *J. Cyst. Fibros.* **2013**, *12*, 206–215. [CrossRef]
75. Werner, T.; Wagner, S.J.; Martinez, I.; Walter, J.; Chang, J.S.; Clavel, T.; Kisling, S.; Schuemann, K.; Haller, D. Depletion of luminal iron alters the gut microbiota and prevents Crohn's disease-like ileitis. *Gut* **2011**, *60*, 325–333. [CrossRef] [PubMed]
76. Zhou, H.; Huang, D.; Sun, Z.; Chen, X. Effects of intestinal *Desulfovibrio* bacteria on host health and its potential regulatory strategies: A review. *Microbiol. Res.* **2024**, *284*, 127725. [CrossRef]
77. Ding, H.; Yu, X.; Chen, L.; Han, J.; Zhao, Y.; Feng, J. Tolerable upper intake level of iron damages the intestine and alters the intestinal flora in weaned piglets. *Metallomics* **2020**, *12*, 1356–1369. [CrossRef] [PubMed]
78. Rieg, T.; Xue, J.; Stevens, M.; Thomas, L.; White, J.R.; Dominguez Rieg, J.A. Intravenous ferric carboxymaltose and ferric derisomaltose alter the intestinal microbiome in female iron deficient anemic mice. *Biosci. Rep.* **2023**, *43*, BSR20231217. [CrossRef]
79. van den Bogert, B.; Erkus, O.; Boekhorst, J.; de Goffau, M.; Smid, E.J.; Zoetendal, E.G.; Kleerebezem, M. Diversity of human small intestinal *Streptococcus* and *Veillonella* populations. *FEMS Microbiol. Ecol.* **2013**, *85*, 376–388. [CrossRef]
80. Council, S.E. *Prevotella melaninogenica, an Oral Anaerobic Bacterium, Prevalent in Cystic Fibrosis Chronic Lung Infection*; University of North Carolina at Chapel Hill: Chapel Hill, NC, USA, 2013.
81. Althani, A.A.; Marei, H.E.; Hamdi, W.S.; Nasrallah, G.K.; El Zowalaty, M.E.; Al Khodor, S.; Al-Asmakh, M.; Abdel-Aziz, H.; Cenciarelli, C. Human Microbiome and its Association With Health and Diseases. *J. Cell Physiol.* **2016**, *231*, 1688–1694. [CrossRef]
82. Clemente, J.C.; Ursell, L.K.; Parfrey, L.W.; Knight, R. The impact of the gut microbiota on human health: An integrative view. *Cell* **2012**, *148*, 1258–1270. [CrossRef]
83. Round, J.L.; Mazmanian, S.K. The gut microbiota shapes intestinal immune responses during health and disease. *Nat. Rev. Immunol.* **2009**, *9*, 313–323. [CrossRef] [PubMed]
84. Baumgart, D.C.; Carding, S.R. Inflammatory bowel disease: Cause and immunobiology. *Lancet* **2007**, *369*, 1627–1640. [CrossRef] [PubMed]
85. Tremaroli, V.; Backhed, F. Functional interactions between the gut microbiota and host metabolism. *Nature* **2012**, *489*, 242–249. [CrossRef]
86. Vijay-Kumar, M.; Aitken, J.D.; Carvalho, F.A.; Cullender, T.C.; Mwangi, S.; Srinivasan, S.; Sitaraman, S.V.; Knight, R.; Ley, R.E.; Gewirtz, A.T. Metabolic syndrome and altered gut microbiota in mice lacking Toll-like receptor 5. *Science* **2010**, *328*, 228–231. [CrossRef]
87. Comini, L.; Pasini, E.; Porta, R.; Olivares, A.; Testa, C.; Scalvini, S.; Vitacca, M. Dysbiosis and leaky gut in hyper-inflated COPD patients: Have smoking and exercise training any role? *Respir. Med. Res.* **2023**, *83*, 100995. [CrossRef]
88. Stolfi, C.; Maresca, C.; Monteleone, G.; Laudisi, F. Implication of Intestinal Barrier Dysfunction in Gut Dysbiosis and Diseases. *Biomedicines* **2022**, *10*, 289. [CrossRef]
89. Zhang, Y.G.; Wu, S.; Xia, Y.; Sun, J. Salmonella infection upregulates the leaky protein claudin-2 in intestinal epithelial cells. *PLoS ONE* **2013**, *8*, e58606. [CrossRef] [PubMed]
90. Candelli, M.; Franza, L.; Pignataro, G.; Ojetti, V.; Covino, M.; Piccioni, A.; Gasbarrini, A.; Franceschi, F. Interaction between Lipopolysaccharide and Gut Microbiota in Inflammatory Bowel Diseases. *Int. J. Mol. Sci.* **2021**, *22*, 6242. [CrossRef]
91. Matsuoka, K.; Kanai, T. The gut microbiota and inflammatory bowel disease. *Semin. Immunopathol.* **2015**, *37*, 47–55. [CrossRef]

Disclaimer/Publisher's Note: The statements, opinions and data contained in all publications are solely those of the individual author(s) and contributor(s) and not of MDPI and/or the editor(s). MDPI and/or the editor(s) disclaim responsibility for any injury to people or property resulting from any ideas, methods, instructions or products referred to in the content.

nutrients

MDPI

Article

Association of Dietary Sodium-to-Potassium Ratio with Nutritional Composition, Micronutrient Intake, and Diet Quality in Brazilian Industrial Workers

Anissa Melo Souza [1], Ingrid Wilza Leal Bezerra [1,2,*], Karina Gomes Torres [1], Gabriela Santana Pereira [1], Raiane Medeiros Costa [1] and Antonio Gouveia Oliveira [1,3]

1 Postgraduate Program in Nutrition, Health Sciences Center, Federal University of Rio Grande do Norte, Natal 59078-900, RN, Brazil; anissa.souza.094@ufrn.edu.br (A.M.S.); karina.torres@ufrn.br (K.G.T.); gabriela.pereira.016@ufrn.edu.br (G.S.P.); raiane.costa.092@ufrn.edu.br (R.M.C.); antonio.gouveia@ufrn.br (A.G.O.)

2 Nutrition Department, Health Sciences Center, Federal University of Rio Grande do Norte, Natal 59078-900, RN, Brazil

3 Pharmacy Department, Health Sciences Center, Federal University of Rio Grande do Norte, Natal 59078-900, RN, Brazil

* Correspondence: ingrid.bezerra@ufrn.br

Abstract: Introduction: The sodium-to-potassium (Na:K) ratio in the diet is a critical biomarker for cardiovascular and metabolic health, yet global adherence to recommended levels remains poor. Objectives: The objective of this study was to identify dietary determinants of the dietary Na:K ratio and its associations with micronutrient intake and diet quality. Methods: An observational cross-sectional survey was conducted in a representative sample of manufacturing workers through a combined stratified proportional and two-stage probability sampling plan, with strata defined by company size and industrial sector from the state of Rio Grande do Norte, Brazil. Dietary intake was assessed using 24 h recalls via the Multiple Pass Method, with Na:K ratios calculated from quantified food composition data. Diet quality was assessed with the Diet Quality Index-International (DQI-I). Multiple linear regression was used to analyze associations of Na:K ratio with the study variables. Results: The survey was conducted in the state of Rio Grande do Norte, Brazil, in 921 randomly selected manufacturing workers. The sample mean age was 38.2 ± 10.7 years, 55.9% males, mean BMI 27.2 ± 4.80 kg/m^2. The mean Na:K ratio was 1.97 ± 0.86, with only 0.54% of participants meeting the WHO recommended target (<0.57). Fast food (+3.29 mg/mg per serving, $p < 0.001$), rice, bread, and red meat significantly increased the ratio, while fruits (−0.16 mg/mg), dairy, white meat, and coffee were protective. Higher Na:K ratios were associated with lower intake of calcium, magnesium, phosphorus, and vitamins C, D, and E, as well as poorer diet quality (DQI-I score: −0.026 per 1 mg/mg increase, $p < 0.001$). Conclusions: These findings highlight the critical role of processed foods in elevating Na:K ratios and the potential for dietary modifications to improve both electrolyte balance and micronutrient adequacy in industrial workers. The study underscores the need for workplace interventions that simultaneously address sodium reduction, potassium enhancement, and overall diet quality improvement tailored to socioeconomic and cultural contexts, a triple approach not previously tested in intervention studies. Future studies should further investigate nutritional consequences of imbalanced Na:K intake.

Keywords: sodium-to-potassium ratio; 24 h dietary recall; diet quality; food groups; food serving sizes; micronutrient; manufacturing workers

1. Introduction

Modern dietary patterns, increasingly characterized by the overconsumption of ultra-processed foods, refined sugars, and unhealthy fats, have been strongly linked to adverse health outcomes [1]. Excessive intake of added sugars, particularly fructose and high-fructose corn syrup, contributes to metabolic dysregulation, promoting insulin resistance, non-alcoholic fatty liver disease (NAFLD), and type 2 diabetes [2]. Diets high in trans and saturated fats elevate low-density lipoprotein (LDL) cholesterol, accelerating atherosclerosis and cardiovascular disease [3]. Sodium consumption, primarily from processed and packaged foods, exacerbates hypertension [4], while insufficient dietary fiber disrupts gut microbiota diversity and impairs metabolic and immune function [5]. Emerging evidence underscores the deleterious effects of artificial additives [6], excessive alcohol consumption [7], and imbalanced omega-6 to omega-3 fatty acid ratios [8], which collectively foster chronic low-grade inflammation.

The sodium–potassium (Na:K) ratio in the human diet has emerged as a critical biomarker for assessing cardiovascular and metabolic health [9]. Sodium (Na) and potassium (K) are essential electrolytes that play interdependent roles in maintaining physiological homeostasis, including fluid balance, nerve transmission, and muscle function [10]. However, modern dietary patterns, characterized by excessive sodium intake and inadequate potassium consumption, have disrupted this balance, elevating the Na:K ratio to levels associated with increased risks of hypertension, cardiovascular disease (CVD), stroke, and chronic kidney disease (CKD). The World Health Organization (WHO) recommends a daily sodium intake of less than 2000 mg (5 g of salt) and a potassium intake of at least 3500 mg, translating to a Na:K ratio of approximately 0.57 [11]. Yet, global adherence to these guidelines remains poor, with populations often consuming ratios exceeding 1.5, driven by processed and ultra-processed foods and insufficient fruit and vegetable intake [12]. This deviation underscores the urgent need to understand dietary contributors to the Na:K ratio, particularly in occupational groups such as manufacturing workers, whose demanding schedules and limited access to healthy foods may exacerbate imbalances.

Sodium and potassium exert opposing effects on blood pressure regulation. Sodium promotes fluid retention and vascular resistance, while potassium enhances vasodilation and sodium excretion via renal mechanisms [13]. Epidemiological studies consistently demonstrate that a high Na:K ratio is a stronger predictor of hypertension than either nutrient alone [14]. For instance, the INTERSALT study, involving over 10,000 participants across 32 countries, revealed that populations with higher urinary Na:K ratios had significantly elevated systolic blood pressure [15]. Similarly, a meta-analysis by Ma et al. [16] found that every unit increase in the Na:K ratio was associated with a 24% greater risk of CVD mortality. These findings highlight the ratio's clinical relevance, positioning it as a target for dietary interventions.

Despite clear guidelines, global sodium consumption averages 3950 mg/day, nearly double the recommended limit, while potassium intake lags at 2300 mg/day, well below the 3500 mg threshold [17]. This imbalance is particularly pronounced in industrialized nations, where processed and convenience foods, which are major sodium sources, dominate diets. Conversely, potassium-rich foods like fruits, vegetables, legumes, and dairy are under consumed, often due to cost, accessibility, or cultural preferences. The resultant elevated Na:K ratios are linked not only to cardiovascular outcomes but also to poorer diet quality. Diets high in sodium often correlate with low micronutrient density, as processed foods typically lack essential vitamins and minerals [18]. This duality suggests that addressing the Na:K ratio could simultaneously improve both electrolyte balance and overall nutritional adequacy.

Food groups contributing to the Na:K ratio have been explored in diverse populations. Processed meats, canned goods, bread, and savory snacks are primary sodium sources, while leafy greens, bananas, potatoes, and beans are key potassium contributors [19]. However, dietary patterns vary by socioeconomic and occupational factors. For example, shift workers, including manufacturing employees, often rely on ready-to-eat meals and vending machine snacks due to time constraints and limited cafeteria options, potentially exacerbating sodium intake [20]. A study by Leone et al. [21] identified that low-income populations, akin to many manufacturing workers, consume fewer fresh product items due to cost barriers, further skewing their Na:K ratios. Yet, few studies have specifically examined food group associations with the Na:K ratio in blue-collar workers, a gap this study aims to address.

Existing research on dietary determinants of the Na:K ratio offers mixed insights. A cross-sectional analysis of NHANES data found that 40% of sodium consumed came from the top 10 food categories, which included prepared foods with sodium added, such as deli meat sandwiches and pizza, and 43% of potassium consumed was from other 10 food categories, which included foods naturally low in sodium, like unflavored milk, fruit, vegetables, and prepared foods [22]. Conversely, a Japanese cohort study emphasized soups and sauces as major sodium sources, with potassium intake heavily dependent on vegetables and fruits, fish, and milk [23]. These discrepancies underscore the influence of cultural and regional dietary habits, suggesting that interventions must be context specific. Furthermore, while some studies correlate high Na:K ratios with lower intakes of calcium, magnesium, and vitamin C, nutrients abundant in fruits and vegetables, the relationship between Na:K ratios and micronutrient consumption remains underexplored. Understanding these associations could inform multifaceted dietary guidelines that target both electrolyte balance and micronutrient sufficiency.

The interaction between a high dietary Na:K ratio and poor diet quality contributes to adverse health outcomes through synergistic pathways involving electrolyte imbalance, chronic inflammation, and metabolic dysregulation. Elevated Na:K ratios impair renal sodium excretion and reduce potassium-mediated vasodilation, increasing blood pressure and arterial stiffness [24]. Poor diet quality further aggravates this by promoting oxidative stress and low-grade inflammation, key contributors to atherosclerosis [25]. Epidemiological studies demonstrate that populations with high Na:K ratios and low diet quality scores exhibit higher cardiovascular mortality, with sodium overload potentiating endothelial damage while potassium deficiency exacerbates cardiac strain [15,26]. Additionally, diets low in magnesium and calcium, common in poor-quality diets, worsen sodium retention, further increasing hypertension and chronic kidney disease risk [27]. High Na:K ratios correlate with insulin resistance and dyslipidemia, particularly when diets lack fiber and antioxidants [28]. Ultra-processed foods often displace whole foods rich in polyphenols and short-chain fatty acids, disrupting gut microbiota composition and increasing systemic inflammation [29]. This microbiota imbalance may further impair electrolyte regulation, creating a vicious cycle of metabolic dysfunction [30].

This study investigated the dietary determinants of the Na:K ratio among manufacturing workers and explored how this ratio relates to micronutrient consumption. Beyond a simple nutrient analysis, the study investigated the contribution of traditional food groups to the Na:K ratio in the diet, estimating the net contribution per serving. In addition, the study investigated whether there is an association of the Na:K ratio with diet quality. Manufacturing workers may be particularly at risk of electrolyte imbalance due to occupational stressors that may compromise dietary habits. Long work shifts, physical labor, and irregular mealtimes often lead to reliance on convenience foods, which are typically energy-dense but nutrient-poor [28]. Such patterns likely elevate Na:K ratios while depriving individuals of essential micronutrients. However, no study has concurrently examined food group contributions to the Na:K ratio and its relationship with micronutrient intake in this population. This dual focus is critical, as sodium reduction and potassium enhancement strategies must not inadvertently compromise overall dietary quality. While previous studies have established the association between dietary Na:K ratio and cardiovascular risk factors, several critical gaps remain unaddressed. By focusing specifically on Brazilian industrial workers, we provide the first comprehensive analysis of how workplace-related dietary patterns influence Na:K ratios. Unlike previous studies that examined Na:K ratios in isolation, we employ the Diet Quality Index-International (DQI-I) to simultaneously evaluate how electrolyte imbalance interacts with overall diet quality, which contributed to elucidating whether high Na:K ratios serve as a marker for broader nutritional deficiencies. Moving beyond nutrient-level analyses, we examined traditional food groups to identify specific groups (e.g., fast food, dairy, fruits) that most significantly impact Na:K ratios in this population.

By identifying specific food groups associated with elevated ratios, the study aims to inform targeted interventions tailored to this population's unique challenges. Additionally, elucidating links between the Na:K ratio and micronutrients may reveal synergies for public health strategies that address both electrolyte imbalance and nutritional deficiencies. Given the paucity of data on occupational dietary patterns, this research contributes novel insights into how workplace environments shape nutritional health, offering a foundation for policies that promote accessible, balanced diets in industrial settings.

2. Materials and Methods

2.1. Survey Design

This was an observational, cross-sectional study based on a representative probability sample of manufacturing workers in the state of Rio Grande do Norte, Brazil. The manufacturing industry stands out as one of the most important segments for the Brazilian economy. According to data from the Ministry of Development, Industry, Trade, and Services, it represents 11.3% of Brazil's GDP and, in 2024, recorded record exports of US$181.9 billion, justifying the interest of this study in this important market segment. The sampling strategy employed a stratified proportional two-stage design, with stratification based on company size (categorized as small for firms with fewer than 50 workers, medium for 50 to 500 workers, and large for more than 500 workers) and sector of activity (food and beverages, non-metallic minerals, and textiles), which represented the most prominent industries in the region.

Ethical approval was obtained from the Research Ethics Committee of the Hospital Universitário Onofre Lopes (CAAE 2.198.545/2017), and the study adhered to the principles of the Declaration of Helsinki. All participants provided written informed consent prior to data collection.

In the first sampling stage, manufacturing companies were selected from each stratum through simple random sampling, with the number of selected companies proportional to their representation in the state's industrial landscape. The sampling frame with the identification of all companies in the state from each of the three defined sectors of activity was provided by the Federation of Industries of the State of Rio Grande do Norte (FIERN). In the second stage, a predetermined number of workers were randomly selected from each participating company using computer-generated random numbers, drawn from employee lists supplied by human resources departments.

Eligibility for the study required companies from the three specified sectors to provide written consent to participate. Workers included in the study were male and female workers aged 18 years or older with permanent employment contracts, while exclusions applied to pregnant women, temporary workers, interns, and employees in probationary periods.

2.2. Data Collection

Fieldwork was conducted through scheduled visits by the research team between Tuesdays and Saturdays. Randomly selected workers were approached during lunch breaks and invited to participate. Following consent, researchers collected biodemographic information, including age, sex, marital status, education level, and income. Anthropometric measurements included body mass index (BMI), calculated from weight and height measurements taken with a digital scale (Inner Scan, Tanita Corp., Tokyo, Japan) and a stadiometer (Sanny, São Bernardo do Campo, SP, Brazil), respectively, and waist circumference (WC), measured at the midpoint between the lower rib and iliac crest using a measuring tape (Cescorf, Porto Alegre, RS, Brazil). All procedures followed the guidelines of the Brazilian Food and Nutrition Surveillance System–SISVAN [31].

Dietary intake was assessed using the 24 h dietary recall (24HR) method, administered through the Multiple Pass Method (MPM) to enhance recall accuracy and completeness [32]. The 24HR is a widely employed method in nutritional epidemiology for estimating individual food and nutrient intakes. Participants were interviewed by trained personnel to recall all foods and beverages consumed during the previous 24 h period, including detailed descriptions, preparation methods, and quantities. The MPM structures the interview into distinct passes, starting with a quick list of consumed items, followed by detailed probing for forgotten foods (e.g., beverages, snacks, condiments), then a collection of detailed descriptions and portion sizes using household measures, and finally, a review of the entire recall for verification and additions.

Diet quality was assessed with the Diet Quality Index-International (DQI-I), a comprehensive tool designed to evaluate overall diet quality across diverse populations, facilitating cross-national comparisons [33]. Developed to address the limitations of region-specific indices, the DQI-I incorporates four key dimensions (variety, adequacy, moderation, and overall balance in nutrient intake). The index assesses dietary patterns by scoring food group diversity (food consumption across major categories), adequacy of nutrient-rich foods (such as fruits, vegetables, and whole grains), moderation of less desirable components (e.g., saturated fats, cholesterol, and empty calories), and macronutrient balance (evaluating the proportion of energy derived from carbohydrates, fats, and proteins). The scale total score ranges from 0 to 100 points, the components score from 0 to 20, 40, 30, and 10 points, respectively, with higher scores indicating better diet quality.

2.3. Nutritional Analysis

Food consumption data from the 24HR were quantified by weight and volume using standardized references, including direct weighing, photographic records, and technical manuals [34,35]. Nutritional analysis relied on Food Composition Tables [36–38] supplemented with information from food labels when necessary.

2.4. Nutritional Composition

Foods were classified according to the Brazilian Food Guide's grouping system, which categorizes foods by biological, nutritional, and culinary characteristics. These groups were further subdivided into 44 subgroups to accommodate the diversity of reported foods and preparations. Mixed dishes were disaggregated into individual ingredients using Technical Preparation Sheets, while traditional Brazilian mixed foods, primarily composed of fresh or minimally processed ingredients, were classified according to their dominant component. For example, the dish *baião-de-dois*, typically prepared with a mixture of rice and beans, but containing a higher proportion of rice according to traditional culinary recipes, was classified in the rice group. Similarly, traditional corn-based dishes such as *canjica*, which includes culinary ingredients such as sugar and coconut milk, were classified in the corn group [39]. The serving size estimates established for each food group were based on specific bibliographic references, which use protocols related to food consumption, such as directly weighing foods to standardize their size (small, medium, large) and/or defining them in household measurements (one cup, one tablespoon, one glass, one ladle, one plate, one handful, etc.) [35,40].

2.5. Statistical Analysis

For sample size calculations, the sampling plan aimed to select 30 companies for the first stage and 30 workers sampled within each company, to achieve a total sample of 900 participants, according to the widely used 30-by-30 two-stage cluster sampled design method in population surveys.

Descriptive data are presented as mean ± standard deviation or n (%). Univariate and multivariate linear regression was used to investigate associations of Na:K ratio with food groups, micronutrient intake, and diet quality, with results presented as partial regression coefficients with 95% confidence intervals (CI). Statistical significance was assumed for $p < 0.05$. Stata 15 (Stata Corp., College Station, TX, USA) was used for statistical analysis.

3. Results

Between September 2017 and July 2018, the study enrolled 921 workers from 33 manufacturing companies across three key industrial sectors: food and beverages (14 companies), non-metallic minerals (6 companies), and textiles (13 companies). The participating companies represented a range of sizes, with 13 classified as small, 14 as medium, and 6 as large companies. Notably, the study achieved complete participation with no refusals and no missing or incomplete data, ensuring a robust dataset. Each worker interview was conducted efficiently, averaging less than 15 min in duration.

The participant demographic profile (Table 1) revealed a mean age of 38.2 ± 10.7 years, with a slight male predominance (55.9%). Most workers (62.7%) reported living with a partner, while the average income stood at 1.46 times Brazil's minimum wage. Anthropometric measurements showed an average body mass index (BMI) of 27.2, with weight distribution categories indicating 34.4% of participants at normal weight, 39.0% overweight, and 25.4% meeting criteria for obesity.

Table 1. Characteristics of the study population.

Variable	Values
Age, years	38.2 ± 10.7
Sex, male	515 (55.9%)
Married/Cohabiting	577 (62.7%)
Education > middle school	527 (57.3%)
Income, minimum wages	1.46 ± 1.61
Body Mass Index, kg/m^2	27.2 ± 4.80
Waist Circumference, cm	89.7 ± 11.9
Underweight	12 (1.30%)
Normal weight	316 (34.4%)
Overweight	359 (39.0%)
Obesity I	178 (19.4%)
Obesity II	40 (4.35%)
Obesity III	15 (1.63%)
Na:K ratio (mg/mg)	1.97 ± 0.86
Na:K ratio > 0.6 mg/mg	5 (0.54%)
Na:K ratio > 1.0 mg/mg	48 (5.21%)

The mean Na:K ratio was 1.97 ± 0.86, with only 0.54% (95% confidence interval (CI) 0.17–1.26%) below the WHO recommended value, and only 5.21% (95% CI 3.87–6.85%) with a Na:K ratio below 1.0.

Table 2 shows which food groups contribute to an increase or decrease in the daily Na:K ratio. Multivariate analysis shows those food groups independently associated with the Na:K ratio. The Na:K ratio decreases 0.16 mg/mg ($p < 0.001$) for each serving of fruit in the diet, 0.04 mg/mg for each serving of dairy products ($p = 0.002$), 0.13 for each serving of white meat ($p < 0.001$), and 0.21 for each serving of coffee ($p < 0.001$). Food groups in the diet that increase the Na:K ratio include rice (0.21, $p < 0.001$), corn (0.12, $p < 0.001$), bread (0.08, $p < 0.001$), pasta (0.06, $p = 0.10$), soup (0.11, $p < 0.001$), red meat (0.12, $p < 0.001$), vegetable oil (0.07, $p = 0.047$) and, above all, fast food (3.29, $p < 0.001$).

Table 2. Changes in Na:K ratios per one serving of each food group in the diet.

	Univariate Analysis				Multivariate Analysis			
Food Group	**Coefficient**	**95% CI**		***p*-Value**	**Coefficient**	**95% CI**		***p*-Value**
Increase Na:K Ratio								
Beans	0.050	0.018	0.082	0.002				
Rice	0.210	0.144	0.276	0.000	0.205	0.151	0.258	<0.001
Corn	0.092	0.043	0.140	0.000	0.115	0.076	0.154	<0.001
Bread	0.086	0.048	0.124	0.000	0.077	0.046	0.107	<0.001
Pasta	0.099	0.051	0.148	0.000	0.057	0.014	0.100	0.010
Soups	0.114	0.096	0.131	0.000	0.109	0.094	0.123	<0.001
Red meat	0.189	0.145	0.232	0.000	0.116	0.077	0.156	<0.001
Processed meat	0.162	0.082	0.243	0.000				
Fast food	3.332	1.644	5.020	0.000	3.294	1.943	4.645	<0.001
Vegetable oil	0.102	0.036	0.168	0.002	0.065	0.001	0.130	0.047
Salt	0.036	0.021	0.052	0.000				
Decrease Na:K Ratio								
Grains, integral	−0.153	−0.247	−0.058	0.002				
Tubercles	−0.105	−0.143	−0.067	0.000				
Fruits	−0.204	−0.254	−0.154	0.000	−0.158	−0.198	−0.117	<0.001
Dairy	−0.056	−0.090	−0.022	0.001	−0.042	−0.069	−0.015	0.002
White meat	−0.151	−0.206	−0.097	0.000	−0.129	−0.179	−0.079	<0.001
Coffee	−0.164	−0.222	−0.106	0.000	−0.208	−0.255	−0.161	<0.001
Vitamin supplements	−0.205	−0.375	−0.036	0.018				

The Na:K ratio of diet is associated with the intake of micronutrients and vitamins. Table 3 shows the average increase or decrease in micronutrient intake per 1 mg/mg increase in Na:K ratio in the diet. Diets with increasing Na:K ratio are associated with decreased intake of several elements (calcium, magnesium, and phosphorus) and increased intake of iron. In the case of vitamins, the increased Na:K ratio in the diet is associated with decreased intake of niacin and vitamins C, D, and E.

Increased Na:K ratio in the diet is also associated with diet quality. Table 4 shows the difference in DQI-I score per 1 mg/mg increase in Na:K ratio in the diet. Total score decreases an average of 0.026 points per unit increase in Na:K ratio ($p < 0.001$), reflecting worse quality of diets with higher Na:K ratios. All DQI-I subscales are also negatively associated with the diet Na:K ratio, except for Overall Balance probably because this dimension primarily assesses macronutrient and fatty acid distribution, without accounting for micronutrient intake or electrolyte ratios.

Table 3. Association of Na:K ratio in the diet with intake of micronutrients.

Micronutrient	Coeficient	95% CI		p-Value
Calcium	−45.774	−76.059	−15.489	0.003
Copper	0.000	−0.209	0.210	0.99
Iron	1.813	1.151	2.476	<0.001
Magnesium	−13.451	−21.820	−5.081	0.002
Manganese	0.070	−0.274	0.413	0.69
Phosphorus	−58.971	−99.339	−18.603	0.004
Selenium	−3.014	−7.212	1.184	0.16
Zinc	1.069	−0.094	2.232	0.07
Vitamin A	−55.447	−267.550	156.656	0.61
Vitamin C	−28.809	−46.282	−11.336	0.001
Vitamin D	−0.379	−0.679	−0.078	0.013
Vitamin E	−0.691	−1.032	−0.350	<0.001
Cobalamin	−0.942	−2.144	0.260	0.13
Niacin	−1.262	−2.211	−0.313	0.009
Pyridoxine	0.011	−0.055	0.078	0.74
Riboflavin	0.047	−0.034	0.129	0.25
Thiamine	0.044	−0.010	0.098	0.11

Table 4. Change in DQI-I score and its components per 1 mg/mg increase in dietary Na:K ratio.

DQI-I Score	Coefficient	95% CI		p-Value
Total score (0–100)	−0.026	−0.032	−0.020	<0.001
Variety (0–20)	−0.046	−0.066	−0.026	<0.001
Adequacy (0–40)	−0.022	−0.034	−0.010	<0.001
Moderation (0–30)	−0.032	−0.040	−0.024	<0.001
Overall Balance (0–10)	0.018	−0.009	0.045	0.19

The range of scores for the DQI-I and each component is shown in brackets, higher scores indicating better diet quality.

4. Discussion

This study provides critical insights into the dietary determinants of the sodium-to-potassium (Na:K) ratio and its broader nutritional implications among manufacturing workers, a population group of interest given the importance of the workforce to a country's economy, with the distinct aspect of assessing the association of the Na:K ratio from a food perspective, based on the servings consumed, which allowed both the analysis of food consumption beyond nutrients and the association to diet quality using an international quality indicator.

4.1. Micronutrient Intake and the Relationship Between Electrolyte Balance and Diet Quality

The mean Na:K ratio observed in this population far exceeds the WHO-recommended threshold of 0.57, with less than 1% of participants meeting this guideline. This disparity mirrors global trends of excessive sodium intake and insufficient potassium consumption, particularly in low- and middle-income settings where processed foods dominate diets [11,17,26]. Notably, the Na:K ratio's association with specific food groups underscores the dual burden of industrialization and occupational constraints on dietary habits. Fast food emerged as the strongest association to elevated Na:K ratios, a finding consistent with studies linking ultra-processed foods to poor electrolyte balance [22,29,30]. Conversely, fruits, dairy, white meat, and coffee were protective, aligning with evidence that potassium-rich foods mitigate sodium's adverse effects [24]. These results highlight the need to contextualize dietary guidelines for blue-collar workers, whose reliance on convenient, energy-dense meals exacerbates nutritional imbalances [41].

The inverse relationship between the Na:K ratio and micronutrient adequacy further underscores the interconnectedness of dietary quality and electrolyte balance. Elevated ratios were associated with reduced intake of calcium, magnesium, and vitamins C, D, and E, which are nutrients abundant in fruits, vegetables, and minimally processed foods. This aligns with findings that sodium-dense diets often displace micronutrient-rich options, compounding cardiovascular risks [25]. The positive association with iron intake may reflect the prevalence of fortified processed foods or red meat consumption, which, while iron-rich, also contribute to sodium overload. Such trade-offs emphasize the complexity of addressing nutritional deficiencies in populations with limited access to fresh produce.

The observed decline in diet quality, as measured by the DQI-I scale, reinforces the Na:K ratio's utility as a proxy for overall dietary patterns. An increase in the Na:K ratio is associated with a reduction in total DQI-I score, driven by poorer variety, adequacy, and moderation, a pattern consistent with studies linking high sodium intake to monotonous, nutrient-poor diets [42].

Numerous studies have investigated the association of high Na:K ratios with arterial hypertension and cardiovascular diseases, but there is limited information on the association of the Na:K ratio with food composition, micronutrient intake, and diet quality. A study based on the Irish Adult Nutrition Survey of a nationally representative sample of the adult population, with sodium and potassium urinary excretion measured with spot urine samples, also investigated foods associated with Na:K ratios with findings similar to those of the present study: fruits, vegetables, potatoes, breakfast cereals, milk, yogurt, and fresh meat contribute to a higher Na:K ratio, while the foods negatively associated with a lower Na:K were breads, cured and processed meats, and butter and fat spreads [43]. Another study conducted in healthy pregnant women also investigated the association of food groups to Na:K ratios, again with findings similar to those in this study, in which soups and sauces, cereals and cereal products, and fats and oils were associated with increased Na:K ratios, while fruit and non-alcoholic beverages were associated with decreased Na:K ratios [44]. However, we have no knowledge of studies investigating the association of Na:K ratios with micronutrient intake. In addition, although the Na:K ratio is being considered a proxy for diet quality because of its association with cardiovascular risk, a direct correlation with a validated measure of diet quality had not been previously studied. Furthermore, we are unaware of previous studies conducted on the population of manufacturing workers.

4.2. Methodological Considerations

For this research, we elected a single 24 h recall for estimating usual nutrient consumption because, although it may not fully capture habitual intake, it offers significant advantages when deployed in large-scale epidemiological surveys such as the present one. First, its administration is relatively quick and cost-effective per participant, compared to methods like food diaries, making it feasible for large samples. Second, it does not require participant literacy, thus reducing non-respondent bias. Third, it minimizes the long-term recall bias inherent in food frequency questionnaires by not relying on memory beyond the immediate past. On the other hand, although a single 24HR per individual captures only a snapshot of intake, which may not reflect habitual dietary patterns due to substantial day-to-day variation, particularly for episodically consumed foods and nutrients, the main consequence of such measurement error is to increase the variance of the distribution of daily intakes. While this may have significant impact on the estimation of individual-level intake, which is particularly worrisome in studies estimating the fraction of the population with usual intake above or below some standard [45], it may not be significant in studies estimating group-level usual intakes. This is supported by several studies in low-income countries, comparing weighted food records with a single 24 h recall, that have found no significant underreporting of intakes estimated at group level [46]. The effect in analyses involving regression is mostly an attenuation of the regression coefficients and the increase in their standard errors, which may lead to loss of statistical power and possibly to missing some significant associations.

Another methodological option was to estimate individual Na:K ratios from 24HR. While 24 h urinary excretion is considered the gold standard for estimating sodium and potassium intake, in large-scale surveys, spot urine measurements and 24HR are often used as alternatives due to practical constraints (such as the challenge of obtaining 24 h urine collections, the dependency on participant compliance, and complex logistics of storage, transportation and analysis), while improving participation rates and decreasing self-selection bias, although its limitations in accuracy are acknowledged.

Regarding the use of the DQI-I to evaluate diet quality, validation studies have demonstrated that the DQI-I correlates with biomarkers of nutritional status and health outcomes, supporting its construct validity [47]. Its adaptability allows for application in different cultural and dietary contexts while maintaining a standardized scoring system, enhancing comparability in epidemiological research [33]. However, like all diet quality indices, the DQI-I relies on accurate dietary intake data, typically obtained through 24HR or food frequency questionnaires, and may be influenced by measurement errors. Despite this limitation, its structured approach provides a useful framework for assessing dietary patterns in relation to public health goals, making it a valuable tool in nutritional surveillance and intervention studies.

4.3. Public Health Implications

The findings of this study carry significant public health implications. Manufacturing workers, often constrained by irregular schedules and limited meal options, face structural barriers to achieving dietary balance. Workplace interventions targeting sodium reduction, such as subsidized healthy meals, nutrition education, and restrictions on fast-food vendors, could mitigate these disparities [48]. Additionally, promoting potassium-rich foods like fruits and dairy through onsite cafeterias may counteract sodium's effects. However, such strategies must account for socioeconomic realities, as cost and accessibility remain key determinants of food choice in this demographic group [49,50]. The regional specificity of dietary patterns, exemplified by the prominence of rice and corn as sodium contributors, further underscores the need for culturally tailored guidelines. For instance, traditional Brazilian dishes like feijoada, though nutrient-dense, may require modifications to reduce hidden sodium without compromising cultural acceptability [51].

4.4. Limitations and Strengths

This study has some limitations. Because it is based on a cross-sectional design, causal relationships between dietary patterns and nutritional outcomes cannot be inferred, but we did not make such inferences. Furthermore, dietary intake was self-reported, which may introduce recall bias, and focusing on the single Brazilian state limits generalizability to other occupational or cultural contexts. As previously discussed, the 24 h recall method has limitations in estimating habitual nutrient intake and sodium and potassium intake. However, to mitigate this limitation, we employed the Multiple Pass Method (MPM), a structured and validated approach known to improve the accuracy and completeness of dietary recall data.

Despite these limitations, the study presents several methodological strengths that increase the robustness and credibility of the results. The use of a probability sampling strategy ensured that the sample was representative of the working adult population across the state. The large sample size increases statistical power and the precision of the estimates. Moreover, data collection was conducted prospectively through in-person interviews conducted by trained nutritionists, minimizing potential measurement errors. It is important to highlight that the absence of non-response and the negligible risk of self-selection bias reinforce the internal validity of the study. Together, these aspects corroborate the reliability of our findings and reinforce their relevance for informing dietary interventions and public health policies focused on workforce nutrition.

5. Conclusions

This study investigated dietary Na:K ratio and its associations with diet composition, micronutrient intake, and overall diet quality among industrial workers. Less than 1% of workers met the WHO-recommended threshold for Na:K ratio. Key dietary contributors to elevated ratios included fast food, rice, bread, and red meat, while fruits, dairy, white meat, and coffee were protective. Higher Na:K ratios were associated with reduced intake of essential micronutrients such as calcium, magnesium, phosphorus, and vitamins C, D, and E, as well as poorer diet quality, as evidenced by lower Diet Quality Index-International scores. These findings underscore the dual burden of excessive sodium and inadequate potassium intake in this population, driven largely by processed and convenience foods. By demonstrating its inverse relationship with micronutrient adequacy and diet quality, this research reinforces the interconnectedness of electrolyte balance and nutritional health.

In conclusion, this study elucidates the multifactorial drivers of elevated Na:K ratios in manufacturing workers and their downstream effects on micronutrient intake and diet quality. By demonstrating its inverse relationship with micronutrient adequacy and diet quality, this research reinforces the interconnectedness of electrolyte balance and nutritional health.

Methodologically, estimating the change in Na:K ratio per serving of a food group (rather than per gram or kilocalorie) represents a meaningful methodological innovation that enhances the practical evaluation of dietary contributions to electrolyte balance. By linking Na:K ratios to standardized serving sizes, the findings become more interpretable and actionable for both public health messaging and individual dietary guidance. In addition, estimating the change in Na:K ratio per serving of a food group captures the combined effects of the multiple ingredients or processing steps in a single serving of a food item that collectively influences its sodium and potassium content. Another benefit of the approach is that it facilitates direct comparisons across food groups in a way that echoes dietary guidelines, which often recommend servings, simplifying the translation of study findings into interventions, such as promoting specific serving substitutions.

By identifying actionable dietary targets, such as reducing fast food and refined grain consumption while increasing fruits and dairy, these findings offer a roadmap for policies aimed at improving both electrolyte balance and nutritional adequacy in high-risk populations. Addressing these dual challenges is essential to curbing the growing burden of diet-related chronic diseases in industrial settings. In addition, to improve the unfavorable Na:K ratios observed among industrial workers, targeted nutritional interventions are needed at both the individual and environmental levels. Strategies could include reformulating commonly consumed processed and convenience foods to reduce sodium content and/or increase potassium density, along with workplace-based programs that promote the consumption of potassium-rich foods such as fruits, vegetables, legumes, and dairy products. In parallel, nutrition education campaigns tailored to this workforce could emphasize the benefits of replacing sodium-rich items (e.g., fast food, processed meats, white bread) with protective alternatives (e.g., whole foods, minimally processed options), using portion-based guidelines to align with the study's methodological framework. These efforts could be further supported by institutional policies that improve the availability and accessibility of healthier options in workplace cafeterias and surrounding food environments, ultimately contributing to improved electrolyte balance and overall diet quality in this population. Future studies should further investigate nutritional consequences of imbalanced Na:K intake, particularly on micronutrient composition of diets, as well as on vascular, renal, hormonal, and metabolic functions, and on gut microbiota composition. Also, additional information is needed on the food groups with greater contribution to Na:K balance, while intervention trials could test the efficacy of workplace-based dietary programs.

Author Contributions: Conceptualization, A.M.S., I.W.L.B. and A.G.O.; data curation, A.M.S., I.W.L.B., K.G.T. and A.G.O.; formal analysis, A.G.O.; funding acquisition, A.G.O.; investigation, A.M.S., I.W.L.B., K.G.T., G.S.P. and R.M.C.; methodology, A.M.S., I.W.L.B., K.G.T. and A.G.O.; project administration, I.W.L.B.; supervision, I.W.L.B.; writing—original draft, A.M.S., I.W.L.B. and A.G.O.; writing—review and editing, A.M.S., I.W.L.B., K.G.T. and A.G.O. All authors have read and agreed to the published version of the manuscript.

Funding: This study was partially financed by the Coordination for the Improvement of Higher Education Personnel—Brazil (CAPES)—Finance Code 001. No additional external funding was received for this study.

Institutional Review Board Statement: The study was conducted in accordance with the Declaration of Helsinki and approved by the Ethics Committee of the Onofre Lopes University Hospital (2.198.545, approved 2 August 2017).

Informed Consent Statement: Informed consent was obtained from all subjects involved in the study.

Data Availability Statement: The study data is available from the corresponding author upon reasonable request. Data are not publicly available due to ethical reasons as data requests will have to be reviewed by the Institutional Review Board.

Acknowledgments: We acknowledge the support of the undergraduate students engaged in scientific initiation activities, whose collaboration was instrumental during the data collection and investigation stages of this research: Andréa Medeiros, Daionara Sousa, Heloísa Gurgel, Rafaella Bezerra, Roberta Carvalho, Nathanael Araújo, Isabel Muniz, Rodrigo Santos, and Noeme Viana.

Conflicts of Interest: The authors declare no conflicts of interest.

Abbreviations

The following abbreviations are used in this manuscript:

Na:K	Sodium-to-potassium
DQI-I	Diet Quality Index-International
BMI	Body Mass Index
24HR	24 h dietary recall
MPM	Multiple Pass Method

References

1. Elizabeth, L.; Machado, P.; Zinöcker, M.; Baker, P.; Lawrence, M. Ultra-Processed Foods and Health Outcomes: A Narrative Review. *Nutrients* **2020**, *12*, 1955. [CrossRef] [PubMed] [PubMed Central]
2. Malik, V.S.; Hu, F.B. The role of sugar-sweetened beverages in the global epidemics of obesity and chronic diseases. *Nat. Rev. Endocrinol.* **2022**, *18*, 205–218. [CrossRef]
3. Mozaffarian, D.; Clarke, R. Quantitative effects on cardiovascular risk factors and coronary heart disease risk of replacing partially hydrogenated vegetable oils with other fats and oils. *Eur. J. Clin. Nutr.* **2021**, *63*, 22–33. [CrossRef]
4. He, F.J.; Tan, M.; Ma, Y.; MacGregor, G.A. Salt reduction to prevent hypertension and cardiovascular disease: JACC state-of-the-art review. *J. Am. Coll. Cardiol.* **2020**, *75*, 632–647. [CrossRef]

5. Sonnenburg, E.D.; Sonnenburg, J.L. The ancestral and industrialized gut microbiota and implications for human health. *Nat. Rev Microbiol.* **2019**, *17*, 383–390. [CrossRef]
6. Suez, J.; Cohen, Y.; Valdés-Mas, R.; Mor, U.; Dori-Bachash, M.; Federici, S.; Zmora, N.; Leshem, A.; Heinemann, M.; Linevsky, R. et al. Personalized microbiome-driven effects of non-nutritive sweeteners on human glucose tolerance. *Cell* **2022**, *185*, 2565–2586 [CrossRef] [PubMed]
7. GBD 2016 Alcohol Collaborators. Alcohol use and burden for 195 countries and territories, 1990–2016: A systematic analysis for the Global Burden of Disease Study 2016. *Lancet* **2018**, *392*, 1015–1035. [CrossRef]
8. Simopoulos, A.P. An increase in the omega-6/omega-3 fatty acid ratio increases the risk for obesity. *Nutrients* **2016**, *8*, 128. [CrossRef] [PubMed]
9. Mosallanezhad, Z.; Jalali, M.; Bahadoran, Z.; Mirmiran, P.; Azizi, F. Dietary sodium to potassium ratio is an independent predictor of cardiovascular events: A longitudinal follow-up study. *BMC Public Health* **2023**, *23*, 705. [CrossRef] [PubMed] [PubMed Central]
10. Gałęska, E.; Wrzecińska, M.; Kowalczyk, A.; Araujo, J.P. Reproductive Consequences of Electrolyte Disturbances in Domestic Animals. *Biology* **2022**, *11*, 1006. [CrossRef]
11. World Health Organization. *Guideline: Sodium Intake for Adults and Children*; World Health Organization: Geneva, Switzerland, 2012.
12. Juul, F.; Vaidean, G.; Parekh, N. Ultra-processed Foods and Cardiovascular Diseases: Potential Mechanisms of Action. *Adv. Nutr.* **2021**, *12*, 1673–1680. [CrossRef] [PubMed] [PubMed Central]
13. Grillo, A.; Salvi, L.; Coruzzi, P.; Salvi, P.; Parati, G. Sodium Intake and Hypertension. *Nutrients* **2019**, *11*, 1970. [CrossRef] [PubMed] [PubMed Central]
14. Perez, V.; Chang, E.T. Sodium-to-potassium ratio and blood pressure, hypertension, and related factors. *Adv. Nutr.* **2014**, *5*, 712–741. [CrossRef] [PubMed] [PubMed Central]
15. Stamler, J.; Rose, G.; Elliott, P.; Dyer, A.; Marmot, M.; Kesteloot, H.; Stamler, R. Findings of the International Cooperative INTERSALT Study. *Hypertension* **1991**, *17*, I9–I15. [CrossRef]
16. Ma, Y.; He, F.J.; Sun, Q.; Yuan, C.; Kieneker, L.M.; Curhan, G.C.; MacGregor, G.A.; Bakker, S.J.L.; Campbell, N.R.C.; Wang, M.; et al. 24-Hour Urinary Sodium and Potassium Excretion and Cardiovascular Risk. *N. Engl. J. Med.* **2022**, *386*, 252–263. [CrossRef] [PubMed]
17. Mozaffarian, D.; Fahimi, S.; Singh, G.M.; Micha, R.; Khatibzadeh, S.; Engell, R.E.; Lim, S.; Danaei, G.; Ezzati, M.; Powles, J.; et al. Global sodium consumption and death from cardiovascular causes. *N. Engl. J. Med.* **2014**, *371*, 624–634. [CrossRef] [PubMed]
18. Louzada, M.L.; Martins, A.P.; Canella, D.S.; Baraldi, L.G.; Levy, R.B.; Claro, R.M.; Moubarac, J.C.; Cannon, G.; Monteiro, C.A. Impact of ultra-processed foods on micronutrient content in the Brazilian diet. *Rev. Saude Publica* **2015**, *49*, 45. [CrossRef]
19. Levings, J.L.; Gunn, J.P. The imbalance of sodium and potassium intake: Implications for dietetic practice. *J. Acad. Nutr. Diet.* **2014**, *114*, 838–841. [CrossRef]
20. Souza, R.V.; Sarmento, R.A.; de Almeida, J.C.; Canuto, R. The effect of shift work on eating habits: A systematic review. *Scand. J. Work Environ. Health* **2019**, *45*, 7–21. [CrossRef]
21. Leone, L.A.; Beth, D.; Ickes, S.B.; Macguire, K.; Nelson, E.; Smith, R.A.; Tate, D.F.; Ammerman, A.S. Attitudes Toward Fruit and Vegetable Consumption and Farmers' Market Usage Among Low-Income North Carolinians. *J. Hunger Environ. Nutr.* **2012**, *7*, 64–76. [CrossRef]
22. Woodruff, R.C.; Zhao, L.; Ahuja, J.K.C.; Gillespie, C.; Goldman, J.; Harris, D.M.; Jackson, S.L.; Moshfegh, A.; Rhodes, D.; Sebastian, R.S.; et al. Top Food Category Contributors to Sodium and Potassium Intake—United States, 2015–2016. *MMWR Morb. Mortal. Wkly. Rep.* **2020**, *69*, 1064–1069. [CrossRef] [PubMed]
23. Okuda, N.; Okayama, A.; Miura, K.; Yoshita, K.; Miyagawa, N.; Saitoh, S.; Nakagawa, H.; Sakata, K.; Chan, Q.; Elliott, P.; et al. Food Sources of Dietary Potassium in the Adult Japanese Population: The International Study of Macro-/Micronutrients and Blood Pressure (INTERMAP). *Nutrients* **2020**, *12*, 787. [CrossRef]
24. Filippini, T.; Naska, A.; Kasdagli, M.I.; Torres, D.; Lopes, C.; Carvalho, C.; Moreira, P.; Malavolti, M.; Orsini, N.; Whelton, P.K.; et al. Potassium Intake and Blood Pressure: A Dose-Response Meta-Analysis of Randomized Controlled Trials. *J. Am. Heart Assoc.* **2020**, *9*, e015719. [CrossRef]
25. Karppanen, H.; Karppanen, P.; Mervaala, E. Why and how to implement sodium, potassium, calcium, and magnesium changes in food items and diets? *J. Hum. Hypertens.* **2005**, *19* (Suppl. 3), S10–S19. [CrossRef]
26. Webster, J.; Waqanivalu, T.; Arcand, J.; Trieu, K.; Cappuccio, F.P.; Appel, L.J.; Woodward, M.; Campbell, N.R.C.; McLean, R. Understanding the science that supports population-wide salt reduction programs. *J. Clin. Hypertens.* **2017**, *19*, 569–576. [CrossRef]
27. Vinson, J.A.; Dabbagh, Y.A.; Serry, M.M.; Jang, J. Plant flavonoids, especially tea flavonols, are powerful antioxidants using an in vitro oxidation model for heart disease. *J. Agric. Food Chem.* **2020**, *43*, 2800–2802. [CrossRef]
28. Previdelli, A.N.; Lipi, M.; Castro, M.A.; Marchioni, D.M. Dietary quality and associated factors among factory workers in the metropolitan region of São Paulo, Brazil. *J. Am. Diet. Assoc.* **2010**, *110*, 786–790. [CrossRef]

29. Steele, E.M.; Popkin, B.M.; Swinburn, B.; Monteiro, C.A. The share of ultra-processed foods and the overall nutritional quality of diets in the US: Evidence from a nationally representative cross-sectional study. *Popul. Health Metr.* **2017**, *15*, 6. [CrossRef]
30. Cediel, G.; Reyes, M.; Corvalán, C.; Levy, R.B.; Uauy, R.; Monteiro, C.A. Ultra-processed foods drive to unhealthy diets: Evidence from Chile. *Public Health Nutr.* **2021**, *24*, 1698–1707. [CrossRef] [PubMed]
31. Brasil, Ministério da Saúde; Secretaria de Atenção à Saúde; Departamento de Atenção Básica. *Orientações Para a Coleta e Análise de Dados Antropométricos em Serviços de Saúde: Norma Técnica do Sistema de Vigilância Alimentar e Nutricional SISVAN*; Série G; Estatística e Informação em Saúde; Ministério da Saúde: Brasília, Brazil, 2011.
32. Raper, N.; Perloff, B.; Ingwersen, L.; Steinfeldt, L. An overview of USDA's dietary intake data system. *J. Food Compos. Anal.* **2004**, *17*, 545–555. [CrossRef]
33. Kim, S.; Haines, P.S.; Siega-Riz, A.M.; Popkin, B.M. The Diet Quality Index-International (DQI-I) provides an effective tool for cross-national comparison of diet quality as illustrated by China and the United States. *J. Nutr.* **2003**, *133*, 3476–3484. [CrossRef]
34. Moshfegh, A.J.; Rhodes, D.G.; Baer, D.J.; Murayi, T.; Clemens, J.C.; Rumpler, W.V.; Paul, D.R.; Sebastian, R.S.; Kuczynski, K.J.; Ingwersen, L.A.; et al. The US Department of Agriculture automated multiple-pass method reduces bias in the collection of energy intakes. *Am. J. Clin. Nutr.* **2008**, *88*, 324–332. [CrossRef]
35. Fisberg, R.M., Marchioni, D.M.L.; Providelli, A.N.; Carvalho, A.M.; Mendes, A.; Timm, A.S.; Gorgulho, B.M.; Verly Junior, E.; Steluti, J.; Brunacio, K.H.; et al. Manual de Avaliação do Consumo Alimentar em Estudos Populacionais: A Experiência do Inquérito de Saúde em São Paulo (ISA); FSP/USP: São Paulo 2012. Available online: http://colecoes.sibi.usp.br/fsp/files/original/8b36141af4e756cbfb889b895b541890.pdf (accessed on 24 June 2025).
36. Núcleo de Estudos e Pesquisas em Alimentação-NEPA; Universidade Estadual de Campinas. *Tabela Brasileira de Composição de Alimentos–TACO. 4. Ed. Rev. e Ampl. Campinas: NEPA-UNICAMP 2011.* Available online: https://www.cfn.org.br/wp-content/uploads/2017/03/taco_4_edicao_ampliada_e_revisada.pdf (accessed on 24 June 2025).
37. Brasil. Ministério do Planejamento, Orçamento e Gestão; Instituto Brasileiro de Geografia e Estatística (IBGE); Diretoria de Pesquisas.; Coordenação de Trabalho e Rendimento. *Pesquisa de Orçamentos Familiares (POF) 2008–2009: Tabelas de Composição Nutricional dos Alimentos Consumidos no Brasil*; IBGE: Rio de Janeiro, Brazil, 2011.
38. United States Department of Agriculture. Agricultural Research Service, Nutrient Data 2024. Available online: https://fdc.nal.usda.gov/ (accessed on 24 June 2025).
39. Souza, A.M.; Bezerra, I.W.L.; Pereira, G.S.; Torres, K.G.; Costa, R.M.; Oliveira, A.G. Relationships between motivations for food choices and consumption of food groups: A prospective cross-sectional survey in manufacturing workers in Brazil. *Nutrients* **2020**, *12*, 1490. [CrossRef] [PubMed]
40. Philippi, S.T. *Nutrição e Técnica Dietética*, 4th ed.; Manole: Barueri, Brazil, 2019.
41. Wronska, M.D.; Coffey, M.; Robins, A. Determinants of nutrition practice and food choice in UK construction workers. *Health Promot. Int.* **2022**, *1*, 37. [CrossRef] [PubMed]
42. Centers for Disease Control and Prevention (CDC). Vital signs: Food categories contributing the most to sodium consumption—United States, 2007–2008. *MMWR Morb. Mortal. Wkly Rep.* **2012**, *61*, 92–98.
43. Morrissey, E.; Giltinan, M.; Kehoe, L.; Nugent, A.P.; McNulty, B.A.; Flynn, A.; Walton, J. Sodium and Potassium Intakes and Their Ratio in Adults (18–90 y): Findings from the Irish National Adult Nutrition Survey. *Nutrients* **2020**, *12*, 938. [CrossRef]
44. Vulin, M.; Magušić, L.; Metzger, A.-M.; Muller, A.; Drenjančević, I.; Jukić, I.; Šijanović, S.; Lukić, M.; Stanojević, L.; Davidović Cvetko, E.; et al. Sodium-to-Potassium Ratio as an Indicator of Diet Quality in Healthy Pregnant Women. *Nutrients* **2022**, *14*, 5052. [CrossRef]
45. French, C.D.; Arsenault, J.E.; Arnold, C.D.; Haile, D.; Luo, H.; Dodd, K.W.; Vosti, S.A.; Slupsky, C.M.; Engle-Stone, R.; Variance Components of Nutrient Intakes Data Working Group. Within-Person Variation in Nutrient Intakes Across Populations and Settings: Implications for the Use of External Estimates in Modeling Usual Nutrient Intake Distributions. *Adv. Nutr.* **2021**, *12*, 429–451. [CrossRef]
46. Gibson, R.S.; Charrondiere, U.R.; Bell, W. Measurement Errors in Dietary Assessment Using Self-Reported 24-Hour Recalls in Low-Income Countries and Strategies for Their Prevention. *Adv. Nutr.* **2017**, *8*, 980–991. [CrossRef]
47. Najibi, N.; Jahromi, M.K.; Teymoori, F.; Farhadnejad, H.; Salehi-Sahlabadi, A.; Mirmiran, P. International diet quality index and revised diet quality index relationship with non-alcoholic fatty liver disease: A case-control study. *BMC Gastroenterol.* **2023**, *23*, 441. [CrossRef]
48. Lassen, A.D.; Beck, A.; Leedo, E.; Andersen, E.W.; Christensen, T.; Mejborn, H.; Thorsen, A.V.; Tetens, I. Effectiveness of offering healthy labelled meals in improving the nutritional quality of lunch meals eaten in a worksite canteen. *Appetite* **2014**, *75*, 128–134. [CrossRef] [PubMed]
49. Ares, G.; Machín, L.; Girona, A.; Curutchet, M.R.; Giménez, A. Comparison of motives underlying food choice and barriers to healthy eating among low medium income consumers in Uruguay. *Cad. Saude Publica* **2017**, *33*, e00213315. [CrossRef] [PubMed]

50. Darmon, N.; Drewnowski, A. Contribution of food prices and diet cost to socioeconomic disparities in diet quality and health: A systematic review and analysis. *Nutr. Rev.* **2015**, *73*, 643–660. [CrossRef] [PubMed]
51. Oliveira, M.S.D.S.; Santos, L.A.D.S. Dietary guidelines for Brazilian population: An analysis from the cultural and social dimensions of food. *Cienc. Saude Colet.* **2020**, *25*, 2519–2528. [CrossRef] [PubMed]

Disclaimer/Publisher's Note: The statements, opinions and data contained in all publications are solely those of the individual author(s) and contributor(s) and not of MDPI and/or the editor(s). MDPI and/or the editor(s) disclaim responsibility for any injury to people or property resulting from any ideas, methods, instructions or products referred to in the content.

nutrients

MDPI

Article

The Effect of Copper Nanoparticles on Liver Metabolism Depends on the Type of Dietary Fiber

Aleksandra Marzec [1], Bartosz Fotschki [2,*], Dorota Napiórkowska [2], Joanna Fotschki [2], Ewelina Cholewińska [1], Piotr Listos [3], Jerzy Juśkiewicz [2] and Katarzyna Ognik [1]

1 Department of Biochemistry and Toxicology, Faculty of Animal Sciences and Bioeconomy, University of Life Sciences in Lublin, Akademicka 13, 20-950 Lublin, Poland; aleksandra.marzec@up.lublin.pl (A.M.); ewelina.cholewinska@up.lublin.pl (E.C.); katarzyna.ognik@up.lublin.pl (K.O.)

2 Division of Food Science, Institute of Animal Reproduction and Food Research, Polish Academy of Sciences, Tuwima 10, 10-748 Olsztyn, Poland; d.napiorkowska@pan.olsztyn.pl (D.N.); j.fotschki@pan.olsztyn.pl (J.F.); j.juskiewicz@pan.olsztyn.pl (J.J.)

3 Sub-Department of Pathomorphology and Forensic Medicine, Faculty of Veterinary Medicine, University of Life Sciences in Lublin, Głęboka 30, 20-612 Lublin, Poland; piotr.listos@up.lublin.pl

* Correspondence: b.fotschki@pan.olsztyn.pl

Abstract: Background/Objectives: A diet enriched with copper nanoparticles (CuNPs) exhibits a wide range of effects on liver metabolism, both positive and negative. Dietary fibers are the key components that may affect the absorption of minerals, including copper, and change their impact on organisms. Methods: Therefore, this study investigated whether and how supplementation with different sources of dietary fiber (cellulose, pectin, inulin, and psyllium) affects the function of CuNPs in the liver of male Wistar rats. Results: The results showed that CuNPs at different doses had varying effects on lipid metabolism and inflammation in the liver. Specifically, higher doses of CuNPs were associated with increased lipid accumulation and the activation of pro-inflammatory mechanisms. However, combining CuNPs with dietary fibers, such as psyllium and inulin, was beneficial in mitigating the effects of the examined nanoparticles, leading to reduced fat, cholesterol, and triglycerides in the liver. Combining psyllium with CuNPs showed the most substantial effect on liver metabolism and inflammation parameters. Furthermore, hepatic histology analyses showed that adding psyllium to the diet with CuNPs reduces changes associated with fat accumulation and mononuclear cell infiltration. The observed beneficial changes in the liver may have been related to a reduction in the gene expression level of sterol regulatory element-binding protein 1 and peroxisome proliferator-activated receptor gamma and cyclooxygenase-2. Conclusions: In conclusion, enriching the diet with dietary fibers such as psyllium can regulate the action of CuNPs, thereby improving lipid metabolism and reducing inflammation in the liver.

Keywords: copper nanoparticles; psyllium; dietary fiber; triglycerides; liver metabolism; aminotransferases

1. Introduction

Due to their small size and high surface-to-volume ratio, copper nanoparticles (CuNPs) exhibit unique properties that make them useful in various applications, including antimicrobial agents, electronics, and agriculture [1–3]. Another way to use these nano molecules is as an ingredient in food products. These compounds show a wide range of effects on the consumer's body, including both positive and negative effects [4].

The liver plays a crucial role in processing compounds absorbed from the digestive tract, making CuNPs in the diet a vital component, with the potential to influence liver metabolism [5]. Using CuNPs as a therapeutic agent for liver metabolic disorders presents promising opportunities as well as significant risks. CuNPs have shown potential in treating non-alcoholic fatty liver disease (NAFLD) by scavenging reactive oxygen species

(ROS) and reducing oxidative stress, which are critical factors in the progression of liver diseases [6,7]. Specifically, ultrasmall copper-based nanoparticles combined with resveratrol have demonstrated efficacy in targeting liver tissues and treating inflammation in NAFLD, suggesting a novel approach for managing such conditions [6]. Other in vitro studies have shown the potential of CuNPs as anticancer agents, particularly in targeting colorectal cancer cells [8]. Additionally, the low-dose administration of various inorganic nanoparticles, including CuNPs, has been found to promote lipid degradation and alleviate liver steatosis without causing adverse effects, indicating a potential therapeutic benefit for metabolic regulation [9]. Moreover, the study involving obese individuals with liver steatosis showed that the level of copper in serum played a significant role in predicting the onset of atherosclerosis [10]. However, the hepatotoxicity of CuNPs cannot be overlooked. Studies have shown that high doses of CuNPs can induce significant oxidative stress, inflammation, and histopathological changes in the liver, leading to severe hepatic damage and impaired drug metabolism [11,12]. Furthermore, repeated exposure to CuNPs has been associated with profibrotic changes and immunosuppressive effects [13]. Recent nutritional studies on mice have shown that copper deficiency is strongly associated with metabolic dysfunction-associated steatotic liver disease (MASLD), and CuNPs can exacerbate this condition by disrupting copper metabolism and inducing oxidative stress [14]. The size of the nanoparticles also plays a crucial role, with smaller nanoparticles showing reversible toxicity and better clearance compared to larger ones, which tend to accumulate in the liver and cause prolonged damage [15].

One effective method to modulate the absorption of CuNPs and their impact on the body might be to enhance the diet with fiber. This dietary component can impact the bioavailability of CuNPs through various mechanisms, including mineral binding, fermentation, and interaction with other dietary components. Dietary fibers can bind minerals and form complexes that are difficult to absorb in the intestinal lumen, potentially decreasing mineral bioavailability [16]. However, the type of dietary fiber plays a crucial role in this interaction. For instance, inulin has been shown to have a beneficial effect on DNA repair mechanisms. At the same time, pectin can inhibit inflammatory processes in the small intestine, thereby protecting against the oxidative effects of CuNPs [17]. The physicochemical properties of fibers, such as solubility and fermentability, also influence their interaction with minerals. Soluble fibers are rapidly fermented, which can liberate bound minerals and promote colonic absorption, potentially offsetting any negative effects on mineral bioavailability [18]. Additionally, the interaction between dietary fibers and copper ions can be complex, as demonstrated by the use of Electron Paramagnetic Resonance (EPR) techniques to study barley β-glucan and copper ions, revealing weak interactions under physiological conditions [19].

Previous nutritional research has shown the protective role of dietary fiber supplementation against the potentially harmful oxidative effects of CuNPs in the small intestine [17]. Furthermore, the dietary combination with dietary fiber affects the absorption of CuNPs from the gastrointestinal tract, thus modulating their levels in the liver [20]. However, there is no information on how the type of dietary fiber modulates the effect of CuNPs on liver function. Therefore, the main objective of this study was to evaluate the dosage of CuPNs and dietary combinations of four different sources of dietary fiber (cellulose, pectin, inulin, and psyllium) with CuPNs on the metabolism and inflammation in the liver of rats. As far as animal models are concerned, there is considerable similarity between rodent and human internal organs, including liver functioning. Although in vivo data on metal nanoparticles obtained from rats may not be entirely transferred to humans, it must be stressed that such new information pushes our knowledge forward.

2. Materials and Methods

The protocol containing the research questions, experimental schema with the in vivo study design, and analysis plan was submitted to the reviewers of the National Science Center (Kraków, Poland) for evaluation, and was subsequently approved and funded.

2.1. Dietary Fibres and Copper Nanoparticles

The copper nanoparticles (99.9% purity) were acquired from SkySpring Nanomaterials, Inc. located in Houston, TX, USA. The detailed physicochemical properties were described by Cholewińska et al. (2023) [17]. For the control dietary sources, $CuCO_3$ was purchased from Merck KGaA in Darmstadt, Germany, and α-cellulose from Sigma in Poznań, Poland. Experimental dietary fiber sources such as pectin (PectinE 440(I) from Brouwland in Beverlo, Belgium), inulin (Frutafit Tex from Sensus in Roosendaal, the Netherlands), and psyllium (Psyllium husk powder from NaturaleBio in Rome, Italy) were also used to design the experimental diets.

2.2. In Vivo Experiment

A total of 100 nine-week-old healthy outbred male Wistar rats (Cmdb:Wi) were part of the research study. The animals were originated from the Medical University of Bialystok, Poland. After a two-week acclimatization period, they were randomly divided into ten groups, with each group consisting of ten rats. Random numbers were created using Microsoft Excel's standard = RAND() function (version 15.0.5589.1000). The rats were housed individually in metabolic cages in a carefully controlled environment. The environment included a 12 h light–dark cycle, a temperature maintained at 21 ± 1 °C, a relative humidity ranging from 50% to 70%, and 20 air changes per hour. Throughout a six-week period, each group was given tap water and a modified version of the semipurified casein diet, as recommended for laboratory rodents by the American Institute of Nutrition. The experimental diets in this study consisted of two levels of copper nanoparticles (CuNPs); the recommended level and the double that level were 6.5 and 13 mg/kg, respectively. These levels of CuNPs were paired with various types of dietary fiber. The control diet included a mineral mixture containing standard and high levels of $CuCO_3$ (6.5 and 13 mg/kg diet). In contrast, the diets with CuNPs utilized a mineral mixture without $CuCO_3$. The control dietary fiber, α-cellulose, was incorporated at 8% of the diet, while the experimental fibers, including inulin with a prebiotic effect, psyllium with a bulking effect, and pectin with a viscous effect, were added at 6% of the diet in place of cellulose. The CuNPs were administered in oil to prevent oxidation, and the experimental diet was kept in the freezer (−70 °C) before administration. The diets provided to the rats throughout the entire experimental period consisted of a detailed composition, which can be found in Table 1. The animals were given unrestricted access to these diets. All animal procedures strictly followed the guidelines outlined in the European Union Directive (2010/63/EU) for animal experiments. Furthermore, the experiment received approval from the local Institutional Animal Care and Use Committee under Permission No. 19/2021 (17 March 2021) in Olsztyn, Poland. The study was conducted in accordance with the ARRIVE guidelines, and all possible measures were taken to minimize the suffering of the animals used in the experiment. During the period of experimental feeding, in the event of adverse effects related to humane endpoints, i.e., cessation of diet intake for more than 2 days, making specific sounds as a pain signal for more than 1 h, the appearance of neurological symptoms (e.g., ataxia, impairment in maintaining a favorable body position), and the presence of blood in the feces for more than 1 day, a veterinarian (employed for these purposes at the institute) could make a decision of humane euthanasia, using the method of gradually filling the chamber of the animal with carbon dioxide or the method of dislodging the cervical vertebrae of a previously sedated animal. To the best of our knowledge, based on literature and our previous experiments, the above symptoms should not be related to the experimental factors (fiber, nanoparticles) and there is only a minimal risk of their occurrence.

Table 1. The experimental diets given to the rats for six weeks.

	C	CH	CN	CNH	PN	PNH	JN	JNH	SN	SNH
					%					
Casein [1]	14.8	14.8	14.8	14.8	14.8	14.8	14.8	14.8	14.8	14.8
DL-methionine	0.2	0.2	0.2	0.2	0.2	0.2	0.2	0.2	0.2	0.2
Cellulose [2]	8.0	8.0	8.0	8.0	2.0	2.0	2.0	2.0	2.0	2.0
Pectin					6	6				
Inulin							6	6		
Psyllium									6	6
Other components [3]	77.0	77.0	77.0	77.0	77.0	77.0	77.0	77.0	77.0	77.0
Calculation:										
Cu from, mg/kg										
$CuCO_3$	6.5	13	0	0	0	0	0	0	0	0
CuNPs	0	0	6.5	13	6.5	13	6.5	13	6.5	13

C, control diet with a standard Cu content in the mineral mix and 8% of cellulose as a dietary fiber source; CH, control diet with an enhanced Cu content in the mineral mix and 8% of cellulose as a dietary fiber; CN, diet with a standard CuNPs content in the mineral mix and 8% of cellulose as a dietary fiber source; CNH, diet with enhanced CuNPs content in the mineral mix and 8% of cellulose as a dietary fiber; PN, diet with a standard CuNPs content in the mineral mix and 2% of cellulose and 6% of pectin as a dietary fiber source; PNH, diet with enhanced CuNPs content in the mineral mix and 2% of cellulose and 6% of pectin as a dietary fiber source; JN, diet with a standard CuNPs content in the mineral mix and 2% of cellulose and 6% of inulin as a dietary fiber source; JNH, diet with enhanced CuNPs content in the mineral mix and 2% of cellulose and 6% of inulin as a dietary fiber source; SN, diet with a standard CuNPs content in the mineral mix and 2% of cellulose and 6% of psyllium as a dietary fiber source; SNH, diet with enhanced CuNPs content in the mineral mix and 2% of cellulose and 6% of psyllium as a dietary fiber source. [1] Casein preparation: crude protein 89.7%, crude fat 0.3%, ash 2.0%, and water 8.0%. [2] α-Cellulose (SIGMA, Poznan, Poland), main source of dietary fiber. [3] 0.2% choline chloride, 8.0% rapeseed oil, 64.0% maize starch, 1.0% vitamin mix (AIN-93G-VM), and 3.5% mineral mix were added to each diet. In the experimental treatments involving copper nanoparticles (CuNPs), the mineral mix was intentionally devoid of copper carbonate ($CuCO_3$). To ensure the safety of the operator during the preparation of the experimental diets, the CuNPs preparation was incorporated into an emulsion, along with dietary rapeseed oil. This approach, which has been successfully employed in the previous study, helped to mitigate potential risks associated with handling CuNPs and ensured the effective administration of the treatment.

2.3. Collection of Biological Material and Analytical Procedures

None of the animals were excluded from the experiment. The project manager was the only person who was aware of the animals' allocation to a particular study group. None of the analysis contractors were acquainted with the treatment animal allocation. During the experiment, the rats were monitored for feed intake and body weight (data published in Cholewińska et al. 2023) [17]. After a period of 6 weeks on the experimental diet, the rats underwent anesthesia with a combination of ketamine and xylazine (100 mg and 10 mg/kg BW, respectively). Subsequently, blood samples were collected from the vena cava into heparinized tubes and then centrifuged at a low speed for 10 min at $350\times g$ and 4 °C. The resulting plasma samples were stored at −70 °C until analysis. Following this, the liver was extracted, weighed, and rapidly frozen using liquid nitrogen. Lipids were then extracted from the liver using the Folch et al. (1957) [21] method. A total of 0.2 g of tissue was used for extraction. The extraction process involved using a mixture of chloroform and methanol in a 2/1 ratio. A total of 4 mL of the mixture was added to the tissue and then homogenized. After centrifugation, the chloroform layer was collected, and the solvent was evaporated under nitrogen. The resulting sample was then used for further determinations. The concentrations of liver cholesterol and triglycerides were determined using spectrophotometric techniques with commercial kits (Cholesterol DST, Triglycerides DST, Alpha Diagnostics, Ltd., San Antonio, TX, USA). The sample size was determined for each analysis based on previous research.

2.4. Blood Plasma Parameters of Metabolism and Inflammation

The plasma concentration of cholesterol (total and its HDL), triglycerides, creatinine, uric acid, urea, glucose, and the plasma activities of aspartate transaminase (AST) and alanine transaminase (ALT) were determined using an automatic biochemical analyzer (Pentra C200, Horiba, Tokyo, Japan). The atherogenic index of plasma was calculated according to the following formula: log(TG/HDL). To measure the concentration of insulin, glucagon, and ghrelin in the plasma, validated rat ELISA kits were used (Shanghai Qayee Biotechnology Co. Ltd., Shanghai, China).

2.5. Liver Gene Expression

Firstly, RNA was extracted from the liver using a TRI Reagent solution (Thermo Fisher Scientific, Waltham, MA, USA), as per the manufacturer's instructions. *β-actin* was chosen as the reference gene. The mRNA expression levels of peroxisome proliferator-activated receptor gamma (*ppar-γ*, Cat# Rn00440945_m1), peroxisome proliferator activated receptor alpha (*ppar-α*, Cat# Rn00566193_m1), sterol regulatory element-binding protein 1 (*srebp1c*, Cat# Rn01495769_m1), and cyclooxygenase-2 (*cox-2*, Cat# Rn01483828_m1) were measured using single tube TaqMan® Gene Expression Assays (Life Technologies, Carlsbad, CA, USA). Amplification was carried out using a 7900HT Fast Real-Time PCR System under the following conditions: initial denaturation for 10 min at 95 °C, followed by 40 cycles of 15 s at 95 °C and 1 min at 60 °C. Finally, the mRNA expression levels of the selected genes were normalized to *β-actin* (Cat# Rn00667869_m1).

2.6. Liver Histopathology

Histopathological examinations, including hematoxylin eosin staining of liver samples from rats, were conducted following the procedure described in our previous study [22].

2.7. Statistical Analysis

The differences among treatment groups were determined using STATISTICA software, version 12.0 (StatSoft Corp., Krakow, Poland). A two-way analysis of variance (ANOVA) was used to evaluate the effects of two main factors, the dosage of copper nanoparticles (CuNPs) (low dose, 6.5 mg/kg and high dose, 13 mg/kg) and the type of dietary fiber (cellulose, pectin, inulin, and psyllium). Following the ANOVA, Duncan's multiple range test was conducted. The data were checked for normality with the aid of the Shapiro–Wilk test. Furthermore, to compare each experimental group fed the low dose of CuNPs with the control group (fed a diet with 6.5 mg/kg Cu from $CuCO_3$ and containing cellulose as the primary dietary fiber source), a *t*-test was employed. Similarly, the *t*-test was used to compare the experimental groups fed diets with the high dose of CuNPs with the control group, CH, in which rats were fed a diet with 13 mg/kg Cu from $CuCO_3$ and containing cellulose as the primary dietary fiber source. Differences with a significance level of $p \leq 0.05$ were considered to be statistically significant.

3. Results

The basic parameters of growth and dietary intake shown in previous work [17] indicate that the experimental diets had no effect on the final weight and body weight gain in rats, while dietary intake significantly decreased when inulin and psyllium were added to the diets with CuNPs. The relative weight of the liver was the highest in rats fed diets with inulin ($p < 0.05$ vs. the treatment with psyllium), regardless of the CuNPs dose (Table 2). The D × F interaction showed the highest fat content was observed in the hepatic tissue in the CHH and PN groups ($p < 0.05$ vs. all other groups), while the lowest fat content was noted in the SNH rats ($p < 0.05$ vs. remaining groups except SN and JN). Irrespective of the nanoparticle dose, psyllium treatment decreased hepatic total cholesterol and triglycerides concentration, as well as the activity of plasma ALT, as compared to other treatments with cellulose, pectin, or inulin. Additionally, we observed reduced ALT activity in the pectin treatment in comparison to the cellulose one ($p < 0.05$). The CuNPs dose by fiber interaction showed the highest plasma AST activity occurred in the CNH and PNH rats ($p < 0.05$ vs. remaining groups except CN), while the lowest AST was noted in the JN and JNH groups ($p < 0.05$ vs. CN, CNH, PNH).

Table 2. Liver function indicators and the levels of AST and ALT in the blood plasma of rats that were fed experimental diets (n = 10 per group) *.

	Weight g/100 g BW	Fat %	TC mg/g	TG mg/g	AST U/L	ALT U/L
Control C	2.80 ± 0.04	14.8 ± 0.74	10.9 ± 1.03	13.8 ± 1.08	57.5 ± 1.55	17.6 ± 0.93
Control CH	2.83 ± 0.07	13.8 ± 0.97	10.8 ± 0.68	13.3 ± 0.93	59.6 ± 1.14	19.0 ± 0.99
Two-way ANOVA:						
CN	2.83 ± 0.08	12.2 ± 0.96 [b#]	6.42 ± 0.58 [#]	10.8 ± 0.48 [#]	63.1 ± 0.67 [ab#]	19.1 ± 0.66
CNH	2.78 ± 0.06	18.8 ± 1.39 [a&]	8.71 ± 1.04	11.1 ± 1.15	66.7 ± 1.88 [a&]	21.8 ± 1.10
PN	2.67 ± 0.04 [#]	18.3 ± 1.13 [a#]	6.40 ± 0.79 [#]	11.4 ± 1.07	59.2 ± 1.82 [bc]	18.2 ± 0.83
PNH	2.73 ± 0.06	12.5 ± 0.81 [b]	7.11 ± 0.88 [&]	12.8 ± 0.98	68.1 ± 2.07 [a&]	18.5 ± 0.45
JN	2.78 ± 0.07	8.26 ± 0.85 [cd#]	8.25 ± 1.09	10.8 ± 0.98 [#]	55.7 ± 1.65 [c]	19.1 ± 0.96
JNH	2.93 ± 0.18	8.49 ± 0.84 [c&]	7.43 ± 0.47 [&]	10.4 ± 0.85 [&]	55.5 ± 1.88 [c]	19.3 ± 0.93
SN	2.58 ± 0.06 [#]	7.90 ± 0.57 [cd#]	3.67 ± 0.23 [#]	9.11 ± 0.56 [#]	60.0 ± 1.91 [bc]	16.6 ± 0.78
SNH	2.65 ± 0.05	5.56 ± 067 [d&]	2.97 ± 0.55 [&]	6.87 ± 0.57 [&]	60.1 ± 1.91 [bc]	16.4 ± 0.73 [&]
SEM	0.026	0.511	0.344	0.333	0.654	0.297
CuNPs dose (D)						
L (6.5 mg/kg)	2.72	11.6	6.18	10.5	59.5	18.2
H (13 mg/kg)	2.76	11.4	6.55	10.3	62.3	18.9
p value	0.509	0.656	0.487	0.697	0.025	0.243
Fiber type (F)						
C (cellulose)	2.78 [ab]	15.5	7.56 [a]	11.0 [a]	64.9	20.5 [a]
P (pectin)	2.70 [ab]	15.4	6.76 [a]	12.1 [a]	63.7	18.3 [b]
J (inulin)	2.85 [a]	8.38	7.84 [a]	10.6 [a]	55.0	19.1 [ab]
S (psyllium)	2.62 [b]	6.73	3.32 [b]	7.99 [b]	60.0	16.5 [c]
p value	0.041	<0.001	<0.001	<0.001	<0.001	<0.001
Interaction D × F						
p value	0.504	<0.001	0.146	0.212	0.022	0.256

* The experimental feeding period involved different dietary treatments: groups C, control diet with a standard Cu content in the mineral mix and 8% of cellulose as a dietary fiber source; group CH, control diet with an enhanced Cu content in the mineral mix and 8% of cellulose as a dietary fiber; group CN, diet with a standard CuNPs content in the mineral mix and 8% of cellulose as a dietary fiber source; group CNH, diet with enhanced CuNPs content in the mineral mix and 8% of cellulose as a dietary fiber; group PN, diet with a standard CuNPs content in the mineral mix and 2% of cellulose and 6% of pectin as a dietary fiber source; group PNH, diet with enhanced CuNPs content in the mineral mix and 2% of cellulose and 6% of pectin as a dietary fiber source; group JN, diet with a standard CuNPs content in the mineral mix and 2% of cellulose and 6% of inulin as a dietary fiber source; group JNH, diet with enhanced CuNPs content in the mineral mix and 2% of cellulose and 6% of inulin as a dietary fiber source; group SN, diet with a standard CuNPs content in the mineral mix and 2% of cellulose and 6% of psyllium as a dietary fiber source; group SNH, diet with enhanced CuNPs content in the mineral mix and 2% of cellulose and 6% of psyllium as a dietary fiber source. Mean values within a column with different superscript letters differ significantly ($p < 0.05$). Differences among the groups (CN, CNH, PN, PNH, JN, JNH, SN, SNH) are indicated with superscripts only if there is a statistically significant interaction D × F ($p < 0.05$). Each experimental group fed CuNPs 6.5 mg/kg (CN, PN, JN, SN) was compared with the control C group using a t-test (# indicates a significant difference versus the C group). Similarly, each experimental group fed CuNPs 13 mg/kg (CNH, PNH, JNH, SNH) was compared with the control CH group using a t-test (& indicates a significant difference versus the CH group). The results are presented as the mean ± standard error of the mean (SEM). ALT, alanine aminotransferase; AST, aspartate aminotransferase; TC, total cholesterol; TG, triglycerides.

The t-test revealed a diminished relative liver weight in the PN and SN rats compared to the C control ($p < 0.05$). Hepatic fat concentration was significantly enhanced in the CNH and PN groups vs. their respective controls, i.e., the CH and C, respectively (t-test; $p < 0.05$). In the remaining groups fed diets with copper nanoparticles, except PNH, the liver fat concentration was decreased in comparison to the respective controls fed diets with $CuCO_3$ (t-test; $p < 0.05$). As compared to the respective controls, the hepatic total cholesterol was diminished in all experimental groups with CuNPs, except CNH and JN.

In the case of hepatic triglycerides, such decreases were noted in the CN, JN, JHNH, SN, and SNH rats ($p < 0.05$; t-test). The t-test showed significantly enhanced blood plasma AST activity in the CN (vs. C), CNH (vs. CH), and PNH (vs. CH) groups. The plasma ALT activity was lower in SNH rats than CH ones (t-test; $p < 0.05$).

The two-way ANOVA showed that irrespective of the CuNPs dose, the treatments with functional fiber, i.e., pectin, inulin, and psyllium, caused a significant increase in the plasma HDL concentration, thus decreasing the TC/HDL ratio compared to the cellulose treatment (Table 3). The lowest AIP index value was noted in the psyllium treatment, regardless of the CuNPs dose ($p < 0.05$ vs. C and P treatments). Additionally, the AIP noted in the inulin treatment was lower than in the C one ($p < 0.05$). The t-test showed that in comparison to a respective control, the SNH rats had decreased creatinine concentrations, and the PNH and SN animals had decreased uric acid levels in their blood plasma. As compared to the C control, plasma urea was significantly enhanced in the JN group (t-test). The plasma HDL concentration was enhanced in the PN and SN rats (vs. control C; t-test), as well as in the JNH and SNH animals (vs. control CH; t-test). The TC/HDL ratio decreased in the PN, JN, JNH, SN, and SNH rats in comparison to their respective controls ($p < 0.05$; t-test). The AIP index decreased in rats JNH vs. CH ($p < 0.05$; t-test).

Table 3. Blood plasma parameters in rats fed experimental diets (n = 10 per group) *.

	Creat. µmol/L	UA µmol/L	Urea mmol/L	HDL mmol/L	TC mmol/L	TG mmol/L	GL mmol/L	AIP	TC/HDL
Control C	23.6 ± 0.95	35.4 ± 1.85	3.99 ± 0.22	0.397 ± 0.026	1.58 ± 0.09	0.927 ± 0.08	12.4 ± 0.63	0.361 ± 0.049	4.04 ± 0.18
Control CH	24.0 ± 0.72	34.2 ± 2.28	4.20 ± 0.09	0.417 ± 0.022	1.69 ± 0.08	1.05 ± 0.07	13.9 ± 0.71	0.399 ± 0.033	4.08 ± 0.13
Two-way ANOVA:									
CN	25.0 ± 1.03	33.4 ± 1.82	4.20 ± 0.11	0.442 ± 0.023	1.75 ± 0.08	1.22 ± 0.12	13.3 ± 0.85	0.427 ± 0.045	4.01 ± 0.16
CNH	23.5 ± 0.71	34.8 ± 2.54	4.35 ± 0.17	0.417 ± 0.030	1.66 ± 0.08	1.01 ± 0.05	13.8 ± 0.57	0.387 ± 0.039	4.06 ± 0.19
PN	22.7 ± 0.95	29.7 ± 2.59	4.28 ± 0.09	0.491 ± 0.022 [#]	1.64 ± 0.07	1.16 ± 0.08	13.7 ± 0.50	0.367 ± 0.034	3.35 ± 0.08 [#]
PNH	22.7 ± 0.91	28.2 ± 1.60 [&]	4.39 ± 0.15	0.467 ± 0.027	1.66 ± 0.08	1.16 ± 0.10	13.7 ± 0.42	0.389 ± 0.041	3.61 ± 0.18
JN	22.5 ± 0.89	30.1 ± 2.56	4.84 ± 0.19 [#]	0.462 ± 0.031	1.56 ± 0.07	0.901 ± 0.08	13.2 ± 0.76	0.283 ± 0.046	3.46 ± 0.19 [#]
JNH	23.7 ± 1.00	33.2 ± 3.15	4.57 ± 0.20	0.499 ± 0.019 [&]	1.61 ± 0.09	0.960 ± 0.05	14.2 ± 0.63	0.283 ± 0.026 [&]	3.26 ± 0.23 [&]
SN	23.3 ± 1.63	28.9 ± 1.96 [#]	4.56 ± 0.33	0.499 ± 0.022 [#]	1.67 ± 0.09	0.926 ± 0.14	12.4 ± 0.37	0.200 ± 0.100	3.35 ± 0.14 [#]
SNH	21.8 ± 0.74 [&]	29.5 ± 2.22	4.28 ± 0.17	0.504 ± 0.019 [&]	1.73 ± 0.08	1.12 ± 0.12	12.7 ± 0.64	0.326 ± 0.046	3.44 ± 0.10 [&]
SEM	0.309	0.740	0.060	0.008	0.025	0.030	0.198	0.016	0.059
CuNPs dose (D)									
L (6.5 mg/kg)	23.4	30.5	4.47	0.474	1.66	1.05	13.1	0.319	3.54
H (13 mg/kg)	22.7	31.5	4.42	0.472	1.67	1.06	13.6	0.346	3.60
p value	0.382	0.548	0.692	0.932	0.841	0.864	0.294	0.470	0.622
Fibre type (F)									
C (cellulose)	24.2	34.1	4.27	0.430 [b]	1.70	1.11	13.5	0.407 [a]	4.04 [a]
P (pectin)	22.7	29.0	4.33	0.479 [a]	1.65	1.16	13.7	0.378 [ab]	3.48 [b]
J (inulin)	22.8	31.9	4.75	0.481 [a]	1.59	0.933	13.7	0.284 [bc]	3.37 [b]
S (psyllium)	22.5	29.2	4.42	0.502 [a]	1.70	1.02	12.6	0.263 [c]	3.40 [b]
p value	0.285	0.105	0.070	0.033	0.511	0.112	0.221	0.017	<0.001
Interaction D × F p value	0.687	0.766	0.594	0.551	0.719	0.232	0.931	0.435	0.662

* Description same as in Table 2. The results are presented as the mean ± standard error of the mean (SEM). Mean values within a column with different superscript letters differ significantly ($p < 0.05$). Differences among the groups (CN, CNH, PN, PNH, JN, JNH, SN, SNH) are indicated with superscripts only if there is a statistically significant interaction D × F ($p < 0.05$). Each experimental group fed CuNPs 6.5 mg/kg (CN, PN, JN, SN) was compared with the control C group using a t-test (# indicates a significant difference versus the C group). Similarly, each experimental group fed CuNPs 13 mg/kg (CNH, PNH, JNH, SNH) was compared with the control CH group using a t-test (& indicates a significant difference versus the CH group). TC, total cholesterol (mmol/L); HDL, high density lipoprotein (mmol/L); TG, triglycerides (mmol/L); AIP, atherogenic index of plasma [log(TG/HDL)]; UA, uric acid (µmol/L); Urea, mmol/L; Creat., creatinine (µmol/L); GL, glucose (mmol/L).

A significant CuNPs by fiber interaction showed that insulin concentration in the plasma of rats fed diets PN, PNH, JN, and SNH was enhanced as compared to JNH rats ($p < 0.05$; Table 4). The highest and lowest plasma ghrelin concentration was noted in the JNH and SNH groups, respectively (in both cases $p < 0.05$ vs. remaining groups). The highest glucagon concentration in the plasma followed feeding with diet CN ($p < 0.05$ vs. all other groups, except CNH), while the lowest glucagon levels were noted in the PN,

JNH, SN, and SNH rats ($p < 0.05$ vs. other groups; see significant D×F interaction). The *t*-test revealed that all groups fed diets with copper nanoparticles were characterized by diminished plasma insulin levels when compared to their respective controls ($p < 0.05$ vs. C or CH). The ghrelin concentration in the plasma was enhanced in JNH rats vs. CH rats, while in the SNH rats, ghrelin levels were lower compared to the CH control ($p < 0.05$; *t*-test). The glucagon content in the plasma was decreased in the groups fed diets with copper nanoparticles, except CN, CNH, and PNH in comparison to the respective controls without CuNPs ($p < 0.05$; *t*-test).

Table 4. Blood levels of insulin, ghrelin, and glucagon in rats fed experimental diets.(n = 10 per group) *.

	Insulin µIU/mL	Ghrelin pg/mL	Glucagon ng/mL
Control C	16.6 ± 0.71	30.4 ± 1.46	9.99 ± 0.36
Control CH	15.6 ± 0.74	28.7 ± 1.61	8.37 ± 0.58
Two-way ANOVA:			
CN	10.7 ± 0.92 $^{ab\#}$	30.0 ± 1.94 ^{b}	9.70 ± 0.92 ^{a}
CNH	10.6 ± 0.30 $^{ab\&}$	29.2 ± 1.78 ^{b}	9.48 ± 0.51 ^{ab}
PN	11.6 ± 0.46 $^{a\#}$	29.5 ± 1.05 ^{b}	5.53 ± 0.38 $^{d\#}$
PNH	12.1 ± 0.72 $^{a\&}$	30.6 ± 1.18 ^{b}	7.07 ± 0.60 ^{c}
JN	12.0 ± 0.50 $^{a\#}$	30.7 ± 2.08 ^{b}	8.03 ± 0.31 $^{bc\#}$
JNH	9.71 ± 0.52 $^{b\&}$	35.1 ± 0.64 $^{a\&}$	5.52 ± 0.50 $^{d\&}$
SN	10.9 ± 0.41 $^{ab\#}$	29.9 ± 1.46 ^{b}	5.23 ± 0.44 $^{d\#}$
SNH	11.7 ± 0.52 $^{a\&}$	20.4 ± 1.15 $^{c\&}$	4.99 ± 0.33 $^{d\&}$
SEM	0.279	0.568	0.246
CuNPs dose (D)			
L (6.5 mg/kg)	11.3	30.0	7.12
H (13 mg/kg)	11.0	28.8	6.76
p value	0.489	0.258	0.342
Fiber type (F)			
C (cellulose)	10.7	29.6	9.59
P (pectin)	11.8	30.0	6.30
J (inulin)	10.8	32.9	6.78
S (psyllium)	11.3	25.2	5.11
p value	0.181	<0.001	<0.001
Interaction D × F			
p value	0.042	<0.001	0.004

* Description same as in Table 2. The results are presented as the mean ± standard error of the mean (SEM). Mean values within a column with different superscript letters differ significantly ($p < 0.05$). Differences among the groups (CN, CNH, PN, PNH, JN, JNH, SN, SNH) are indicated with superscripts only if there is a statistically significant interaction D × F ($p < 0.05$). Each experimental group fed CuNPs 6.5 mg/kg (CN, PN, JN, SN) was compared with the control C group using a *t*-test (# indicates a significant difference versus the C group). Similarly, each experimental group fed CuNPs 13 mg/kg (CNH, PNH, JNH, SNH) was compared with the control CH group using a *t*-test (& indicates a significant difference versus the CH group).

The hepatic *srebp-1c* expression was significantly enhanced in the CNH group, while the lowest expression of *srebp-1c* in the liver was observed in the SNH rats (Figure 1; in both cases $p < 0.05$ vs. all other groups). The CNH rats were also characterized by the highest hepatic *cox-2* expression ($p < 0.05$ vs. PN, JN, SN, SNH). The SNH liver had decreased *ppar-*γ expression as compared to other groups ($p < 0.05$). The *t*-test showed a decrease in *srebp-1c* and *ppar-*γ hepatic expression in SNH rats vs. control CH ($p < 0.05$). The *cox-2* expression was significantly higher in the CNH group than in the CH one (*t*-test).

The histopathological examination revealed minor changes in the liver structure of rats from the C and CH groups (Figure 2). A physiological structure with small clusters of mononuclear cell infiltration and slight congestion of the liver was observed in the CH group. The CN group showed a normal liver structure without any pathological changes. In the SN group, minor histological changes were observed, including minor fatty changes and mononuclear cell infiltration. Additionally, the SNH, JN, and JNH groups, which were administered with inulin and psyllium, exhibited mild fatty degeneration and single clusters of mononuclear cell infiltration. The most severe changes, characterized by extensive fatty degeneration, were observed in the CNH, PN, and PNH groups.

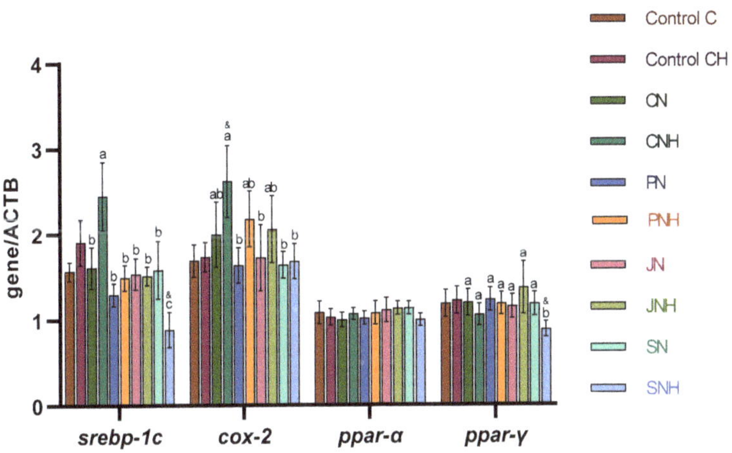

Figure 1. Hepatic mRNA expression (normalized to β-actin) of metabolic and inflammatory factors in rats fed experimental diets (n = 10 per group). Description of the groups same as in Table 2. *cox-2*, cyclooxygenase-2; *ppar-α*, peroxisome proliferator-activated receptor alpha; *ppar-γ*, peroxisome proliferator-activated receptor gamma; *srebp-1c*, sterol regulatory element-binding protein 1. Mean values within a column with different superscript letters differ significantly ($p < 0.05$). Similarly, each experimental group fed CuNPs 13 mg/kg (CNH, PNH, JNH, SNH) was compared with the control CH group using a *t*-test (& indicates a significant difference versus the CH group).

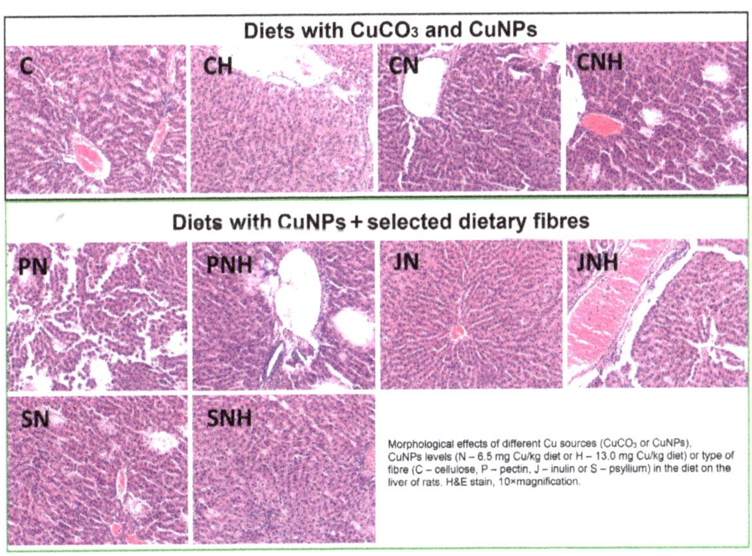

Figure 2. Hepatic histology in rats fed experimental diets. Description of the groups same as in Table 2. C, single foci of fatty degeneration of a physiological nature combined with the presence of small clusters of mononuclear cell infiltration; CH, physiological structure combined with slight congestion of the liver; CN, physiological structure of the liver and the presence of multiple small foci

of fatty degeneration; CNH, extensive multiple-multifocal foci of fatty degeneration of the liver; PN, extensive singular fatty degeneration of the liver; PNH, extensive multifocal fatty degeneration of the liver; JN, physiological structure of the liver and presence of multiple small foci of fatty degeneration, a single cluster of mononuclear cell infiltration; JNH, multiple small fatty degeneration of the liver, combined with multiple clusters of mononuclear cell infiltration; SN, presence of multiple small foci of fatty degeneration, a single cluster of mononuclear cell infiltration; SNH, single-focal fatty degeneration of the liver, combined with single clusters of mononuclear cell infiltration.

4. Discussion

The present experiment involves Wistar rats (Cmdb:Wi) because the rat model is an established host model for nutritional and metabolic studies, including the gastrointestinal response to nutritional interventions and the systemic (metabolic) changes observed with test dietary supplementation. Of course, there are limitations of applying findings from an animal study to humans. These limitations stem from differences in species, fiber, and copper content of the diet, and the inability to test dietary nanoparticles on humans. However, sufficient literature supports the statement that the rat model provides important strengths for the study of human health and disease in relation to human dietary habits and environment. Copper is essential for numerous metabolic activities, including lipid metabolism, redox balance, and iron mobilization, and its homeostasis is crucial for maintaining cellular function [23,24]. Dysregulation of copper levels can lead to oxidative stress, which is implicated in developing many liver metabolic disorders. Recently, there has been much interest in using nanocompounds in food. Due to their size and surface interaction area, these compounds may affect organisms differently than their native form. Indeed, in this study, replacing $CuCO_3$ with CuNPs at a dose of 6.5 mg/kg significantly reduced fat, cholesterol, and triglycerides in the liver, while increasing the dose to 13 mg/kg showed the opposite effect. Due to the enhanced interaction of nano Cu with the body, they have a more significant impact, which may be comparable to an excess of copper in the diet. Other studies also showed that excess Cu could induce mitochondrial dysfunction, promoting lipid deposition and lipogenesis [25]. Furthermore, histological analysis confirmed the observed changes regarding the effect of CuNPs dose on hepatic lipid metabolism. Numerous multifocal foci of lipid degeneration were noted in the group with an increased dose of CuNPs. Irrespective of the dose, CuNPs increased plasma aspartate aminotransferase activity, which may indicate the activation of inflammatory mechanisms in the liver. These changes were most likely related to the increased expression levels of *srebp-1c* and *cox-2* in groups with CuNPs, which are involved in lipid accumulation mechanisms and liver inflammation development [26,27]. Srebp-1c is intricately linked to insulin signaling and plays a pivotal role in lipid and glucose metabolism. Insulin stimulates the transcription of the *srebp-1c* gene, activating the transcription of the genes necessary for fatty acid synthesis, thereby promoting de novo lipogenesis in the liver [26,28]. Interestingly, in this study, blood insulin levels decreased significantly, while glucose levels did not change when the dietary source of Cu was replaced with CuNPs. Other studies showed that srebp-1c in the liver represses insulin receptor substrate-2 (IRS-2) transcription by displacing forkhead proteins from the insulin response element on the IRS-2 promoter, contributing to hepatic insulin resistance [29]. This indicates that an increased dose of CuNPs can significantly interfere with the regulation of glucose and lipid metabolism in the liver. This may disrupt the mechanism linking the interaction of *srebp-1c* and insulin, thus increasing liver fat accumulation and developing inflammation.

Our previous nutritional study with CuNPs suggests that adding dietary fiber to rats' diets decreases Cu intake, affecting Cu bioavailability [20]. Based on these results, it was assumed that combining CuNPs with dietary fibers may regulate the intensity of the nanocompound's action on hepatic metabolism. Indeed, this study confirmed that the beneficial effects of combining CuNPs with dietary fiber depend on the type of fiber. The combination of psyllium and CuNPs in the diet, regardless of dose, was most beneficial in reducing fat, total cholesterol, and triglycerides in the liver. In addition, a comparable effect was observed when inulin was combined with an increased dose of CuNPs in the diet. While inulin and psyllium offer significant health benefits, their nutritional and functional properties differ. Inulin is primarily valued for its prebiotic effects and its role in enhancing mineral absorption, metabolic health, and reducing inflammation [30]. In contrast, psyllium is prized for its ability to form a gel-like substance in the gut that helps slow digestion, manage blood sugar levels, and lower cholesterol [31]. Therefore, adding fibers like psyllium and inulin to the diet could be an explanation for the mitigation of the CuNPs effect by reducing the activity of aminotransferases in the blood. Additionally, higher blood HDL levels and lower atherogenicity factor values were observed in the groups with CuNPs supplemented with psyllium and inulin. The same was observed in the histological analysis, where adding inulin and psyllium to a diet with a higher dose of CuNPs reduced the foci of fatty degeneration of the liver. The beneficial effects of combining psyllium with CuNPs on lipid metabolism and activating pro-inflammatory mechanisms in the liver are most likely related to the observed downregulation of *srebp1c*, *ppar* γ, and *cox-2* expression levels. These molecule indicators are involved in many mechanisms regulating lipids, glucose metabolism, and inflammation in the liver [26,32]. Particularly interesting are the changes in *ppar-*γ expression. Activation of this receptor in the liver can lead to the increased storage of fatty acids in the form of triglycerides, which is a protective mechanism against lipotoxicity, preventing the accumulation of free fatty acids that can cause cellular damage [33,34]. This mechanism may explain why liver triglyceride levels were reduced in the group with psyllium.

Moreover, *ppar-*γ plays a crucial role in glucose metabolism. It enhances insulin sensitivity by modulating the expression of genes involved in glucose uptake and utilization. It is essential in developing NAFLD and type 2 diabetes, where insulin resistance is a standard disorder. By improving insulin sensitivity, *ppar-*γ activation can help reduce hepatic glucose production and improve overall glucose homeostasis [35,36]. Interestingly, when the examined fibers were added to the diets, CuPNs' insulin-lowering effect with unchanged blood glucose levels was also observed. However, the examined dietary fibers reduced glucagon levels. Glucagon is produced by the alpha cells of the pancreas and serves to increase blood glucose levels, acting as a counter-regulatory hormone to insulin [37]. Consequently, the natural mechanism for lowering blood insulin levels involves a decrease in glucagon secretion. This mechanism was disrupted when CuNPs were added to the diet. However, adding dietary fibers (specifically psyllium) reversed the action of CuNPs and lowered glucagon levels.

Another mechanism that can indirectly influence blood insulin levels is related to ghrelin secretion. Ghrelin has been shown to inhibit insulin secretion, which can lead to increased blood glucose levels and potentially contribute to insulin resistance over time [38]. When comparing the dietary fibers tested, psyllium showed the most significant reduction in blood ghrelin levels. However, inulin increases the level of this hormone. This effect may also explain the lowest insulin level observed in the group with CuNPs and inulin. One of the mechanisms regulating ghrelin secretion is associated with the feeling of satiety. Therefore, the characteristic gelling effect of psyllium [31], which can make it easier to feel full, may also reduce the secretion of this hormone. The effect of a decrease in ghrelin levels matches the reduced dietary intake in the psyllium and CuNPs group described in an earlier paper [17]. For inulin, the mechanism of increasing ghrelin levels in blood is probably associated with its prebiotic properties and the production of short-chain fatty acids (SCFAs). The fermentation of inulin by gut bacteria increases the production of SCFAs, which can enhance the secretion of ghrelin by stimulating the enteroendocrine cells in the stomach and the small intestine. Moreover, the increase in SCFAs can also affect the expression of genes involved in ghrelin production [39–41].

5. Conclusions

In conclusion, the gathered data show that CuNPs can significantly impact hepatic lipid metabolism and inflammatory mechanisms in the liver. The study revealed that replacing $CuCO_3$ with CuNPs at different doses had distinct, unfavorable effects on liver fat, cholesterol, and triglyceride levels. High doses of CuNPs induced lipid degeneration and activated inflammatory mechanisms, as indicated by the increased plasma aspartate aminotransferase activity and the elevated expression levels of specific genes associated with lipid accumulation and liver inflammation. Interestingly, the study also demonstrated that combining CuNPs with dietary fibers, such as psyllium and inulin, modulates the effects of CuNPs on hepatic metabolism. Combining CuNPs with these dietary fibers showed beneficial effects in reducing fat, cholesterol, and triglycerides, while also decreasing the activity of aminotransferases in the blood. Moreover, it was observed that adding psyllium and inulin resulted in higher blood HDL levels and lower liver fatty degeneration foci when combined with CuNPs. These findings underscore the intricate relationship between CuNPs, dietary fibers, and hepatic metabolism. The addition of psyllium to the diet with CuNPs downregulated specific gene expression levels, such as *srebp-1c*, *ppar-γ*, and *cox-2*, responsible for the regulation of lipids, glucose metabolism, and inflammatory pathways in the liver. The study highlights the potential of dietary fibers in reducing the negative effects of CuNPs on liver function, opening the door for further research into the combined effects of nanocompounds and dietary interventions on metabolic health. However, it is important to note that the potential toxicity of CuNPs still poses a significant barrier to clinical translation.

Author Contributions: Conceptualization, J.J. and K.O.; methodology, A.M., B.F., D.N., E.C., and P.L.; software, B.F.; validation, B.F., J.F., and J.J.; formal analysis, B.F. and J.J.; investigation, J.J., K.O., B.F., and A.M.; resources, J.J. and K.O.; data curation, J.J. and K.O.; writing—original draft preparation, B.F., J.J., and A.M.; writing—review and editing, B.F., J.J., and A.M.; visualization, B.F. and J.F.; supervision, J.J., B.F., and K.O.; project administration, J.J.; funding acquisition, J.J. and K.O. All authors have read and agreed to the published version of the manuscript.

Funding: This research was funded by National Science Center, Poland (Grant No. 2021/41/B/NZ9/01104).

Institutional Review Board Statement: The animal study protocol was approved by the National Ethics Committee for Animal Experiments (Permission No. 19/2021; Olsztyn, Poland, 17 March 2021). The study was conducted in accordance with the ARRIVE guidelines, and every attempt was made to minimize the animals' suffering during the experiment.

Data Availability Statement: The original contributions presented in the study are included in the article, further inquiries can be directed to the corresponding author.

Conflicts of Interest: The authors declare no conflicts of interest. The funders had no role in the design of the study; in the collection, analyses, or interpretation of data; in the writing of the manuscript; or in the decision to publish the results.

References

1. Chibber, S.; Shanker, R. Can CuO Nanoparticles Lead to Epigenetic Regulation of Antioxidant Enzyme System? *J. Appl. Toxicol.* **2017**, *37*, 84–91. [CrossRef] [PubMed]
2. Pérez-Hernández, H.; García-Mayagoitia, S.; Torres-Gómez, P.A.; Campos-Montiel, R.G.; Fernández-Luqueño, F. Ecological Effects of Copper NPs: Advantages and Drawbacks Regarding Current and Potential Applications. In *Copper Nanostructures: Next-Generation of Agrochemicals for Sustainable Agroecosystems*; Elsevier: Amsterdam, The Netherlands, 2022; pp. 719–750, ISBN 978-0-12-823833-2.
3. Ravi, S.; Kishore, I.R. Consumer Nanoproducts in Antimicrobial Application. In *Handbook of Consumer Nanoproducts*; Springer Nature: Singapore, 2022; pp. 1499–1513. ISBN 9789811686979.
4. Adetunji, C.O.; Michael, O.S.; Rathee, S.; Singh, K.R.; Ajayi, O.O.; Adetunji, J.B.; Ojha, A.; Singh, J.; Singh, R.P. Potentialities of Nanomaterials for the Management and Treatment of Metabolic Syndrome: A New Insight. *Mater. Today Adv.* **2022**, *13*, 100198. [CrossRef]
5. Dardevet, D.; Moore, M.C.; Remond, D.; Everett-Grueter, C.A.; Cherrington, A.D. Regulation of Hepatic Metabolism by Enteral Delivery of Nutrients. *Nutr. Res. Rev.* **2006**, *19*, 161–173. [CrossRef] [PubMed]
6. Guo, X.; Huang, Z.; Chen, J.; He, K.; Lin, J.; Zhang, H.; Zeng, Y. Synergistic Delivery of Resveratrol and Ultrasmall Copper-Based Nanoparticles by Aptamer-Functionalized Ultrasound Nanobubbles for the Treatment of Nonalcoholic Fatty Liver Disease. *Front. Physiol.* **2022**, *13*, 950141. [CrossRef] [PubMed]

7. Padmanaban, S.; Pully, D.; Samrot, A.V.; Gosu, V.; Sadasivam, N.; Park, I.; Radhakrishnan, K.; Kim, D.-K. Rising Influence of Nanotechnology in Addressing Oxidative Stress-Related Liver Disorders. *Antioxidants* **2023**, *12*, 1405. [CrossRef] [PubMed]
8. Ghasemi, P.; Shafiee, G.; Ziamajidi, N.; Abbasalipourkabir, R. Copper Nanoparticles Induce Apoptosis and Oxidative Stress in SW480 Human Colon Cancer Cell Line. *Biol. Trace Elem. Res.* **2023**, *201*, 3746–3754. [CrossRef]
9. Cai, J.; Peng, J.; Feng, J.; Li, R.; Ren, P.; Zang, X.; Wu, Z.; Lu, Y.; Luo, L.; Hu, Z.; et al. Antioxidant Hepatic Lipid Metabolism Can Be Promoted by Orally Administered Inorganic Nanoparticles. *Nat. Commun.* **2023**, *14*, 3643. [CrossRef]
10. Tarantino, G.; Porcu, C.; Arciello, M.; Andreozzi, P.; Balsano, C. Prediction of Carotid Intima–Media Thickness in Obese Patients with Low Prevalence of Comorbidities by Serum Copper Bioavailability. *J. Gastroenterol. Hepatol.* **2018**, *33*, 1511–1517. [CrossRef]
11. Tang, H.; Xu, M.; Shi, F.; Ye, G.; Lv, C.; Luo, J.; Zhao, L.; Li, Y. Effects and Mechanism of Nano-Copper Exposure on Hepatic Cytochrome P450 Enzymes in Rats. *Int. J. Mol. Sci.* **2018**, *19*, 2140. [CrossRef]
12. Tang, H.; Xu, M.; Luo, J.; Zhao, L.; Ye, G.; Shi, F.; Lv, C.; Chen, H.; Wang, Y.; Li, Y. Liver Toxicity Assessments in Rats Following Sub-Chronic Oral Exposure to Copper Nanoparticles. *Environ. Sci. Eur.* **2019**, *31*, 30. [CrossRef]
13. Lee, I.-C.; Ko, J.-W.; Park, S.-H.; Shin, N.-R.; Shin, I.-S.; Moon, C.; Kim, S.-H.; Yun, W.-K.; Kim, H.-C.; Kim, J.-C. Copper Nanoparticles Induce Early Fibrotic Changes in the Liver via TGF-β/Smad Signaling and Cause Immunosuppressive Effects in Rats. *Nanotoxicology* **2018**, *12*, 637–651. [CrossRef] [PubMed]
14. Li, C.; Liu, Z.; Wei, W.; Chen, C.; Zhang, L.; Wang, Y.; Zhou, B.; Liu, L.; Li, X.; Zhao, C. Exploring the Regulatory Effect of LPJZ-658 on Copper Deficiency Combined with Sugar-Induced MASLD in Middle-Aged Mice Based on Multi-Omics Analysis. *Nutrients* **2024**, *16*, 2010. [CrossRef] [PubMed]
15. Xia, Y.; Zu, H.; Guo, H.; Jiang, T.; Yang, S.; Yu, H.; Zhang, S.; Ding, H.; Li, X.; Wang, Y.; et al. Preclinical Safety and Hepatotoxicity Evaluation of Biomineralized Copper Sulfide Nanoagents. *J. Nanobiotechnol.* **2022**, *20*, 185. [CrossRef] [PubMed]
16. Kishida, T. Beneficial Effect of Dietary Fiber and Other Luminacoids on Mineral Bioavailability. *J. Jpn. Assoc. Diet. Fiber Res.* **2008**, *12*, 1–7. [CrossRef]
17. Cholewińska, E.; Marzec, A.; Sołek, P.; Fotschki, B.; Listos, P.; Ognik, K.; Juśkiewicz, J. The Effect of Copper Nanoparticles and a Different Source of Dietary Fibre in the Diet on the Integrity of the Small Intestine in the Rat. *Nutrients* **2023**, *15*, 1588. [CrossRef]
18. Fardet, A. Do the Physical Structure and Physicochemical Characteristics of Dietary Fibers Influence Their Health Effects. In *Dietary Fibre Functionality in Food & Nutraceuticals: From Plant to Gut*; John Wiley & Sons Ltd: Hoboken, NJ, USA, 2016; pp. 1–19.
19. Syryamina, V.N.; Yulikov, M.; Nyström, L. The Cu(Ii)—Dietary Fibre Interactions at Molecular Level Unveiled via EPR Spectroscopy. *RSC Adv.* **2022**, *12*, 19901–19916. [CrossRef]
20. Marzec, A.; Cholewińska, E.; Fotschki, B.; Juśkiewicz, J.; Stepniowska, A.; Ognik, K. Are the Biodistribution and Metabolic Effects of Copper Nanoparticles Dependent on Differences in the Physiological Functions of Dietary Fibre? *Ann. Anim. Sci.* **2024**. [CrossRef]
21. Folch, J.; Lees, M.; Stanley, G.H.S. A Simple Method for the Isolation and Purification of Total Lipides from Animal Tissues. *J. Biol. Chem.* **1957**, *226*, 497–509. [CrossRef]
22. Cholewińska, E.; Juśkiewicz, J.; Majewski, M.; Smagieł, R.; Listos, P.; Fotschki, B.; Godycka-Kłos, I.; Ognik, K. Effect of Copper Nanoparticles in the Diet of WKY and SHR Rats on the Redox Profile and Histology of the Heart, Liver, Kidney, and Small Intestine. *Antioxidants* **2022**, *11*, 910. [CrossRef]
23. Chen, J.; Jiang, Y.; Shi, H.; Peng, Y.; Fan, X.; Li, C. The Molecular Mechanisms of Copper Metabolism and Its Roles in Human Diseases. *Pflüg. Arch. Eur. J. Physiol.* **2020**, *472*, 1415–1429. [CrossRef]
24. Morrell, A.; Tallino, S.; Yu, L.; Burkhead, J.L. The Role of Insufficient Copper in Lipid Synthesis and Fatty-Liver Disease. *IUBMB Life* **2017**, *69*, 263–270. [CrossRef] [PubMed]
25. Zhong, C.-C.; Zhao, T.; Hogstrand, C.; Chen, F.; Song, C.-C.; Luo, Z. Copper (Cu) Induced Changes of Lipid Metabolism through Oxidative Stress-Mediated Autophagy and Nrf2/PPARγ Pathways. *J. Nutr. Biochem.* **2022**, *100*, 108883. [CrossRef] [PubMed]
26. Tian, J.; Goldstein, J.L.; Brown, M.S. Insulin Induction of SREBP-1c in Rodent Liver Requires LXRα-C/EBPβ Complex. *Proc. Natl. Acad. Sci. USA* **2016**, *113*, 8182–8187. [CrossRef] [PubMed]
27. Hu, Y.; Yang, X.; Wu, S.; Xiao, J. COX-2 in Liver Fibrosis. *Clin. Chim. Acta* **2020**, *506*, 196–203. [CrossRef]
28. Dong, Q.; Majumdar, G.; O'Meally, R.N.; Cole, R.N.; Elam, M.B.; Raghow, R. Insulin-Induced de Novo Lipid Synthesis Occurs Mainly via mTOR-Dependent Regulation of Proteostasis of SREBP-1c. *Mol. Cell. Biochem.* **2020**, *463*, 13–31. [CrossRef]
29. Ide, T.; Shimano, H.; Yahagi, N.; Matsuzaka, T.; Nakakuki, M.; Yamamoto, T.; Nakagawa, Y.; Takahashi, A.; Suzuki, H.; Sone, H.; et al. SREBPs Suppress IRS-2-Mediated Insulin Signalling in the Liver. *Nat. Cell Biol.* **2004**, *6*, 351–357. [CrossRef]
30. Qin, Y.-Q.; Wang, L.-Y.; Yang, X.-Y.; Xu, Y.-J.; Fan, G.; Fan, Y.-G.; Ren, J.-N.; An, Q.; Li, X. Inulin: Properties and Health Benefits. *Food Funct.* **2023**, *14*, 2948–2968. [CrossRef]
31. Gunn, D.; Abbas, Z.; Harris, H.C.; Major, G.; Hoad, C.; Gowland, P.; Marciani, L.; Gill, S.K.; Warren, F.J.; Rossi, M.; et al. Psyllium Reduces Inulin-Induced Colonic Gas Production in IBS: MRI and in Vitro Fermentation Studies. *Gut* **2022**, *71*, 919–927. [CrossRef]
32. Wang, Y.; Nakajima, T.; Gonzalez, F.J.; Tanaka, N. PPARs as Metabolic Regulators in the Liver: Lessons from Liver-Specific PPAR-Null Mice. *Int. J. Mol. Sci.* **2020**, *21*, 2061. [CrossRef]
33. Pan, Y.; Li, Y.; Fan, H.; Cui, H.; Chen, Z.; Wang, Y.; Jiang, M.; Wang, G. Roles of the Peroxisome Proliferator-Activated Receptors (PPARs) in the Pathogenesis of Hepatocellular Carcinoma (HCC). *Biomed. Pharmacother.* **2024**, *177*, 117089. [CrossRef]
34. Singh, S.; Kumar, A.; Gupta, S.; Agrawal, R. Curative Role of Natural PPARγ Agonist in Non-Alcoholic Fatty Liver Disease (NAFLD). *Tissue Barriers* **2024**, *12*, 2289830. [CrossRef] [PubMed]

35. Changizi, Z.; Kajbaf, F.; Moslehi, A. An Overview of the Role of Peroxisome Proliferator-Activated Receptors in Liver Diseases. *J. Clin. Transl. Hepatol.* **2023**, *11*, 1542–1552. [CrossRef] [PubMed]
36. Qiu, Y.-Y.; Zhang, J.; Zeng, F.; Zhu, Y.-Z. Roles of the Peroxisome Proliferator-Activated Receptors (PPARs) in the Pathogenesis of Nonalcoholic Fatty Liver Disease (NAFLD). *Pharmacol. Res.* **2023**, *192*, 106786. [CrossRef] [PubMed]
37. Robertson, R.P. Brief Overview: Glucagon History and Physiology. *J. Endocrinol.* **2023**, *258*, e220224. [CrossRef]
38. Price, M.L.; Ley, C.D.; Gorvin, C.M. The Emerging Role of Heterodimerisation and Interacting Proteins in Ghrelin Receptor Function. *J. Endocrinol.* **2022**, *252*, R23–R39. [CrossRef]
39. Cani, P.D.; Dewever, C.; Delzenne, N.M. Inulin-Type Fructans Modulate Gastrointestinal Peptides Involved in Appetite Regulation (Glucagon-like Peptide-1 and Ghrelin) in Rats. *Br. J. Nutr.* **2004**, *92*, 521–526. [CrossRef]
40. Delzenne, N.M.; Cani, P.D.; Daubioul, C.; Neyrinck, A.M. Impact of Inulin and Oligofructose on Gastrointestinal Peptides. *Br. J. Nutr.* **2005**, *93*, S157–S161. [CrossRef]
41. Delzenne, N.M.; Cani, P.D.; Neyrinck, A.M. Modulation of Glucagon-like Peptide 1 and Energy Metabolism by Inulin and Oligofructose: Experimental Data. *J. Nutr.* **2007**, *137*, 2547S–2551S. [CrossRef]

Disclaimer/Publisher's Note: The statements, opinions and data contained in all publications are solely those of the individual author(s) and contributor(s) and not of MDPI and/or the editor(s). MDPI and/or the editor(s) disclaim responsibility for any injury to people or property resulting from any ideas, methods, instructions or products referred to in the content.

nutrients MDPI

Article

Short-Term Effects of Escalating Doses of Cholecalciferol on FGF23 and 24,25(OH)$_2$ Vitamin D Levels: A Preliminary Investigation

Jessica Pepe [1,*], Luciano Colangelo [1], Roberta Pilotto [1], Viviana De Martino [1], Carla Ferrara [2], Alfredo Scillitani [3], Mirella Cilli [1], Salvatore Minisola [1], Ravinder Singh [4] and Cristiana Cipriani [1]

1 Department of Clinical, Internal, Anesthesiology and Cardiovascular Sciences (SCIAC), "Sapienza" University of Rome, 00161 Rome, Italy; luciano.colangelo@uniroma1.it (L.C.); robertapilotto@virgilio.it (R.P.); viviana.demartino@uniroma1.it (V.D.M.); mirella.cilli@uniroma1.it (M.C.); salvatore.minisola@uniroma1.it (S.M.); cristiana.cipriani@gmail.com (C.C.)

2 Istituto Superiore di Sanità, 00161 Rome, Italy; carla.ferrara@iss.it

3 Unit of Endocrinology, Ospedale Casa Sollievo della Sofferenza, IRCCS, San Giovanni Rotondo, 71013 Foggia, Italy; alfredo.scillitani@gmail.com

4 Department of Laboratory Medicine and Pathology, Mayo Clinic, Rochester, MN 55905, USA; singh.ravinder@mayo.edu

* Correspondence: jessica.pepe@uniroma1.it

Abstract: Background: There are few and controversial results on 24,25(OH)$_2$D and FGF23 acute changes following supplementation with cholecalciferol. Methods: Twenty-seven subjects with 25(OH)D < 30 ng/mL were randomized into three groups to receive a single oral dose of 25,000 I.U. or 600,000 I.U. of cholecalciferol or placebo, respectively. We measured 25(OH)D, 1,25(OH)$_2$D, 24,25(OH)$_2$D, and FGF23 levels at baseline and after 72 h. The 1,25(OH)$_2$D/25(OH)D, 1,25(OH)$_2$D/24,25(OH)$_2$D, and 24,25(OH)$_2$D/25(OH)D ratios were also calculated. Results: There was an increase in 25(OH)D and 1,25 (OH)$_2$D following both doses of cholecalciferol. In the group administered 600,000 I.U., there was a significant increase in the delta changes in 25(OH)D and 1,25(OH)$_2$D compared to the placebo and in the delta 24,25(OH)D$_2$ compared to the placebo and 25,000 I.U. groups (all $p < 0.05$). A decrease in both the 1,25(OH)$_2$D/25(OH)D and 1,25(OH)$_2$D/24,25(OH)$_2$D ratio (all $p < 0.05$) was observed in the 600,000 I.U. group. FGF23 values significantly increased only in the group administered 600,000 I.U. Conclusions: 25(OH)D and 1,25(OH)D levels significantly increased following 600,000 IU cholecalciferol administration compared to 25,000 I.U. and placebo. Following the massive administration of cholecalciferol, the CYP24A1 enzyme is actively involved in catabolism, thus, avoiding toxic effects.

Keywords: vitamin D; 24,25(OH)$_2$D; FGF23; cholecalciferol

1. Introduction

There are many ways to replenish the human body with vitamin D. The most physiological one is represented by ultraviolet B (UVB) radiation. UVB is in the region of the ultraviolet spectrum, which extends from about 280 to 320 nm in wavelength. Everything below 290 is absorbed by the stratosphere; therefore, for the natural light range the string from 290 is relevant. Erythemally weighted UVB above 315 nm is irrelevant in D3 formation, thus, the biological effect of this spectrum is through a different mechanism. The peak of production is around 295 nm. It has been shown that many biological effects of UVB cannot be replaced by oral delivery [1]. However, some concerns have been raised, since UVB radiation is primarily responsible for sunburn, aging of the skin, and the development of skin cancer. In this context, it is important to highlight the discovery of novel pathways of vitamin D3 initiation, for example, by CYP11A1 and/or CYP27A1. The generated metabolites among other things, enhance the protection of skin against DNA damage and oxidative stress, thus, protecting its integrity [2]. Furthermore, it has also been hypothesized that these hydroxy-metabolites might play a role in some so-called non-skeletal effects of vitamin D, including systemic diseases such as cancer, autoimmune and inflammatory disorders, atherosclerosis, and many metabolic diseases. Under most living conditions, approximately 75 to 90% of this pro-hormone is produced in the skin after UVB exposure. However, many factors can influence production, the most important being latitude, season, and time of day as well as ozone pollution in the atmosphere. Concerning pollution, it has been shown that the more an area is polluted the more the passage of UVB is diminished, consequently impairing the synthesis of vitamin D. Another physiological possibility is represented by foods and other nutritional sources [3]. However, only a small amount of vitamin D is available from nutritional sources since very few foods naturally contain vitamin D. The flesh of fatty fish (including, trout, salmon, tuna, and mackerel) and fish oils are among the best sources. Beef liver, egg yolks, and cheese have small amounts of vitamin D, primarily in the form of vitamin D3 and its metabolite 25(OH)D3. Mushrooms provide variable amounts of vitamin D2 [4]. The third option is represented by enriching foods with vitamin D or 25(OH)D. Several countries (for example, USA, India, Finland, Canada, and other nations in the Middle East and North African regions) mandate the addition of vitamin D to various foods. However, many other countries are reluctant to accept this approach, even though this could substitute pharmacological vitamin D supplementation without any harmful effects. Finally, the last option is represented by pharmacological supplementation [5]. Concerning the last option, there are a number of commercial products that can be utilized that basically recapitulate the biosynthetic pathway of vitamin D from its origin to the final metabolite. The most utilized of these products are ergocalciferol, cholecalciferol, calcifediol, 1α(OH)D, and calcitriol, even though their utilization is markedly different in European, Asian, and American countries. Other compounds, for example, paricalcitol, are utilized in specific clinical conditions, such as in chronic kidney disease [4].

Cholecalciferol (vitamin D3) is the most commercialized product in Europe. Its supplementation has been proposed as one of the modalities to overcome the widespread prevalence of vitamin D insufficiency in the general population [6,7]. To date, most studies have mainly focused on the ability of different schedules of cholecalciferol administration (daily, weekly, fortnightly, monthly, and even every six months) to reach the vitamin D sufficiency evaluated by measuring circulating 25(OH)D levels, mainly on long-term basis. However, there is little data on the effect of different doses of cholecalciferol on their final active [i.e., 1,25(OH)$_2$D] and catabolic [i.e., 24,25(OH)$_2$D] derivatives [8]. Furthermore, the majority of the studies evaluate the production of 24,25(OH)D over the long term, i.e., weeks or months after vitamin D supplementation of vitamin D deficient population [9–13]. However, to the best of our knowledge, there are no data on the early changes in 24,25(OH)$_2$D nor of 24,25(OH)$_2$D/25(OH)D ratio (a marker of vitamin D metabolism) after different doses of cholecalciferol supplementation. In this context, it should be noted that some data seem to suggest that 24,25(OH)$_2$D cannot be considered only as an inert metabolite. Indeed, it has been implicated in many biological processes independent of the action of 1,25(OH)$_2$D.

CYP24A1, the enzyme needed for the catabolism of both 25(OH)D and 1,25(OH)$_2$D into presumably inactive carboxylic acids, is also influenced by fibroblast growth factor (FGF23) [13]. In particular, FGF23 suppresses the expression of the 1α-hydroxylase enzyme in the kidney and induces the expression of 25-hydroxyvitamin D 24-hydroxylase enzyme, resulting in the decreased conversion of 25(OH)D into 1,25(OH)$_2$D and increased formation of inactive carboxylic acids [14]. A recent meta-analysis showed that vitamin D supplementation leads to a significant increase in serum intact FGF23 among vitamin D-deficient patients [15]. Authors considered studies employing both daily doses and boluses of vitamin D up to 300,000 international units (I.U.), with a follow-up from eight weeks to three months. Finally, according to some authors, a more extensive metabolome profile of vitamin D, including different vitamin D metabolic ratios, could offer insights to both evaluate genetic and acquired diseases of calcium and phosphorus metabolism and better elucidate metabolic pathways [16,17].

Vitamin D metabolism strictly regulates not only calcium but also phosphate metabolism. For example, if the dietary intake of phosphate is low, absorption is mostly active and mediated through 1,25-dihydroxyvitamin D. This final metabolite is able to induce the expression of NPT2b (sodium-phosphate transport protein 2B), which is a sodium-phosphate cotransporter on the apical membrane of intestinal epithelial cells. In the kidneys, 1,25$(OH)_2$D can stimulate NPT2a to increase phosphate reabsorption. NPT2c is typically increased during a low-phosphate diet but reduced in response to a high-phosphate diet. 1,25$(OH)_2$D stimulates both the functional role and biosynthesis of NPT2c [18].

Therefore, the aim of this study was to evaluate the short-term effect (i.e., 72 h) of a small and large oral dose of cholecalciferol on FGF23 and 24,25$(OH)_2$D in vitamin D deficient patients, together with the evaluation of three vitamin D metabolic ratios (1,25$(OH)_2$D/25(OH)D, 1,25$(OH)_2$D/24,25(OH)2D, and 24,25$(OH)_2$D/25(OH)D).

2. Materials and Methods

2.1. Study Design, Recruitment, Inclusion and Exclusion Criteria

Between January and March 2015, we initially investigated the first 70 Caucasian post-menopausal women and men (age range 40–90 years), routinely followed as outpatients at our Department of Internal Medicine. Potentially eligible individuals were subjected to an assessment visit, consisting of medical history, clinical examination (vitals, weight, height, and physical examination, including blood pressure and pulse rate), and biochemical evaluations. The aim was the enrollment of 27 patients, satisfying the inclusion [25(OH)D levels < 30 ng/mL] and exclusion criteria, as a preliminary investigation to define future protocols on the basis of the results obtained. Therefore, as an exploratory analysis, a sample size calculation was not estimated. The exclusion criteria were as follows: estimated creatinine clearance by Cockcroft–Gault formula < 60 mL/min, not suffering from diseases such as neoplasia, hepatitis C and/or B virus co-infection and other liver diseases, skeletal diseases (such as primary hyperparathyroidism, Paget's disease or osteoporosis under treatment), severe water and electrolyte disturbances, malabsorption, acute or chronic renal failure, or any serious systemic diseases that could affect the interpretation of the data. Among 70 patients screened, we report the data from 27 enrolled patients whose vitamin D levels were below 30 ng/mL. None of them were taking drugs affecting bone metabolism or calcium or vitamin D supplements. Patients were randomized into three groups of 9 subjects each to receive a single oral dose of 25,000 or 600,000 I.U. of cholecalciferol or the placebo in an open-label modality. All participants were randomly assigned by a computer-based randomization system. Fasting blood samples were collected at the baseline and 72 h after different doses of cholecalciferol or placebo administration, in all participants. One subject in the 25,000 IU was lost in the follow-up. There were no changes in dietary habits that could have influenced the results during the period of observation.

2.2. Ethical Considerations

The study was approved by our local Ethic Committees and written informed consents have been preliminarily obtained by each subject.

2.3. Biochemistry

Following the immediate analysis of serum ionized calcium, the blood samples were centrifuged at a speed of 3000 rpm for 10 min and the serum was frozen at −80 °C. Assays of the analytes were performed in a single batch at the end of the study. We measured the serum intact parathyroid hormone (PTH), 25-hydroxyvitamin D [25(OH)D], 1,25-dihydroxyvitamin D [1,25$(OH)_2$D], and intact FGF23 (FGF23), as previously described [19,20]. Briefly, the serum ionized calcium (Ca^{2+}) was determined using an ion-selective electrode with the fully automated biochemical analyzer NOVA 8 (Nova Biomedical, Waltham, MA, USA). Serum 25-hydroxyvitamin D [25(OH)D] levels and

1,25-dihydroxy-vitamin D [1,25(OH)$_2$D] were analyzed by radioimmunoassay (RIA, Immunodiagnostic Systems Inc. (IDS Inc.), Fountain Hills, AZ, USA). The sensitivity of the 25(OH)D assay is 1.5 ng/mL, and the intra-assay and inter-assay CVs are below 10.8% and 9.4%, respectively, and the sensitivity of the 1,25(OH)$_2$D assay is 4.6 pg/mL, and the intra-assay and inter-assay CVs are below 6.3% and 9.5%, respectively. The serum levels of PTH were analyzed by immunoradiometric assay (IRMA N-TACT PTH SP, Diasorin, Saluggia, Italy) according to the manufacturer's instructions. The sensitivity of this assay is 0.07 pg/mL, and the intra-assay and inter-assay CVs are below 2.6% and 4.3%, respectively. FGF23 was determined by a human intact FGF23 ELISA kit (Kainos Laboratories, Tokyo, Japan), and the intra- and inter-assay CVs were less than 8% and sensitivity was 3 pg/mL [21]. 24,25(OH)$_2$ vitamin D levels were measured by liquid chromatography–tandem mass spectrometry (LC–MS/MS) at the Mayo Clinic. More specifically, 10 g/mL solution of 24R,25(OH)$_2$D3 and 24R,25(OH)$_2$D2 (Medical Isotopes) in 100% ethanol was prepared in the laboratory. Concentrations of 0, 0.1, 0.5, 1, 5, 10, and 25 ng/mL [0, 0.2, 1.2, 2.4, 12.0, 24.0, and 60.0 nmol/L for 24,25(OH)$_2$D3 and 0, 0.2, 1.2, 2.3, 11.7, 23.3, and 58.3 nmol/L for 24,25(OH)$_2$D2] for calibrators; QC samples (prepared in stripped serum) were at a concentration of 0.6, 5.5, and 14 ng/mL [1.4, 13.2, and 33.6 nmol/L for 24,25(OH)$_2$D3; 1.4, 12.8, and 32.7 nmol/L for 24,25(OH)$_2$D2]. The internal standard of 50 µL [1.25 ng (0.003 nmol)] deuterated 24,25(OH)$_2$D3 and deuterated 24,25(OH)$_2$D2 (Toronto Research Chemicals, North York, ON, Canada) to 500 µL serum followed by 500 µL hydrochloric acid (0.2 N) 15 min later. Extraction was performed with BondElut (C18, 250 mg, 6 mL) cartridge (Varian Instruments, Palo Alto, CA, USA), which was washed once with 2 mL of 70:30 methanol:water (1.4 mL methanol:0.6 mL water) and then with 2 mL of 90:10 hexane:methylene chloride (1.8 mLhexane:0.2 mL methylene chloride). Samples were eluted with 2 mL of 90:10 hexane:isopropyl alcohol (1.8 mL hexane:0.2 mL isopropyl alcohol). Eluents were dried and derivatized with 4-phenyl-1,2,4,triazoline-3,5-dione (PTAD) [Sigma, Kawasaki, Japan; 250 µL of 200 µg/mL (285 µmol/L) solution in acetonitrile]. Analytes were separated by column chromatography at a flow rate of 0.25 mL/min on an Agilent XDB-C8 (Agilent, Santa Clara, CA, USA), 2.1 50 mm column over 7.25 min with a methanol-H_2O-ammonium formate (1 mM) linear gradient (60–95%). Mass spectrometry was performed in the multiple reaction monitoring mode on a AB Sciex 5500 mass spectrometer with Analyst 1.6.2 software (AB Sciex, Framingham, MA, USA) for data acquisition and analysis. We also calculated the following derived vitamin D metabolic ratios: (a) 1,25(OH)$_2$D/25(OH)D ratio, mainly representing the 1α-hydroxylase activity, an enzyme encoded by *CYP27B1* gene in the kidney; (b) 1,25(OH)$_2$D/24,25(OH)$_2$D ratio as an indicator of vitamin D status; (c) 24,25(OH)$_2$D/25(OH)D as a ratio mediated by the *CYP24A1* gene that encodes the enzyme 24-hydroxylase, which catalyzes the conversion of 25(OH)D into 24,25(OH)$_2$D, thus, representing the catabolic activity.

2.4. Safety Endpoints

Safety and tolerability assessments consisted of adverse events (AEs), clinical laboratory (clinical chemistry, haematology, coagulation, and urinalysis), vital signs (systolic and diastolic blood pressure, heart rate, and body temperature), 12-lead electrocardiogram, body weight, falls, and physical examination.

2.5. Statistical Analyses

Patient's biochemical measures according to time from vitamin D or placebo administration are reported as means ± SD. The Kolmogorov–Smirnov test was used to evaluate the normality of the distributions. All parameters were normally distributed with the exception of ionized calcium, BMI, 1,25(OH)$_2$D/25(OH)D, 1,25(OH)$_2$D/24,25(OH)$_2$D, 24,25(OH)$_2$D/25(OH)D, and FGF23. Since the above-mentioned distribution parameters did not follow the Gaussian curve, non-parametric tests were carried out as follows: Kruskal–Wallis ANOVA for comparison between groups (Table 1) and pairwise comparisons using the Wilcoxon test was made in each group (Table 2). For the other variables,

ANOVA was performed to evaluate the delta changes in indices in the groups throughout the study (Figure 1). Adjustments for multiple comparisons (Bonferroni) were also made when the ANOVA test was statistically significant. Fisher's exact two-tailed test were used for categorical variables comparisons between groups. Bonferroni correction was applied to account for multiple tests. The significance was set at a p value of <0.05. The statistical analyses were carried out using the SSPS version 21.

Table 1. Anthropometric and biochemical parameters of three groups of subjects investigated.

	Controls (n = 9)	25,000 I.U. (n = 8)	600,000 I.U. (n = 9)	p Values
Age (years)	70.4 ± 17.1	75.9 ± 10.6	75.2 ± 9.7	0.63
BMI (Kg/m^2)	25.4 ± 5.2	26.2 ± 2.0	24.4 ± 4.6	0.90
Ca^{++} (mmol/L)	1.21 ± 0.08	1.24 ± 0.04	1.24 ± 0.03	0.50
25(OH)D (ng/mL)	11.5 ± 8.6	12.2 ± 8.8	16.55 ± 13.1	0.50
1,25(OH)$_2$D (pg/mL)	34.8 ± 12.6	27.8 ± 11.3	39.8 ± 13.55	0.13
24,25(OH)$_2$D (ng/mL)	0.50 ± 0.45	0.72 ± 0.76	0.72 ± 0.75	0.91
PTH (pg/mL)	55.2 ± 36.0	74.0 ± 57.7	45.0 ± 13.1	0.25
FGF23 (ng/mL)	40.0 ± 13.8	39.7 ± 26.9	48.2 ± 18.7	0.59
1,25(OH)$_2$D/25(OH)D	0.004 ± 0.001	0.002 ± 0.001	0.003 ± 0.001	0.21
1,25(OH)$_2$D/24,25(OH)$_2$D	0.14 ± 0.12	0.16 ± 0.20	0.13 ± 0.10	0.60
24,25(OH)$_2$D/25(OH)D	0.06 ± 0.08	0.11 ± 0.12	0.04 ± 0.03	0.31

Table 2. Absolute changes in the biochemical parameters at 72 h following cholecalciferol or placebo supplementation. Time 0 vs. time 72 h in each group * $p < 0.05$.

	25,000 I.U.		600,000 I.U.		Controls	
	0	72	0	72	0	72
Ca^{++} (mmol/L)	1.24 ± 0.04	1.27 ± 0.04	**1.24 ± 0.03**	**1.28 ± 0.16 ***	1.21 ± 0.08	1.22 ± 0.72
25(OH)D (ng/mL)	**12.2 ± 8.8**	**19.9 ± 11.3 ***	**16.55 ± 13.1**	**58.6 ± 29.4 ***	11.5 ± 8.6	15.5 ± 11.9
1,25(OH)$_2$D (pg/mL)	**27.8 ± 11.3**	**38.6 ± 15.8 ***	**39.8 ± 13.55**	**70.5 ± 17.1 ***	34.8 ± 12.6	30.3 ± 10.25
24,25(OH)$_2$D (ng/mL)	0.72 ± 0.76	0.95 ± 0.98	0.72 ± 0.75	2.56 ± 1.52	0.50 ± 0.45	0.59 ± 0.32
PTH (pg/mL)	74.0 ± 57.7	71.7 ± 61.1	46.0 ± 13.1	38.7 ± 16.6	55.2 ± 36.0	59.7 ± 61.3
FGF23 (ng/mL)	39.7 ± 26.9	38.9 ± 18.08	**48.2 ± 18.7**	**68.9 ± 34.2 ***	40.0 ± 13.85	37.5 ± 27.3
1,25(OH)$_2$D/25(OH)D	0.002 ± 0.001	0.002 ± 0.001	**0.003 ± 0.001**	**0.001 ± 0.001 ***	0.004 ± 0.001	0.004 ± 0.004
1,25(OH)$_2$D/24,25(OH)$_2$D	0.16 ± 0.20	0.07 ± 0.08	**0.13 ± 0.10**	**0.06 ± 0.06 ***	0.14 ± 0.12	0.06 ± 0.03
24,25(OH)$_2$D/25(OH)D	0.11 ± 0.12	0.07 ± 0.04	0.04 ± 0.03	0.03 ± 0.01	0.06 ± 0.08	0.07 ± 0.06

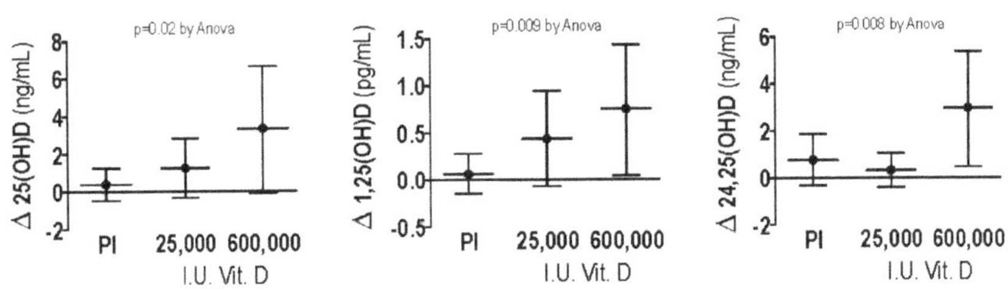

Figure 1. Delta changes in 25(OH)D, 1,25(OH)$_2$D, and 24,25(OH)$_2$D, after 72 h, in the groups studied. Pl: placebo.

3. Results

We enrolled 12 men and 14 women in the study. The mean (± SD) age of the sample studied was 73.8 ± 12.7 years, while the mean body mass index was 25.3 ± 4.1 kg/m^2. Table 1 reports the main anthropometric and biochemical parameters of the patients investigated. There were no significant differences in the distribution of females or males among the three groups investigated, as well as in the anthropometric and basal biochemical parameters.

There was a significant increase in the mean 25(OH)D levels with respect to the basal values both in the 25,000 and 600,000 I.U. groups but not in the placebo group on the third day following cholecalciferol or placebo supplementation, respectively (Table 2). If we consider 30 ng/mL value as a sufficiency threshold for vitamin D, none of the subjects in the group supplemented with 25,000 I.U. reached this target after 72 h, while seven out of nine patients in the 600,000 I.U. group reached this target. The number of patients reaching sufficiency in the last group (i.e., 600,000 I.U.) was significantly higher compared to those supplemented with 25,000 I.U. of cholecalciferol (p = 0.004).

In the 600,000 I.U. group, the mean 25(OH)D values increased from 16.55 ± 13.1 to 58.6 ± 29.4 ng/mL ($p < 0.05$), while in the 25.000 I.U. group mean 25(OH)D values increased from 12.2 ± 8.8 to 19.9 ± 11.3 ng/mL (all $p < 0.05$). A significant difference was observed among the three groups (ANOVA p = 0.02) (Figure 1 left panel) when comparing the delta changes in 25(OH)D from the baseline after vitamin D supplementation. The difference in the delta changes in 25(OH)D between the controls and 600,000 I.U. group persists also after Bonferroni correction.

Delta changes (72 h value–basal value) of 25(OH)D vitamin D (left panel), 1,25(OH)$_2$D dihydroxy vitamin D (middle panel), and 24,25(OH)$_2$D dihydroxy vitamin D (right panel) in the controls and in the patients supplemented with 25,000 and 600,00 I.U. of cholecalciferol after 72 h are reported in Figure 1.

In the case of 1,25(OH)$_2$D serum levels, we found a significant absolute increase (all $p < 0.05$) after 72 h both in the groups supplemented with 25,000 I.U. and 600,000 I.U. of cholecalciferol. Differences in the mean delta changes among groups were significant (ANOVA p = 0.009), as shown in Figure 1 middle panel. After Bonferroni correction, a statistically significance difference in the mean delta increase was observed only in the group supplemented with 600.000 I.U. compared to the control group.

The administration of 600,000 I.U. of cholecalciferol led to an increase in the mean delta 24,25(OH)$_2$D changes that were significantly higher compared to the other groups, (ANOVA p = 0.008), as shown in Figure 1 right panel. In particular, the differences between the delta changes in those supplemented with 600,000 I.U. were 2.11 ± 0.73 compared to the controls (Bonferroni corrected p = 0.03) and 2.56 ± 0.76 compared to those supplemented with 25,000 (Bonferroni corrected p = 0.01).

We found a significant mean decrease in the ratio of 1,25(OH)$_2$D/25(OH)D and 1,25(OH)$_2$D/24,25(OH)$_2$D (both $p < 0.05$), but not of 24,25(OH)$_2$D/25(OH)D (Table 2) ratio at 72 h, only in the group supplemented with 600,000 I.U. of cholecalciferol.

Considering the subjects investigated as a whole, we found a significant correlation between the delta 25(OH)D and delta 24,25(OH)$_2$D (r = 0.47, p = 0.02), and a significant positive correlation between the absolute value of 25(OH)D at baseline and 24,25(OH)$_2$D after 72 h (r = 0.56, p = 0.04). Moreover, a significant positive correlation between the absolute value of 1,25(OH)$_2$D at baseline and 24,25(OH)$_2$D after 72 h was observed (r = 0.56, p = 0.005).

Concerning PTH, we did not find significant mean changes after 72 h following cholecalciferol supplementation, even though there was a decreasing trend in those supplemented with 600,000 I.U. Interestingly, in the last group (i.e., in those administered 600,000 I.U) a significant increase in the mean serum ionized calcium (1.24 ± 0.03 vs. 1.28 ± 0.16 mmol/L, $p < 0.05$) and mean serum FGF23 were observed (48.2 ± 18.7 vs. 68.9 ± 34.2, $p < 0.05$). However, delta changes among the three groups for these two variables did not reach statistical significance.

No case of severe hypercalcemia was observed, nor any clinical symptom was reported by the subjects during the 3 days after the administration of cholecalciferol or placebo.

4. Discussion

To the best of our knowledge, this is the first investigation showing early changes in $24,25(OH)_2D$ and FGF23 values after two different dosing regimens of cholecalciferol in ambulatory patients with vitamin D insufficiency attending an internal medicine department. As expected [22–24], we found that the mean increase in 25(OH)D and $1,25(OH)_2D$ was more pronounced following 600,000 I.U. administration. This last dose also determined a very mild increase in the mean values of the serum ionized calcium.

Concerning changes in $24,25(OH)_2D$ values, we found that there were significant mean increases following the bolus dose (i.e., 600,000 I.U.) with respect to both the control subjects and those administered 25,000 I.U. of cholecalciferol. Previous studies have shown changes in $24,25(OH)_2D$ on a long-term basis, i.e., 30 days after a single bolus (by the oral or intramuscular route) of 600,000 I.U. [25], 4 weeks following an oral bolus dose of 100,000 I.U. [12], 8 weeks following 28,000 I.U. once per week [11], or 12 weeks following a daily dose of 20 mcg [26]. However, in a study carried out on healthy young subjects administered different doses of cholecalciferol (10,000 I.U./day for 8 weeks followed by 1000 I.U./day for 4 weeks; or 50,000 I.U./week for 12 weeks, or 100,000 I.U. every other week for 12 weeks), $24,25(OH)_2D$ values did not change over time when measured at 28, 53, 84 and 112 days [27]. These inconsistencies may be ascribed to many factors, including assay procedure, dose and time interval of cholecalciferol administration, age, and comorbidities.

Our investigation fills a gap in knowledge in this field. Indeed, as outlined in a recent editorial [28], it is important to measure the levels of vitamin D metabolites in patients administered bolus doses of vitamin D3. This is particularly important in view of some previous studies warning against the use of bolus doses of cholecalciferol in vitamin D-repleted subjects [29]. These huge doses would favor falls with consequent increased risk of fractures, so that they should be avoided [30]. Even though the mechanisms linking bolus cholecalciferol administration to falls have not been completely elucidated, a potential negative effect of increased $24,25(OH)_2D$ has been called into question. We did not register any falls during the three-day observation period; however, the number of subjects investigated was very small so that we cannot draw any definitive conclusion. In addition, our study was carried out in vitamin D insufficient subjects that is a completely different clinical condition. In this context, the role of $24,25(OH)_2D$ with respect to skeletal health has not been completely clarified. For example, some studies show that the administration of 10,000 I.U. of cholecalciferol for 36 months has negative effects on radial total bone mineral density [31]. However, previous studies have shown that the administration of huge doses of 24,25(OH)D to mice with X-linked hypophosphatemia [32] or patients with kidney failure [33] did not result in an increased osteoclast surface area when evaluated from a histological point of view.

The changes we found following the administration of the largest dose of cholecalciferol (600,000 I.U), are in line with our current understanding of vitamin D metabolism. Indeed, the administration of a huge dose of cholecalciferol results in a more consistent increase in 25(OH)D than $1,25(OH)_2D$ (and, therefore, a decrease in the 1,25/25 ratio). The massive increase in both metabolites triggers the expression of mitochondrial CYP24A1, the principal enzyme that catalyzes the metabolic clearance of 25(OH)D into 24,25-dihydroxyvitamin D, (thus, also determining the reduction of 1,25/24,25 ratio). As a result, the ratio between 24,25/25 does not change.

In our study, we found no differences in $24,25(OH)_2$ D/25(OH)D ratio, in line with what was observed in previous studies [8,10] but contrary to what was observed in Tang's study [34].

Changes in the serum calcium and PTH values following the bolus dose are in line with our previous paper showing similar biochemical findings [25]. However, also in this case, we must acknowledge inconsistent results in the literature regarding parathyroid hormone behavior following cholecalciferol supplementation [22,34,35]. The same confounding factors previously addressed might underlie different results.

Concerning serum FGF23 levels, we showed significant mean increases after 72 h of 600,000 I.U. of cholecalciferol, while no changes were detected following 25,000 I.U. Indeed, our results are in line with data from a recent meta-analysis documenting an associated increase in serum FGF23 with cholecalciferol supplementation after a longer period of time following vitamin D supplementation with different dosages [15]. A number of possible mechanisms can underlie this increase. Among these, is the possibility that repletion of vitamin D might lead to an increase in intracellular 1,25(OH)$_2$D concentration in the osteocyte, which would then increase the synthesis and secretion of FGF23. Also, the systemic increase in 1,25(OH)$_2$D may be counteracted by the increase in FGF23 by downregulating the expression of the vitamin D-activating enzyme cytochrome P450 subfamily 27 polypeptide 1 (CYP27B1; also known as mitochondrial 25-hydroxyvitamin D1 α-hydroxylase) and by upregulating the vitamin D-degrading enzyme CYP24A1.

Strengths and Limitations

Our study has some limitations worth mentioning. First, we only focused on acute changes, since this information is still missing in the literature, as previously reported [28]. A longer treatment period may have revealed additional information on vitamin D metabolism. A follow-up investigation is ongoing in order to meet these needs. Second, we might also have considered the effects of other doses of vitamin D. However, at the time we carried out this investigation, the doses we employed were the most frequently used in clinical practice in our country. However, following a paper by Sanders and coworkers [29] that utilized 500,000 I.U. annually, the utilization of huge bolus doses of vitamin D has been abandoned for the possible increased risk of falls and fractures. In this context, it should be emphasized that, even though there has been a large amount of scientific investigation, the causal link between the administration of huge doses of vitamin D and the increased risk of falls has never been elucidated. Third, we did not measure other vitamin D metabolites such as 3-epi-25(OH)D3; however, the role of this metabolite seems to be marginal in adults. Other derivatives (such as CYP111A1 derivate vitamin D) might give more information if assayed in future studies. In future research, we are also focusing on serum phosphate and possible sex differences following the administration of different doses of cholecalciferol. Finally, considering the ongoing debate on adverse events linked to bolus doses of vitamin D, we might have measured other parameters such as, for example, the biochemical markers of skeletal turnover. However, this was not the end point of our investigation.

5. Conclusions

In conclusion, both a single dose of 25,000 and 600,000 I.U. of cholecalciferol increase 25(OH) levels even though to a different degree. Moreover, the 600,000 I.U. dose results in a higher magnitude of conversion to the 24,25(OH)$_2$D compared to the lower dosage of vitamin D supplementation. Furthermore, the highest dose results in immediate homeostatic adjustments, although they are slower with the more physiological dose. These adjustments are mainly represented by increased FGF23 secretion and CYP24A1 activity. Our study was preliminary in nature since we investigated three small groups of subjects. However, if our results can be confirmed in larger populations with different doses and compounds, they will give important clues about homeostatic adjustments following vitamin D supplementation and treatment. This is especially important in light of the discovery of new metabolic functions of previously disregarded vitamin D compounds.

Author Contributions: J.P., L.C., R.P. and C.C. contributed to data collection and analysis, and the drafting and revision of the manuscript; J.P., L.C., V.D.M., A.S., M.C., R.S., S.M. and C.C. contributed to data collection and revision of the manuscript; C.F. contributed to data analysis and revision of the manuscript and carried out the statistical analysis; J.P., R.S., S.M. and C.C. contributed to the study design. All authors have read and agreed to the published version of the manuscript.

Funding: This study was partly funded by an unrestricted grant from Nycomed pharmaceutical company.

Institutional Review Board Statement: This study was conducted according to the guidelines of the Declaration of Helsinki and approved by the Ethics Committee of Ospedale Casa Sollievo della Sofferenza, IRCCS, San Giovanni Rotondo (protocol code 07/CE, 31 January 2013) and Policlinico Umberto I (protocol code 3267, 26 June 2014).

Informed Consent Statement: Informed consent was obtained from all subjects involved in the study.

Data Availability Statement: Some or all datasets generated during the current study are not publicly available due to privacy but are available from the corresponding author upon reasonable request.

Conflicts of Interest: S.M. served as a speaker for Abiogen, Beijing Society Biomedical Engineering, Geopharma, Sandoz, UCB. He also served on the advisory board of Kyowa Kirin and Faes Farma. C.C. served as speaker for Abiogen and on the advisory board of IBSA. The other authors have no conflicts to declare.

References

1. Slominski, R.M.; Chen, J.Y.; Reman, C.; Slominski, A.T. Photo-neuro-immune endocrinology: How the ultraviolet radiation regulates the body, brain and immune system. *Proc. Natl. Acad. Sci. USA* **2024**, *121*, e2308374121. [CrossRef] [PubMed]
2. Slominski, A.T.; Kim, T.K.; Janjetovic, Z.; Slominski, R.M.; Li, W.; Jetten, A.M.; Indra, A.K.; Mason, R.S.; Tuckey, R.C. Biological effects of CYP11A1-derived vitamin D and lumisterol metabolites in the skin. *J. Invest. Dermatol.* **2024**, *144*, 2145–2161. [CrossRef] [PubMed]
3. Kim, T.K.; Slominski, R.M.; Pyza, E.; Kleszczynski, K.; Tuckey, R.C.; Reiter, R.J.; Holick, M.F.; Slominski, A.T. Evolutionary formation of melatonin and vitamin D in early life forms: Insects take centre stage. *Biol. Rev. Camb. Philos. Soc.* **2024**, *99*, 1772–1790. [CrossRef] [PubMed]
4. Occhiuto, M.; Pepe, J.; Colangelo, L.; Lucarelli, M.; Angeloni, A.; Nieddu, L.; De Martino, V.; Minisola, S.; Cipriani, C. Effect of 2 years of monthly calcifediol administration in postmenopausal women with vitamin D insufficiency. *Nutrients* **2024**, *16*, 1754. [CrossRef] [PubMed]
5. Ebeling, P.R.; Adler, R.A.; Jones, G.; Liberman, U.A.; Mazziotti, G.; Minisola, S.; Munns, C.F.; Napoli, N.; Pittas, A.G.; Giustina, A.; et al. Management Of Endocrine Disease: Therapeutics of vitamin D. *Eur. J. Endocrinol.* **2018**, *179*, R239–R259. [CrossRef] [PubMed]
6. Giustina, A.; Bouillon, R.; Binkley, N.; Sempos, C.; Adler, R.A.; Bollerslev, J.; Dawson-Hughes, B.; Ebeling, P.R.; Feldman, D.; Heijboer, A.; et al. Controversies in vitamin D: A statement from the third international conference. *J. Bone Miner. Res. Plus* **2020**, *4*, e10417. [CrossRef]
7. Minisola, S.; Pepe, J.; Donato, P.; Vigna, E.; Occhiuto, M.; Ferrone, F.; Biamonte, F.; Cecchetti, V.; Danese, V.C.; Sonato, C.; et al. Replenishment of vitamin D status: Theoretical and practical considerations. *Hormones* **2019**, *18*, 3–5. [CrossRef]
8. Francic, V.; Ursem, S.R.; Dirks, N.F.; Keppel, M.H.; Theiler-Schwetz, V.; Trummer, C.; Pandis, M.; Borzan, V.; Grübler, M.R.; Verheyen, N.D.; et al. The effect of vitamin d supplementation on its metabolism and the vitamin D metabolite ratio. *Nutrients* **2019**, *11*, 2539. [CrossRef]
9. Ketha, H.; Thacher, T.D.; Oberhelman, S.S.; Fischer, P.R.; Singh, R.J.; Kumar, R. Comparison of the effect of daily versus bolus dose maternal vitamin D3 supplementation on the 24,25-dihydroxyvitamin D3 to 25-hydroxyvitamin D3 ratio. *Bone* **2018**, *110*, 321–325. [CrossRef]
10. Aloia, J.; Fazzari, M.; Shieh, A.; Dhaliwal, R.; Mikhail, M.; Hoofnagle, A.N.; Ragolia, L. The vitamin D metabolite ratio (VMR) as a predictor of functional biomarkers of bone health. *Clin. Endocrinol.* **2017**, *86*, 674–679. [CrossRef]
11. Wagner, D.; Hanwell, H.E.; Schnabl, K.; Yazdanpanah, M.; Kimball, S.; Fu, L.; Sidhom, G.; Rousseau, D.; Cole, D.E.; Vieth, R. The ratio of serum 24,25-dihydroxyvitamin D (3) to 25-hydroxyvitamin D (3) is predictive of 25-hydroxyvitamin D (3) response to vitamin D (3) supplementation. *J. Steroid Biochem. Mol. Biol.* **2011**, *126*, 72–77. [CrossRef] [PubMed]
12. Saleh, L.; Tang, J.; Gawinecka, J.; Boesch, L.; Fraser, W.D.; von Eckardstein, A.; Nowak, A. Impact of a single oral dose of 100,000 IU vitamin D3 on profiles of serum 25(OH)D3 and its metabolites 24,25(OH)2D3, 3-epi-25(OH)D3, and 1,25(OH)2D3 in adults with vitamin D insufficiency. *Clin. Chem. Lab Med.* **2017**, *55*, 1912–1921. [CrossRef]
13. Minisola, S.; Fukumoto, S.; Xia, W.; Corsi, A.; Colangelo, L.; Scillitani, A.; Pepe, J.; Cipriani, C.; Thakker, R.V. Tumor-induced Osteomalacia: A Comprehensive Review. *Endocr. Rev.* **2023**, *44*, 323–353. [CrossRef] [PubMed]
14. Shimada, T.; Hasegawa, H.; Yamazaki, Y.; Muto, T.; Hino, R.; Takeuchi, Y.; Fujita, T.; Nakahara, K.; Fukumoto, S.; Yamashita, T. FGF-23 is a potent regulator of vitamin d metabolism and phosphate homeostasis. *J. Bone Miner. Res.* **2003**, *19*, 429–435. [CrossRef] [PubMed]
15. Charoenngam, N.; Rujirachun, P.; Holick, M.F.; Ungprasert, P. Oral vitamin D3 supplementation increases serum fibroblast growth factor 23 concentration in vitamin D-deficient patients: A systematic review and meta-analysis. *Osteoporos. Int.* **2019**, *30*, 2183–2193. [CrossRef]

16. Jones, G.; Kaufmann, M. Diagnostic aspects of vitamin D: Clinical utility of vitamin D metabolite profiling. *J. Bone Miner. Res Plus* **2021**, *5*, e10581. [CrossRef]
17. Alonso, N.; Zelzer, S.; Eibinger, G.; Herrmann, M. Vitamin D metabolites: Analytical challenges and clinical relevance. *Calcif. Tissue Int.* **2023**, *112*, 158–177. [CrossRef]
18. Florenzano, P.; Cipriani, C.; Roszko, K.L.; Fukumoto, S.; Collins, M.T.; Minisola, S.; Pepe, J. Approach to patients with hypophosphataemia. *Lancet Diabetes Endocrinol.* **2020**, *8*, 163–174. [CrossRef]
19. Pepe, J.; Romagnoli, E.; Nofroni, I.; Pacitti, M.T.; De Geronimo, S.; Letizia, C.; Tonnarini, G.; Scarpiello, A.; D'Erasmo, E.; Minisola, S. Vitamin D status as the major factor determining the circulating levels of parathyroid hormone: A study in normal subjects *Osteoporos. Int.* **2005**, *16*, 805–812. [CrossRef]
20. Ferrone, F.; Pepe, J.; Danese, V.C.; Fassino, V.; Cecchetti, V.; De Lucia, F.; Biamonte, F.; Colangelo, L.; Ferrazza, G.; Panzini, E.; et al The relative influence of serum ionized calcium and 25-hydroxyvitamin D in regulating PTH secretion in healthy subjects. *Bone* **2019**, *125*, 200–206. [CrossRef]
21. Yamazaki, Y.; Okazaki, R.; Shibata, M.; Hasegawa, Y.; Satoh, K.; Tajima, T.; Takeuchi, Y.; Fujita, T.; Nakahara, K.; Yamashita, T.; et al. Increased circulatory level of biologically active full-length FGF-23 in patients with hypophosphatemic rickets/osteomalacia. *J. Clin. Endocrinol. Metab.* **2002**, *87*, 4957–4960. [CrossRef] [PubMed]
22. Kearns, M.D.; Alvarez, J.A.; Tangpricha, V. Large, single-dose, oral vitamin D supplementation in adult populations: A systematic review. *Endocr. Pract.* **2014**, *20*, 341–351. [CrossRef] [PubMed]
23. Romagnoli, E.; Mascia, M.L.; Cipriani, C.; Fassino, V.; Mazzei, F.; D'erasmo, E.; Carnevale, V.; Scillitani, A.; Minisola, S. Short and long-term variations in serum calciotropic hormones after a single very large dose of ergocalciferol (vitamin D2) or cholecalciferol (vitamin D3) in the elderly. *J. Clin. Endocrinol. Metab.* **2008**, *93*, 3015–3020. [CrossRef] [PubMed]
24. Pepe, J.; Mezzaroma, I.; Fantauzzi, A.; Falciano, M.; Salotti, A.; Di Traglia, M.; Diacinti, D.; Biondi, P.; Cipriani, C.; Cilli, M.; et al. An oral high dose of cholecalciferol restores vitamin D status in deficient postmenopausal HIV-1-infected women independently of protease inhibitors therapy: A pilot study. *Endocrine* **2016**, *53*, 299–304. [CrossRef]
25. Cipriani, C.; Romagnoli, E.; Pepe, J.; Russo, S.; Carlucci, L.; Piemonte, S.; Nieddu, L.; McMahon, D.J.; Singh, R.; Minisola, S. Long-term bioavailability after a single oral or intramuscular administration of 600,000 IU of ergocalciferol or cholecalciferol: Implications for treatment and prophylaxis. *J. Clin. Endocrinol. Metab.* **2013**, *98*, 2709–2715. [CrossRef]
26. Lehmann, U.; Riedel, A.; Hirche, F.; Brandsch, C.; Girndt, M.; Ulrich, C.; Seibert, E.; Henning, C.; Glomb, M.A.; Dierkes, J.; et al. Vitamin D3 supplementation: Response and predictors of vitamin D3 metabolites—A randomized controlled trial. *Clin. Nutr.* **2016**, *35*, 351–358. [CrossRef]
27. Fassio, A.; Gatti, D.; Rossini, M.; Benini, C.; Fracassi, E.; Bertoldo, E.; Viapiana, O.; Milleri, S.; Gatti, M.; Adami, G. Pharmacodynamics of oral cholecalciferol in healthy Individuals with vitamin D deficiency: A randomized open-label study. *Nutrients* **2021**, *13*, 2293. [CrossRef] [PubMed]
28. Liu, E.S.; LeBoff, M.S. Editorial: The role of vitamin D metabolites in the evaluation of bone health: Are they physiologically relevant? *J. Bone Miner. Res.* **2024**, *39*, 1–2. [CrossRef]
29. Sanders, K.M.; Stuart, A.L.; Williamson, E.J.; Simpson, J.A.; Kotowicz, M.A.; Young, D.; Nicholson, G.C. Annual high-dose oral vitamin D and falls and fractures in older women: A randomized controlled trial. *JAMA* **2010**, *303*, 1815–1822. [CrossRef]
30. Giustina, A.; Bilezikian, J.P.; Adler, R.A.; Banfi, G.; Bikle, D.D.; Binkley, N.C.; Bollerslev, J.; Bouillon, R.; Brandi, M.L.; Casanueva, F.F.; et al. Consensus statement on vitamin D status assessment and supplementation: Whys, whens, and hows. *Endocr. Rev.* **2024**, *45*, bnae009. [CrossRef]
31. Burt, L.A.; Billington, E.O.; Rose, M.S.; Raymond, D.A.; Hanley, D.A.; Boyd, S.K. Effect of high-dose vitamin D supplementation on volumetric bone density and bone strength: A randomized clinical trial. *JAMA* **2019**, *322*, 736–745. [CrossRef] [PubMed]
32. Ono, T.; Tanaka, H.; Yamate, T.; Nagai, Y.; Nakamura, T.; Seino, Y. 24R,25-dihydroxyvitamin D3 promotes bone formation without causing excessive resorption in hypophosphatemic mice. *Endocrinology* **1996**, *137*, 2633–2637. [CrossRef] [PubMed]
33. Birkenhäger-Frenkel, D.H.; Pols, H.A.; Zeelenberg, J.; Eijgelsheim, J.J.; Schot, R.; Nigg, A.L.; Weimar, W.; Mulder, P.G.; Birkenhäger, J.C. Effects of 24R,25-dihydroxyvitamin D3 in combination with 1 alpha-hydroxyvitamin D3 in predialysis renal insufficiency: Biochemistry and histomorphometry of cancellous bone. *J. Bone Miner. Res.* **1995**, *10*, 197–204. [CrossRef] [PubMed]
34. Tang, J.C.Y.; Team, B.F.; Jackson, S.; Walsh, N.P.; Greeves, J.; Fraser, W.D.; Ball, N. The dynamic relationships between the active and catabolic vitamin D metabolites, their ratios, and associations with PTH. *Sci. Rep.* **2019**, *9*, 6974. [CrossRef]
35. Ilahi, M.; Armas, L.A.; Heaney, R.P. Pharmacokinetics of a single, large dose of cholecalciferol. *Am. J. Clin. Nutr.* **2008**, *87*, 688–691. [CrossRef]

Disclaimer/Publisher's Note: The statements, opinions and data contained in all publications are solely those of the individual author(s) and contributor(s) and not of MDPI and/or the editor(s). MDPI and/or the editor(s) disclaim responsibility for any injury to people or property resulting from any ideas, methods, instructions or products referred to in the content.

nutrients MDPI

Article

The Temporal Change in Ionised Calcium, Parathyroid Hormone and Bone Metabolism Following Ingestion of a Plant-Sourced Marine Mineral + Protein Isolate in Healthy Young Adults

Marta Kozior [1], Olusoji Aboyeji Demehin [2], Michelle Mary Ryan [2], Shane O'Connell [2,3] and Philip Michael Jakeman [1,*]

[1] Food, Diet and Nutrition, Health Research Institute, University of Limerick, V94 T9PX Limerick, Ireland; martakozior@yahoo.com

[2] Marigot Research Centre, Sycamore Court, V92 N6C8 Tralee, Ireland; o.demehin@marigot.ie (O.A.D.); michelle.ryan@marigot.ie (M.M.R.); shane.oconnell@marigot.ie (S.O.)

[3] Shannon Applied Technology Centre, Munster Technological University, Clash, V92 CX88 Tralee, Ireland

* Correspondence: phil.jakeman@ul.ie; Tel.: +353-61-202-800

Abstract: Background: An increase in plant-sourced (PS) nutrient intake is promoted in support of a sustainable diet. PS dietary minerals and proteins have bioactive properties that can affect bone health and the risk of fracture. **Methods:** In a group randomised, cross-over design, this study evaluated the post-ingestion temporal pattern of change in arterialised ionised calcium (iCa), parathyroid hormone (PTH), *C*-terminal crosslinked telopeptide of type I collagen (CTX) and procollagen type 1 amino-terminal propeptide (P1NP) for 4 h following ingestion of a novel supplement (SUPP) containing a PS marine multi-mineral + PS protein isolate. A diurnally matched intake of mineral water was used as a control (CON). **Results:** Compared to baseline, the change in iCa concentration was 0.022 (95% CI, 0.006 to 0.038, $p = 0.011$) mmol/l greater in SUPP than CON, resulting in a −4.214 (95% CI, −8.244 to −0.183, $p = 0.042$) pg/mL mean reduction in PTH, a −0.64 (95% CI, −0.199 to −0.008, $p = 0.029$) ng/mL decrease in the biomarker of bone resorption, CTX, and no change in the biomarker of bone formation, P1NP. **Conclusions:** When used as a dietary supplement, or incorporated into a food matrix, the promotion of PS marine multi-mineral and PS protein isolates may contribute to a more sustainable diet and overall bone health.

Keywords: sustainable diet; plant mineral; plant protein; bone health; post-prandial; young adults

1. Introduction

An increase in plant-sourced (PS) and decrease in animal-sourced (AS) nutrient intake is promoted in support of a sustainable diet [1] and to maintain human and planetary health [2]. However, PS dietary patterns as practised by vegetarians and vegans are associated with a lower bone mineral density (BMD) and increased fracture risk [3]. Therefore, a greater understanding of the functional properties of PS nutrient supplements is required to inform and support the transition to a higher PS nutrient intake for bone health.

Protein and calcium are major components of bone tissue. Protein provides the structural matrix of bone, occupying ~50% of bone volume and one-third of bone mass, and calcium is the dominant mineral within that matrix. Collagen and non-collagenous proteins form the organic matrix of bone, so an adequate dietary protein and calcium intake is essential for optimal acquisition and maintenance of adult bone mass [4]. However, at present, animal-sourced (AS) protein and dairy-based calcium predominate in the Western diet [1,5,6].

Bone remodelling describes the normally balanced rate of bone resorption and bone formation that ensures 5–10% of the skeleton per annum is remodelled for essential maintenance and repair. The fracture risk is increased by high remodelling rates or an imbalance (uncoupling) between rates of formation and resorption [7]. Importantly, in the context of PS nutrient intake, a sufficient intake of minerals and protein is required in support of bone remodelling in the adult skeleton [8]. In this respect, legumes are an example of a low-cost PS food of high nutritional value and an excellent source of good-quality protein (20–45% protein), providing high amounts of the essential amino acids such as lysine and leucine [9]. Natural plant sources of calcium are rare, but the multi-mineral extract from the cytoskeleton of the red seaweed *Lithothamnion* sp. (containing 32.6% calcium) is considered an effective PS of mineral for bone health [10].

In addition to the supply of mineral and protein in support of bone remodelling, the ingestion of protein and calcium exert an active, temporal change in bone metabolism that favours a post-prandial reduction in the rate of bone resorption and potential for longer-term change in the rate of bone formation [4]. These nutrient bioactivities are related to calcium bioavailability and the subsequent increase in serum ionised calcium [11] and increase in enteroendocrine peptides known to modulate osteoclast function [12] via the entero-osseous axis [13]. Of interest to the present study is the augmentation of post-prandial secretion of enteroendocrine peptides by co-ingestion of calcium [14] and the development of a PS calcium + protein supplement of benefit to bone health.

The principal nutrient modulators of post-prandial bone metabolism under consideration in this report are a plant-sourced marine multi-mineral derived from the cytoskeleton of the red seaweed *Lithothamnion* sp. containing 32.6% calcium (Aquamin F, Marigot Ltd., Cork, Ireland) co-ingested with a plant-sourced legume protein isolate (*Vicia faba* L., Marigot Ltd., Cork, Ireland). We hypothesised that a meal-sized bolus (0.33 $g{\cdot}kg^{-1}$) of a plant-sourced protein isolate co-ingested with 800 mg of calcium derived from a plant-sourced marine multi-mineral would stimulate a greater post-ingestion calcitropic response compared to a non-bioactive control, leading to a positive effect on bone metabolism. To this end, a study of the acute (0–4 h) temporal change in ionised calcium (iCa), parathyroid hormone (PTH) and the International Osteoporosis Foundation's (IOF) recommended biomarkers of bone resorption, C-terminal crosslinked telopeptide of type I collagen (CTX), and formation, procollagen type 1 amino-terminal propeptide (P1NP), following co-ingestion of a PS marine multi- mineral + legume protein isolate was undertaken in healthy young adult men and women.

2. Materials and Methods

2.1. Ethical Approval and Participant Recruitment

This study was granted ethical approval by the University of Limerick Education and Health Sciences Research Ethics Committee (2022_03_06_EHS), and it was conducted in accordance with the ethical standards outlined in the most recent version of the Declaration of Helsinki and registered with clinicaltrials.gov identifier NCT5533502. Potential participants were informed of the risks and benefits before providing written informed consent. Eligibility criteria: (i) aged 18 to 35 years; (ii) healthy (i.e., no current injury, illness, medication, history of chronic disease, known allergies and intolerances; normotensive; non-obese; with normal blood chemistry). All participant volunteers were assessed for eligibility and screened using a medical questionnaire, physical examination, dietary intake record, anthropometry, and body composition analysis (Tanita MC, 180-MA, Tanita Europe B.V., Amsterdam, The Netherlands). Supplementary Figure S1 provides a CONSORT flow diagram of participant enrolment, allocation, follow-up and analysis. Ten participants were recruited to this study, five men and five women. Participant characteristics are presented in Table 1.

Table 1. Participant characteristics (n = 10; 5 men, 5 women).

	Mean	SD
Age (year)	26.5	4.4
Height (cm)	173.1	9.6
Body mass (kg)	72.5	14.0
Body mass index (kg·m^{-2})	24.1	3.3
Fat mass (%)	20.5	8.2
Fat-free mass (kg)	57.2	9.7
Bone mass (kg)	2.89	0.5

2.2. Study Design

This study was open to young adult men and women, independent of sex, who satisfied the inclusion criteria. This study had a within-subject, single-blind, block-randomised, repeated-measures, cross-over design comprising two study arms, control (CON) and supplement (SUPP), i.e., subjects acting as their own control. Participant volunteers were randomised at entry to this study. Participants maintained their habitual diet and refrained from purposeful exercise and alcohol consumption for the previous 24 h on two separate occasions at least three days apart. On trial days, participants arrived at the laboratory at 8:00 a.m. following an overnight fast and remained seated throughout the trial. A cannula was inserted into a superficial vein on the dorsal surface of the hand and the hand placed in a heated hand box (air temperature 55 °C) for 15 min to arterialise venous drainage. Following baseline blood draw, participants ingested, in a randomised order, either CON or SUPP within 5 min. Serial samples of arterialised venous blood were then drawn 15, 30, 45, 60, 90, 120, 180 and 240 min post-ingestion.

2.3. Supplement Composition

The supplement (SUPP) was composed of a plant-sourced marine multi-mineral (Aquamin F, Marigot Ltd., Cork, Ireland) + plant-sourced protein isolate (*Vicia faba* L., Marigot Ltd., Cork, Ireland). The final composition of the mineral + protein ingested by participants contained 800 mg of calcium (equivalent to 2.45 g of Aquamin F) and 0.33 g of *Vicia faba* L. protein per kilogram of body mass (Table 2) dissolved in 500 mL of water. CON provided an equal volume of mineral water (<15 mg/dL Ca). A detailed analysis of the supplement constituents is provided in Supplementary Table S1.

Table 2. Composition of plant-sourced mineral and protein supplement based on participants receiving 800 mg calcium and 0.33 g protein per kilogram of body mass [1].

	Aquamin F			*Vicia faba* L.	
	g/100 g	mg/2.45 g		Mean	Range
Calcium	32.63	800	g/kg body mass	0.33	
Magnesium	2.59	65	Powder mass (g)	30.9	19.8
Phosphorous	0.044	1.1	Protein mass (g)	23.9	15.3
Zinc	0.2	0.005	Energy (kcal)	117	75

[1] Note: Full compositional analysis is presented in Supplementary Table S1.

2.4. Blood Sample Collection and Analysis

A portion of the whole-blood sample was analysed immediately after collection for iCa and K concentrations by ion-selective potentiometry (I-STAT®; Abbott Laboratories, Dublin, Ireland). Intra-assay CV using control solutions was <1% for both iCa and K. Based on baseline samples, the inter-assay CV was <1.5% for iCa and <11% for K. The remaining blood was separated by centrifugation at 10,000× g at 4 °C for 5 min and frozen at −80 °C until analysis. In total, 12 of the 180 blood samples were haemolysed and not analysed.

PTH, CTX and P1NP was analysed by 2-site immunometric assay using electrochemiluminescent detection (Roche Cobas e411, Roche Diagnostics, Burgess Hill, UK). The inter-assay CV was <10% for PTH, 5.3% for CTX and 4.5% for P1NP. Active plasma GLP-$1_{(7–36)}$ and total GIP $_{(1–42)}$ were measured using the MSD® metabolic assay (Meso Scale Discovery, Rockville, MD, USA) based on sandwich ELISA and according to the manufacturer's instructions. The inter-assay CVs were 7.3 and 16.1% for active GLP-1 and total GIP, respectively.

2.5. Treatment of Data and Statistical Analyses

Prior to statistical analysis, potential outliers were assessed by a boxplot, and data were assessed for a normal distribution by the Shapiro–Wilk test ($p > 0.05$), homogeneity of variance by Levene's test ($p > 0.05$) and Mauchly's test of sphericity ($p > 0.05$). Data are reported as the mean (SD) unless stated otherwise. An a priori hypothesis for statistical analysis (H_0) assumed the mean response for SUPP was equal to CON (i.e., $\mu_{SUPP} = \mu_{CON}$). Temporal change in serially sampled data was analysed by repeated-measures ANOVA, with the level of significance for post hoc tests subject to Bonferroni correction. The overall change in analyte with respect to baseline over the 4 h period post-ingestion was computed by trapezoidal integration and presented as the area under the curve ($AUC_{0–240}$). The difference between treatment $AUC_{0–240}$ was analysed by a paired t-test. Cohen's effect size was calculated using the standardised formula t-score/$\sqrt{n}$. Statistical significance was set at $p \leq 0.05$. SPSS Version 28 (IBM Corporation, Armonk, NY, USA) was employed for all statistical analyses.

3. Results

3.1. Temporal Pattern of Change in Ionised Calcium (iCa) Following Ingestion

No change in iCa was observed post-ingestion in CON. iCa increased after 60 min in SUPP with peak values attained at 120 min post-ingestion. These data showed a statistically significant two-way interaction between treatment and time, $F(8, 72) = 12.35$, $p < 0.001$, $\eta_p^2 = 0.58$. The mean increase in iCa following ingestion was greater in SUPP than CON at the times of 60, 90, 120, 180 and 240 min (Figure 1A,B). When assessed by the integrated $AUC_{0–240}$, SUPP increased the overall ΔiCa response > 5-fold compared to CON (mean difference 7.4 ($CI_{95\%}$, 3.16 to 11.7) mmol·min/L, $t(9) = 3.94$, $p = 0.002$).

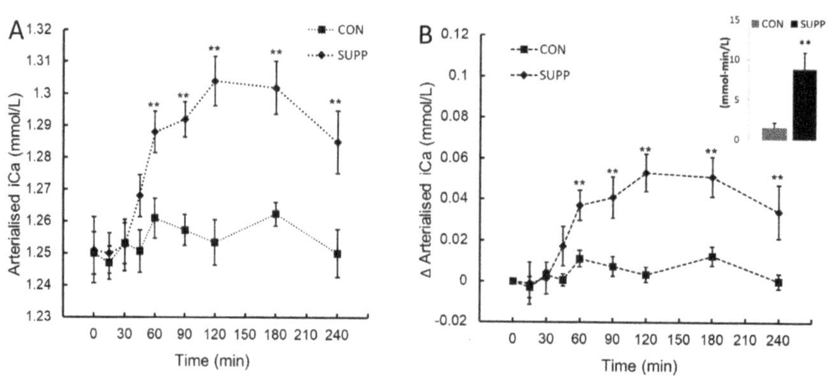

Figure 1. Ionised calcium (iCa; panel (**A**)) and the change in ionised calcium from baseline (ΔiCa; panel (**B**)) following ingestion of plant-sourced marine multi-mineral + plant-sourced protein isolate (SUPP) or mineral water (CON). The insert in panel (**B**) reports the integrated area under the curve ($AUC_{0–240}$) of the change from baseline. Data are the mean (SEM). ** Significant difference between SUPP and CON, $p < 0.05$.

3.2. Temporal Pattern of Change in Parathyroid Hormone (PTH) Following Ingestion

An inverse PTH response to the change in iCa was observed. These data showed a statistically significant two-way interaction between treatment and time, $F(8, 72) = 5.47$, $p < 0.006$, $\eta_p^2 = 0.38$. The mean decrease in PTH following ingestion was greater in SUPP than CON at the times of 30, 45, 60, 90, 120, 180 and 240 min (Figure 2A,B). When assessed by the integrated AUC_{0-240} ΔPTH, the overall ΔPTH response to SUPP was 66% lower than CON (mean difference −1246 (95% CI, −2365 to −126) pg·min/mL, $t(9) = -2.517$, $p = 0.016$).

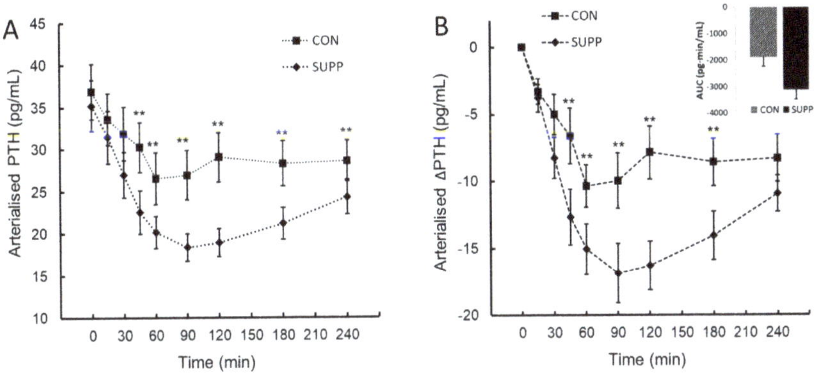

Figure 2. Parathyroid hormone (PTH; panel (**A**)) and the change in parathyroid hormone from baseline (ΔPTH; panel (**B**)) following ingestion of plant-sourced marine multi-mineral + plant-sourced protein isolate (SUPP) or mineral water (CON). The insert in panel (**B**) reports the integrated area under the curve (AUC_{0-240}) of the change from baseline. Data are the mean (SEM). ** Significant difference between SUPP and CON, $p < 0.05$.

3.3. Temporal Change in C-Terminal Peptide of Type I Collagen (CTX) and Procollagen Type 1 Amino-Terminal Propeptide (P1NP) Following Ingestion

CTX followed the normal, mid-morning diurnal decrease in CON. A statistically significant two-way interaction between treatment and time, $F(8, 72) = 6.19$, $p < 0.013$, $\eta_p^2 = 0.41$, was observed post-ingestion. The mean decrease in CTX following ingestion was greater in SUPP than CON at the times of 120, 180 and 240 min (Figure 3A,B). When assessed by the integrated AUC_{0-240} ΔCTX, SUPP decreased the overall ΔCTX response by 70% compared to CON (mean difference −22 (95% CI, −39.7 to −3.30) ng·min/mL, $t(9) = -2.673$, $p = 0.013$.

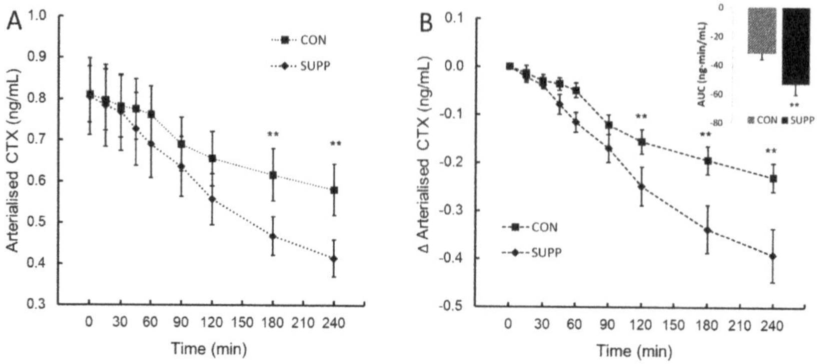

Figure 3. C-terminal peptide of type I collagen (CTX; panel (**A**)) and the change in CTX from baseline (ΔCTX; panel (**B**)) following ingestion of plant-sourced marine multi-mineral + plant-sourced protein isolate (SUPP) or mineral water (CON). The insert in panel (**B**) reports the integrated area under the curve ($AUC_{0–240}$) of the change from baseline. Data are the mean (SEM). ** Significant difference between SUPP and CON, $p < 0.05$.

There was no change in values of procollagen type 1 amino-terminal propeptide (P1NP) at baseline or with respect to the time post-ingestion for CON or SUPP. Baseline values averaged 67.6 (18.7) and 68.1 (2.4) pM for CON and SUPP, respectively.

3.4. Temporal Change in Glucose-Dependent Insulinotropic Peptide (GIP) and Glucagon-like Peptide-1 (GLP-1) Following Ingestion

A statistically significant effect of time was observed post-ingestion for GIP ($F(8, 72) = 4.78$, $p < 0.001$, $\eta_p{}^2 = 0.407$) and GLP-1 ($F(8, 72) = 5.22$, $p < 0.001$, $\eta_p{}^2 = 0.43$). The mean GIP increased 3-fold from baseline, from 6.4 (2.6) pmol/L to a peak concentration of 17.2 (9.1) pmol/L 30 min post-ingestion (Figure 4A,B). Similarly, the mean GLP-1 increased 4-fold from 0.22 (0.12) pg/mL to a peak concentration of 0.87 (0.28) pg/mL 30 min post-ingestion (Figure 4A,B).

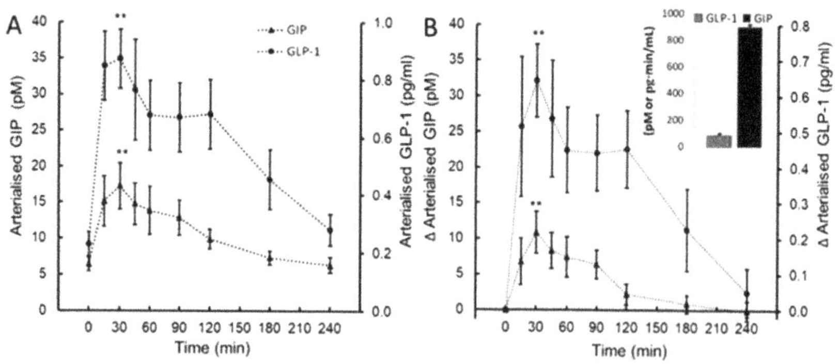

Figure 4. Glucose-dependent insulinotropic peptide (GIP) and glucagon-like peptide-1 (GLP-1; panel (**A**)) and the change in GIP and GLP-1 from baseline (ΔGIP, ΔGLP-1; panel (**B**)) following ingestion of plant-sourced marine multi-mineral + plant-sourced protein isolate (SUPP). The insert in panel (**B**) reports the integrated area under the curve ($AUC_{0–240}$) of the change from baseline. Data are the mean (SEM). ** Significant difference from basal, $p < 0.05$.

4. Discussion

The interaction between dietary calcium and protein intake and its impact on bone health has been reviewed extensively, with reports of a positive association between increased dietary protein (AS or PS) intake and BMD in healthy men and women and a reduction in the rate of bone loss by supplemental calcium intake [15,16]. Novel PS calcium and protein supplements that positively influence Ca absorption and Ca bioavailability and are beneficial to bone health are desirable, particularly in those who fail to achieve the dietary recommended level of Ca and/or protein [11,15–18]. Prompted by the global challenge faced in attaining a sustainable diet and to inform and support an increase in plant-sourced (PS) and decrease in animal-sourced (AS) nutrient intake, this study characterised the bone-related nutrient bioactivity following co-ingestion of a novel PS mineral and protein supplement.

The homeostatic control of Ca absorption, excretion, secretion and storage in bone is governed by the requirement to maintain a plasma concentration of ionised Ca (iCa) within a range of 1.1–1.3 mmol/L. This is achieved by the interaction of the calcitropic parathyroid hormone (PTH), 1,25 dihydroxycholecalciferol (1,25 $(OH)_2D_3$) and calcitonin. Counter-regulatory responses, i.e., negative feedback between PTH and 1,25 $(OH)_2D_3$ when iCa decreases, and calcitonin when iCa increases, maintain a tight regulatory control of calcium in the circulation.

A pilot study comparing the pharmacokinetic and pharmacodynamic responses to the ingestion of an equivalent (720 mg elemental calcium) of plant-sourced marine multi-mineral derived from the cytoskeleton of the red seaweed *Lithothamnion* sp. (Aquamin F) and non-PS (calcium carbonate) reported no difference in ionised or total serum calcium, but a greater lowering of PTH post-ingestion and the calciuric response for PS vs. non-PS calcium in young (mean age 28.8 years), health women [10]. In the present study, the ingestion of a matrix of PS protein and calcium (SUPP) led to an early increase in iCa after 30 min, a rapid rise to a mean peak concentration of 1.30 (Δ 0.053) mmol/L after 120 min post-ingestion, and a 5.3-fold overall 4 h ($AUC_{0–240}$) mean increase in circulating iCa of 8.79 mmol·min/L compared to CON. The counterregulatory effect of an increase in iCa was a suppression of PTH (and presumably renal production of 1,25$(OH)_2D_3$) to reduce the stimulus for osteoclastic bone resorption. Indeed, our observations confirm a temporal pattern of decrease in PTH that mirrored inversely the increase in iCa, resulting in a mean decrease of 66% (−3120 pg·min/mL) in the overall 4 h ($AUC_{0–240}$) circulating PTH compared to CON.

The acute effect of PS Ca + protein supplementation on bone turnover was measured by the temporal change in validated biomarkers of bone resorption, C-terminal peptide of type I collagen (CTX), and formation, procollagen type 1 amino-terminal propeptide (P1NP). Defined by circadian variation in bone turnover markers, bone remodelling exhibits a unimodal diurnal rhythm with a nocturnal peak and daytime nadir [19–21]. The diurnal amplitude of bone resorption (CTX) is greater than bone formation and considered the prominent and sensitive biomarker of an acute change in bone remodelling. CTX in the circulation is derived from osteoclastic degradation of type 1 collagen and is widely employed to assess osteoclastic bone resorption and the dynamic change in bone remodelling, e.g., sensitivity to modulation by feeding and specific nutrient intake [22]. In this study, the diurnally matched acute effect of SUPP invoked a 68% decrease in the overall post-prandial $AUC_{0–240}$ for CTX and a statistically significant temporal reduction in CTX from 120 to 240 min. Though not measured, the trend in the observed temporal response suggests that the reduction in bone resorption would extend beyond 240 min. Although SUPP ingestion evoked a transient inhibition of bone resorption, there was no evidence of a post-prandial change in the biomarker of bone collagen formation, P1NP, suggesting a transient shift toward skeletal deposition of calcium in the post-prandial period

The reduction in the appearance of CTX in the circulation probably reflects the calcium-induced reduction in circulating calcitropic PTH. However, the response of bone remodelling to food ingestion per se is also linked to the secretion of the enterogastric hormones glucose-dependent insulinotropic peptide ($GIP_{1–42}$) and glucagon-like peptide-1 (GLP-$1_{7–36}$). Following intake of a standardised mixed meal (498 kcal) or granola bar (260 kcal), levels of GIP and GLP-1 were found to be positively associated with a decrease in

CTX [23] and to constitute a link between food intake and bone homeostasis acting via an entero-osseous axis to decrease osteoclast activity [24]. These actions are supported by human studies of an intravenous infusion of GIP that report a ~50% reduction in CTX at 90 min [25]. To affirm the magnitude of enteroendocrine response to co-ingestion of a PS protein and mineral supplement that provided 117 (75) kcal total energy, we observed a temporal increase in GIP and GLP-1 rising ~3 and 4-fold, respectively, from baseline to a peak at 60 min post-ingestion. In the development of a nutrient supplement for bone health, co-ingestion of nutrients that act favourably on bone metabolism via a separate, but complimentary, mechanism of action is an attractive option. Though the present study design could not determine whether co-ingestion of 800 mg PS calcium augmented the post-prandial rise in GIP and GLP-1 evoked by ingestion of 0.33 g/kg body mass of PS protein, or vice versa, a previous study found that co-ingestion of calcium (~1240 mg) with a standardised mixed meal (~300 kcal) resulted in a post-prandial increase of 47% in GIP and a 22% increase in GLP-1 [26].

To be considered a good source of Ca, a food or dietary supplement must have a high Ca concentration of high bioavailability, i.e., the proportion of a nutrient present in a food absorbed and utilised. Absorption of the available calcium from foods ranges from <10% to >50% with AS foods, principally dairy, considered to be of high Ca density and bioavailability. In comparison, the Ca density in PS foods is generally lower, and the Ca bioavailability of PS foods or food matrices may be modulated by the presence of anti-nutritional factors (ANFs) such as oxalates and phytates that bind Ca into insoluble and unabsorbable salts by the small intestine, resulting in faecal excretion [27]. However, processing of foods may remove or reduce the presence of ANFs. To this end, the process of isolating the protein from *Vicia faba* L. removed all detectable ANFs present in the native legume. Although the present study primarily targeted PS calcium and protein, the complex relationships among these and other nutrients affecting bone health should be considered. Minerals other than calcium (e.g., magnesium, phosphorous, zinc) are required to optimise bone health. Indeed, in addition to calcium, 2.45 g of SUPP provided ~25% of the DRI for magnesium, <2% of the DRI for phosphorous and <0.1% of the DRI for zinc for adult men and women. Following ingestion, these minerals compete for absorption in the small intestine such that excess consumption of one can impair the absorption of the others and thereby affect the balance of mineral bioavailability. For example, the optimal calcium-to-magnesium ratio from food is 2:1, but it is 12:1 in SUPP (Table 2), raising the possibility of an acute effect on intestinal absorption that could influence the bioavailability of minerals, such as magnesium, contained in SUPP. Calcium and zinc also compete for common absorption sites in the small intestine, though there is limited evidence that supplements containing high levels (~10-fold DRI) of zinc affect calcium absorption when the calcium intake meets the DRI [28], and it is highly unlikely that trace levels of zinc in SUPP would be of concern. Conversely, supplemental calcium (~450 mg/d additional to the DRI) has been shown to reduce net zinc absorption and upset the balance, which may increase the requirement for zinc should a supplemental calcium intake be sustained over prolonged periods [29]. Present only in trace amounts in SUPP, a dietary intake of phosphorous supports the formation and deposition of calcium phosphate within the mineral matrix of bone. Following absorption, phosphorous binds to calcium, reducing the available free calcium within the bloodstream such that an excess phosphorous ingestion is considered hypocalciuric. The complex interplay between minerals highlighted above is further enhanced by the incorporation of minerals with PS protein, or as whole food matrix containing variable amount of minerals, e.g., legumes, nuts and seeds contain high amounts of calcium, while green leafy vegetables, legumes, nuts, seeds and whole grains are rich in magnesium. Furthermore, incorporation into a food matrix results in a slowing of the transit time within the gastrointestinal tract and a change in relative rates of absorption and subsequent bioavailability [11,27,30].

5. Limitations of the Study

The authors acknowledge that these data pertain to a small sample population of young men and women recruited as healthy volunteers. This study makes no prior assumption as to differences that may be attributed to confounding factors such as each participant's sex or lifestyle (e.g., diet, physical activity, etc.). As demonstrated by the present study, and that undertaken previously [31], the temporal change in biomarkers of bone metabolism following single-dose ingestion can only provide an indication of the magnitude and time course of the acute post-prandial bioactive effect on bone metabolism and should not be interpreted as a predictor of the long-term outcome on bone health. Qualifying an interventional change in bone health by using the change in bone mineral density may only be possible following a 3–6-month intervention and is only of relevance to the specific population under investigation [32].

6. Conclusions

Ingestion of the PS mineral + protein supplement evoked a 70% decrease in the diurnally matched 4 h post-prandial rate of bone resorption and no change in the rate of bone formation. PS nutrient intake is promoted in support of a sustainable diet and to maintain human and planetary health. An adequate dietary protein and calcium intake is essential for the optimal acquisition and maintenance of an adult bone mass. Incorporation of PS mineral and protein into foods as an ingredient, or as a supplement to the diet, is considered beneficial to promote adequate Ca and protein intake and encourage individuals to meet their Ca and protein requirements. The temporal pattern of the response to ingestion of the PS mineral + protein supplement indicates that the mineral extract from the marine red algae, *Lithothamnion* sp., contains a readily bioavailable source of calcium that may be enhanced by the co-ingestion of the *Vicia faba* L. protein isolate, meaning that the Ca bioavailability from foods can be optimised. Though a direct comparison with other sources of calcium + protein intake was not undertaken, the appearance and rise in circulating iCa for the PS mineral + protein supplement occurred earlier in comparison to that of a dairy-based protein and calcium matrix. Incorporating a supplement within the food matrix such as a PS protein isolate should be considered as the matrix effects of PS Ca absorption have a positive influence, meaning that Ca bioavailability is optimised.

Supplementary Materials: The following supporting information can be downloaded at https://www.mdpi.com/article/10.3390/nu16183110/s1, Figure S1: CONSORT flow diagram of participant enrolment, allocation, follow-up and analysis; Table S1: Compositional analysis of the PS mineral + protein matrix.

Author Contributions: Conceptualisation: P.M.J. and M.K.; methodology, M.K., O.A.D. and S.O.; validation, M.K. and P.M.J.; formal analysis, M.K., P.M.J. and S.O.; investigation, M.K.; resources, P.M.J.; data curation, M.K., P.M.J., O.A.D. and M.M.R.; writing—original draft preparation, P.M.J.; writing—review and editing, M.K., O.A.D., S.O. and M.K.; supervision, P.M.J. and S.O.; project administration, P.M.J.; funding acquisition, P.M.J. All authors have read and agreed to the published version of the manuscript.

Funding: This research was funded by Marie Curie (M-F2021-0180) and an Enterprise Ireland Innovation Partnership Grant (IP/2019/0870) with Marigot Ltd. (Cork, Ireland).

Institutional Review Board Statement: This study was conducted in accordance with the Declaration of Helsinki and approved by the Education and Health Sciences Research Ethics Committee of the University of Limerick (2022_03_06_EHS). Approved 6 March 2022.

Informed Consent Statement: Written informed consent was obtained from all participants involved in this study.

Data Availability Statement: The raw data supporting the conclusions of this article will be made available by the authors on request.

Acknowledgments: The authors would like to acknowledge Elaine Ahern for contributing to the data collection and thank all volunteers for their participation in this study.

Conflicts of Interest: The authors declare no conflicts of interest. This research was funded by Marie Curie and an Enterprise Ireland Innovation Partnership Grant with Marigot Ltd. (Cork, Ireland). The funders had no role in the design of this study; in the collection, analyses or interpretation of data; in the writing of the manuscript; or in the decision to publish the results.

References

1. Willett, W.; Rockström, J.; Loken, B.; Springmann, M.; Lang, T.; Vermeulen, S.; Garnett, T.; Tilman, D.; Declerck, F.; Wood, A.; et al. Food in the Anthropocene: The EAT–Lancet Commission on healthy diets from sustainable food systems. *Lancet* **2019**, *393*, 447–492. [CrossRef] [PubMed]
2. Tilman, D.; Clark, M. Global diets link environmental sustainability and human health. *Nature* **2014**, *515*, 518–522. [CrossRef] [PubMed]
3. Rizzoli, R.; Chevalley, T. Bone health: Biology and nutrition. *Curr. Opin. Clin. Nutr. Metab. Care* **2024**, *27*, 24–30. [CrossRef] [PubMed]
4. Dawson-Hughes, B. Interaction of dietary calcium and protein in bone health in humans. *J. Nutr.* **2003**, *133*, 852S–854S. [CrossRef]
5. Hone, M.; Nugent, A.P.; Walton, J.; McNulty, B.A.; Egan, B. Habitual protein intake, protein distribution patterns and dietary sources in Irish adults with stratification by sex and age. *J. Hum. Nutr. Diet.* **2020**, *33*, 465–476. [CrossRef]
6. Pasiakos, S.M.; Agarwal, S.; Lieberman, H.R.; Fulgoni, V.L. Sources and Amounts of Animal, Dairy, and Plant Protein Intake of US Adults in 2007–2010. *Nutrients* **2015**, *7*, 7058–7069. [CrossRef]
7. Heaney, R.P. Is the paradigm shifting? *Bone* **2003**, *33*, 457–465. [CrossRef]
8. Reid, D.M.; New, S.A. Nutritional influences on bone mass. *Proc. Nutr. Soc.* **1997**, *56*, 977–987. [CrossRef]
9. *Pulses: Nutritious Seeds for a Sustainable Future*; FAO: Rome, Italy, 2016; pp. 35–37. [CrossRef]
10. Zenk, J.L.; Frestedt, J.L.; Kuskowski, M.A. Effect of Calcium Derived from Lithothamnion sp. on Markers of Calcium Metabolism in Premenopausal Women. *J. Med. Food* **2018**, *21*, 154–158. [CrossRef]
11. Cashman, K.D. Calcium intake, calcium bioavailability and bone health. *J. Nutr.* **2002**, *87* (Suppl. 2), S169–S177. [CrossRef]
12. Zhong, Q.; Itokawa, T.; Sridhar, S.; Ding, K.H.; Xie, D.; Kang, B.; Bollag, W.B.; Bollag, R.J.; Hamrick, M.; Insogna, K.; et al. Effects of glucose-dependent insulinotropic peptide on osteoclast function. *Am. J. Physiol. Endocrinol. Metab.* **2007**, *292*, E543–E548. [CrossRef] [PubMed]
13. Stensen, S.; Gasbjerg, L.S.; Helsted, M.M.; Hartmann, B.; Christensen, M.B.; Knop, F.K. GIP and the gut-bone axis—Physiological, pathophysiological and potential therapeutic implications. *Peptides* **2020**, *125*, 170197. [CrossRef] [PubMed]
14. Gonzalez, J.T.; Stevenson, E.J. Calcium co-ingestion augments postprandial glucose-dependent insulinotropic peptide(1–42), glucagon-like peptide-1 and insulin concentrations in humans. *Eur. J. Nutr.* **2014**, *53*, 375–385. [CrossRef] [PubMed]
15. Shams-White, M.M.; Chung, M.; Du, M.; Fu, Z.; Insogna, K.L.; Karlsen, M.C.; LeBoff, M.S.; Shapses, S.A.; Sackey, J.; Wallace, T.C.; et al. Dietary protein and bone health: A systematic review and meta-analysis from the National Osteoporosis Foundation. *Am. J. Clin. Nutr.* **2014**, *105*, 1528–1543. [CrossRef] [PubMed]
16. Dawson-Hughes, B.; Harris, S.S. Calcium intake influences the association of protein intake with rates of bone loss in elderly men and women. *Am. J. Clin. Nutr.* **2002**, *75*, 773–779. [CrossRef]
17. Weaver, C.M. The growing years and prevention of osteoporosis in later life. *Proc. Nutr. Soc.* **2000**, *59*, 303–306. [CrossRef]
18. British Nutrition Foundation. *Calcium*; British Nutrition Foundation: London, UK, 1989.
19. Redmond, J.; Fulford, A.J.; Jarjou, L.; Zhou, B.; Prentice, A.; Schoenmakers, I. Diurnal Rhythms of Bone Turnover Markers in Three Ethnic Groups. *J. Clin. Endocrinol. Metab.* **2016**, *101*, 3222–3230. [CrossRef]
20. Qvist, P.; Christgau, S.; Pedersen, B.J.; Schlemmer, A.; Christiansen, C. Circadian variation in the serum concentration of C-terminal telopeptide of type I collagen (serum CTx): Effects of gender, age, menopausal status, posture, daylight, serum cortisol, and fasting. *Bone* **2002**, *31*, 57–61. [CrossRef]
21. Pedersen, B.J.; Schlemmer, A.; Rosenquist, C.; Hassager, C.; Christiansen, C. Circadian rhythm in type I collagen formation in postmenopausal women with and without osteopenia. *Osteoporos. Int.* **1995**, *5*, 472–477. [CrossRef]
22. Heaney, R.P. The bone remodeling transient: Interpreting interventions involving bone-related nutrients. *Nutr. Rev.* **2001**, *59*, 327–334. [CrossRef]
23. Jensen, N.W.; Clemmensen, K.K.B.; Jensen, M.M.; Pedersen, H.; Færch, K.; Diaz, L.J.; Quist, J.S.; Størling, J. Associations between Postprandial Gut Hormones and Markers of Bone Remodeling. *Nutrients* **2021**, *13*, 3197. [CrossRef] [PubMed]
24. Nissen, A.; Christensen, M.; Knop, F.K.; Vilsboll, T.; Holst, J.J.; Hartmann, B. Glucose-dependent insulinotropic polypeptide inhibits bone resorption in humans. *J. Clin. Endocrinol. Metab.* **2014**, *99*, E2325–E2329. [CrossRef] [PubMed]
25. Christensen, M.B.; Lund, A.; Calanna, S.; Jørgensen, N.R.; Holst, J.J.; Vilsbøll, T.; Knop, F.K. Glucose-Dependent Insulinotropic Polypeptide (GIP) Inhibits Bone Resorption Independently of Insulin and Glycemia. *J. Clin. Endocrinol. Metab.* **2018**, *103*, 288–294. [CrossRef] [PubMed]
26. Amigo-Benavent, M.; Power-Grant, O.; FitzGerald, R.J.; Jakeman, P. The insulinotropic and incretin response to feeding a milk-based protein matrix in healthy young women. *J. Funct. Foods* **2020**, *72*, 1040–1056. [CrossRef]
27. Muleya, M.; Bailey, E.F.; Bailey, E.H. A comparison of the bioaccessible calcium supplies of various plant-based products relative to bovine milk. *Food Res. Int.* **2024**, *175*, 11375. [CrossRef]

28. Spencer, H.; Norris, C.; Osis, D. Further studies of the effect of zinc on intestinal absorption of calcium in man. *J. Am. Coll. Nutr.* **1992**, *11*, 561–566. [CrossRef]
29. Wood, R.J.; Zheng, J.J. High dietary calcium intakes reduce zinc absorption and balance in humans. *Am. J. Clin. Nutr.* **1997**, *65*, 1803–1809. [CrossRef]
30. Shkembi, B.; Huppertz, T. Calcium Absorption from Food Products: Food Matrix Effects. *Nutrients* **2022**, *14*, 180. [CrossRef]
31. Hettiarachchi, M.; Cooke, R.; Norton, C.; Jakeman, P. Temporal Change in Biomarkers of Bone Turnover Following Late Evening Ingestion of a Calcium-Fortified, Milk-Based Protein Matrix in Postmenopausal Women with Osteopenia. *Nutrients* **2019**, *11*, 1413. [CrossRef]
32. Norton, C.; Hettiarachchi, M.; Cooke, R.; Kozior, M.; Kontro, H.; Daniel, R.; Jakeman, P. Effect of 24-Week, Late-Evening Ingestion of a Calcium-Fortified, Milk-Based Protein Matrix on Biomarkers of Bone Metabolism and Site-Specific Bone Mineral Density in Postmenopausal Women with Osteopenia. *Nutrients* **2022**, *14*, 3486. [CrossRef]

Disclaimer/Publisher's Note: The statements, opinions and data contained in all publications are solely those of the individual author(s) and contributor(s) and not of MDPI and/or the editor(s). MDPI and/or the editor(s) disclaim responsibility for any injury to people or property resulting from any ideas, methods, instructions or products referred to in the content.

nutrients

MDPI

Article

Hemoglobin Regeneration Efficiency and Relative Iron Bioavailability of Four Elemental Iron Powders in Rats

James H. Swain [1,*], Ryan C. Nemeth [2], Anshul R. Bethi [2], Caroline J. Jang [2] and Eva L. Zheng [2]

1 Department of Nutrition, School of Medicine, Case Western Reserve University, 10900 Euclid Avenue, Cleveland, OH 44106, USA
2 Scientific Enrichment Opportunity Program, School of Medicine, Case Western Reserve University, 10900 Euclid Avenue, Cleveland, OH 44106, USA
* Correspondence: james.swain@case.edu; Tel.:+1-216-368-8554

Abstract: Effective food fortification strategies using elemental iron powders (EIPs) are needed to combat iron deficiency anemia. The purpose of this study was to determine hemoglobin regeneration efficiency (HRE) and relative iron bioavailability (RBV) of four food-grade EIPs (El-Lyte (EL), Hi-Sol (HS), H-325 (H3), and A-131 (A1)) by treating anemic rats with 14 d iron repletion diets (uncooked and cooked), fortified with a 12, 24, or 36 mg iron/kg diet of the EIPs, ferrous sulfate monohydrate (FS, $FeSO_4 \bullet H_2O$), or no added iron (control), n = 9–12/group. The ability of EL and HS to maintain hemoglobin for 6 weeks on the 6 mg iron/kg diet was also studied. The dissolution rate of iron from the EIPs was measured in hydrochloric acid at pH 1.0. Compared to FS, the EL, HS, and A1 EIPs had >50% overall RBV, with the following order: HS > A1 > EL > H3 ($p \leq 0.05$); the effect of cooking was not significant ($p > 0.05$). Dissolution testing revealed that the mean RBV of the EIPs was positively associated with the percentage of iron solubility. In the 6-week maintenance study, EL and HS maintained hemoglobin as well as FS. Overall, the findings show that at the concentrations of iron tested, these EIPs are effective fortification agents to replenish hemoglobin and correct iron deficiency anemia.

Keywords: elemental iron powder; fortification; absorption; relative bioavailability; hemoglobin regeneration efficiency

1. Introduction

Elemental iron powders (EIPs) are often used to fortify staple foods because they are a cost-effective approach to reducing iron deficiency anemia, the most prevalent micronutrient deficiency worldwide [1]. EIPs are often preferred as fortification agents because they do not promote unfavorable sensory or organoleptic changes in food during storage. However, the ability of many EIPs that are commonly used to fortify foods to improve iron status has not been thoroughly studied [2]. Furthermore, methods used to produce EIPs vary among manufacturers, resulting in physicochemical differences [3], and evidence suggests that such variation may influence iron bioavailability [4,5]. Additional information regarding the comparative nutritional usefulness of EIPs would enhance ongoing food fortification programs worldwide [6].

An expert panel reviewing animal [4–12] and human studies [6,13–21] concluded that there is inadequate information regarding the types of EIPs currently being used to fortify staple foods [22]. Historically, EIPs have comprised the majority of iron used to fortify foods since the late 1980s [23,24]. Prior studies of nutrient efficacy have included the assessment of iron absorption using rats as an in vivo model to determine hemoglobin regeneration efficiency (HRE) and the relative bioavailability (RBV, compared to ferrous sulfate, a highly bioavailable form of iron) of a micronutrient [11], and a previous review study found excellent agreement between human tracer studies and rat hemoglobin repletion [12]. As new EIPs enter the market for use as food fortificants, thorough testing of the ability of these new agents and the impact they may have on improving iron status globally is needed.

The purpose of this study was to determine the HRE and RBV of four food-grade elemental iron powders, the effect of uncooked vs. cooked diets on RBV, and the association between these elemental iron powders' solubility and their RBVs.

2. Materials and Methods

2.1. Elemental Iron Powders

In this study, four food-grade elemental iron powders (El-Lyte (EL, an electrolytic iron powder), Hi-Sol (HS, a novel iron powder), H-325 (H3, a hydrogen-reduced iron powder), and A-131 (A1, a well-known standard food-grade electrolytic iron powder) (North American Höganas, Inc., Somerset County, PA, USA) were tested. All of these elemental iron powders (EIPs) meet Food Chemical Codex (FCC) specifications [25]. The EIPs were incorporated into diets that were uncooked or cooked. Bakery-grade ferrous sulfate monohydrate (FS, $FeSO_4 \bullet H_2O$, Crown Technologies, Inc, Indianapolis, IN, USA), an iron salt known for its highly absorbable non-heme iron, also in uncooked and cooked diets, was used as a positive control and for comparisons to determine relative iron bioavailability (RBV). A "no added iron" diet served as the standard control. EIPs and FS were stored in a desiccator under vacuum at room temperature (72 °F) until use. All four EIPs (EL, HS, H3, and A1) are considered "elemental iron powders", referring to the final chemical form or compound, which is relatively pure elemental iron (>98% iron *w/w*, zero oxidation state, with typically >95% of the particle sizes <45 µm diameter). The apparent densities (g/cm^3) of the EIPs are as follows: EL = 1.96, HS = 1.77, H3 = 1.97, and A1 = 2.03. Although structurally different (i.e., flake-like vs. granular for electrolytic vs. H-reduced, respectively), >98% of all four EIPs' particles pass through a +325 mesh sieve.

2.2. Study Design and Dietary Treatments

An initial 432 weanling male Sprague-Dawley rats (Charles River/SASCO, Wilmington, MA, USA) were used. Rats were housed individually in wire-bottom stainless steel mesh cages in a room controlled for temperature (21 ± 1 °C), humidity, and given a 12 h light:dark cycle. After 24 d of depletion with an iron-deficient diet (1.6 mg iron/kg AIN-93G diet; approximately 1.4 mg iron/kg diet based on analyzed iron content), iron-deficient rats with hemoglobin values between 3 and 6 g/dL (mean ± SEM of 4.1 ± 0.4 g/dL; range 3.1–5.9 g/dL) were then randomly assigned by blocking on hemoglobin to one of 36 different repletion-period diet groups. Rats consumed the repletion diets, uncooked or cooked, for 14 d, fortified with EL, HS, H3, or A1 (at 12, 24, or 36 mg iron/kg diet), FS (at 6, 12, 18, 24, or 36 mg iron/kg diet), or no added iron, n = 9–12/group. During the repletion period, daily and total food consumption was measured, including adjustments for spilled food. All diets and deionized, distilled water were fed ad libitum. The HRE ratio and RBV calculations account for iron intake and body weight. All animal procedures followed the Institutional Animal Care and Use Committee procedures at Case Western Reserve University (CWRU), in accordance with the NIH guidelines.

The four EIPs or FS were incorporated into a phytate-free diet modified to have a very low base iron content, using vitamin-free casein (Harlan Teklad, Madison, WI, USA) and a high-purity cellulose fiber source (Alphacel™; ICN Biomedicals, Irvine, CA, USA). The diet pH was neutral (7.0). The base modified diet [26] (AIN-93G*[M]) composition, from which treatment (repletion) diets were prepared, is shown in Table 1. Without added iron, the diet contained approximately 1.5 mg iron/kg diet according to the analysis. EIPs and FS were added into diets taking into account the baseline amount (analyzed) already present in the control (no added iron) group, uncooked and cooked. The modified mineral mix omitted ferric citrate, and the vitamin mix used considered calcium and ascorbic acid content (AIN-93-VX; Harlan Teklad, Madison, WI, USA).

Mixing of each EIP and FS into repletion-period diets and analysis of each diet's iron content were performed in a similar manner as previously described [27], with modifications. Briefly, before being added to the rest of the diet, each EIP or FS was combined with a 500 g portion of corn starch. After thoroughly mixing as previously described [27], the corn starch/iron powder mixture was then added to the other dry diet ingredients and mixed (D330 mixer; Hobart, Troy, OH, USA) for 30 min. Oil was then added, and the diet was mixed for an additional 30 min, followed by the addition of 10% v/w distilled water; this percentage of added water represents a typical flour–water mixture often used in flour-based prepared foods. A concurrent (matching) diet was prepared in this manner and then cooked in a convection oven (model VSOF7301; Viking Inc., Greenwood, MS, USA) at 375 °F (190 °C) for 25 min. Diets were then vacuum-packed (Pro-2100 Vacuum Sealer; Westin, Inc., Southern Pines, NC, USA) and stored in the dark at 39 °F until use. A small portion of each uncooked and cooked diet was taken for analysis to confirm the iron content, as previously described [27].

Table 1. Baseline diet composition [1], from which treatment diets were prepared by adding iron (Fe, as ferrous sulfate monohydrate or one of the elemental iron powders).

Formula	g/Kg	
Casein, low Cu and Fe	200.0	
Sucrose	314.5	
Corn starch	315.0	
Soybean oil	70.0	
Cellulose, microcrystalline (Alphacel™)	50.0	
Mineral mix modified, no added iron (81062) [2] *	35.0	
Vitamin mix, AIN-93-VX (40077) [3] **	10.0	
L-cysteine	3.0	
Choline bitartrate	2.5	
TBHQ, antioxidant	0.014	
Macronutrients	**% dry weight**	**% kcal**
Protein	17.7	17.8
Carbohydrate	69.8	70.4
Fat	5.2	11.8

[1] Ref: [26]. (The base AIN-93G[M] diet was prepared by Harlan Teklad (Catalog #TD.99397; Harlan Teklad, Madison, WI, USA)). [2] Calcium at 3.5 g/kg diet. * Atomic absorption spectrophotometry, performed in triplicate, was used to confirm iron concentrations of the diets, as previously described [27]). Without added iron, the baseline diet contained approximately 1.5 mg iron/kg diet. [3] Ascorbic acid at 200 mg/kg diet. ** Diet/ingredient catalog numbers shown; all other ingredients obtained from Harlan Teklad, Madison, WI, USA).

2.3. Iron Powder Dissolution (Solubility) Testing

The dissolution (solubility) of iron from the elemental iron powder particles was measured using a method similar to that previously described [6] using analytical-grade 18.0 ΩM-cm water (Nanopure, Model D11900 series; ThermoFisher Scientific, Dubuque, IA, USA). After placing 20 mg of each elemental iron powder in 250 mL of dilute 0.1 mol/L hydrochloric acid at pH 1.0, the solution was immediately subjected to constant orbital stirring (Model ORS 200; Boekel Scientific, Feasterville, PA, USA) at 150 rpm and 37 °C. No magnetic stirring devices or glass beads were used. Iron in the solution was then assayed at 30 min; these conditions (pH of 1.0 and sampling at 30 min) were previously found to be optimal [27]. At 30 min, stirring was stopped, and 1 mL of solution was quickly withdrawn and, within 30 sec, centrifuged (Bio-fugePico, 7500 series model; Heraeus Instruments, South Plainfield, NJ, USA) for 5 min at 11,600× *g*. Immediately after centrifugation, 500 μL of the supernatant was diluted to 5 mL with an additional 0.1 mol/L of hydrochloric acid. The concentration of iron in the solution was then analyzed through inductively coupled plasma–atomic emission spectroscopy (ICP-AES, Model ICPE-9800; Shimadzu Scientific

Instruments, Kyoto, Japan). Solubility values are expressed as the mean ± SEM of triplicate samples determined via separate tests.

2.4. Hemoglobin and Hemoglobin Iron Determinations

Hemoglobin and hemoglobin iron determinations were performed similarly as previously described [27,28] but using tail vein blood collection. The following calculation was used to determine hemoglobin (Hb) iron:

$$\text{Hb Fe (mg)} = \text{BW (kg)} \times 0.067 \times \text{Grams Hb per mL} \times 3.35 \text{ mg Fe}$$

Note: the calculation assumes that 6.7% of body weight (BW; kg) is blood and the iron content of hemoglobin is 3.35 mg/g [8,9]. Therefore, hemoglobin iron was determined on the basis of 3.35 mg iron/g hemoglobin and 0.075 L blood/kg body weight [29].

Following repletion, the rats were anesthetized and sacrificed as previously described [27].

2.5. Hemoglobin Regeneration Efficiency and Relative Iron Bioavailability

The hemoglobin regeneration efficiency (HRE) of the four EIPs (EL, HS, H3, and A1) and FS and the relative biological value (relative iron bioavailability, RBV) of the four EIPs were determined as described previously [30]. The RBV of the EIPs was calculated as a percentage of HRE relative to the HRE of FS; HRE and RBV determinations were made by matching cooked vs. cooked and uncooked vs. uncooked data for the same concentration of iron in the diet (diet-matched) and using the analyzed value of iron for each diet. For RBV determinations, bakery-grade ferrous sulfate monohydrate (FS) was selected as the reference standard (positive control) because of its known and higher bioavailability and the prevalence of data in the literature [5,6]. HRE and RBV were calculated based on the following formulas:

$$\text{HRE ratio} = [\text{Final Hb Fe (mg)} - \text{Initial Hb Fe (mg)}]/\text{Fe intake (mg total consumed; analyzed diet value)}$$

$$\text{RBV} = \text{Percentage (\%) HRE relative to ferrous sulfate monohydrate (FS): HRE ratio of elemental iron powder/HRE ratio of the FS group} \times 100 \text{ (diet-matched)}$$

2.6. Hemoglobin Maintenance Study Arm

In a separate arm of the study, EL and HS were added to uncooked and cooked diets to provide 6 mg Fe/kg diet to determine how well these two EIPs maintained iron status (hemoglobin) during a 6-week period. Hemoglobin measures were taken at baseline, mid-treatment (week 3), and at the end of the treatment period (week 6). Ninety-six rats (with normal hemoglobin, not anemic) were randomly assigned to one of eight treatment groups: control (no added iron), FS (positive control), EL, or HS, uncooked or cooked (n = 9–12/group).

2.7. Statistical Analyses

Power analyses were performed based on previous published data [27]. Data on repletion of hemoglobin concentration and change in total hemoglobin iron relative to dietary iron concentration (analyzed) from each iron source as well as absolute iron intake (mg/d) were analyzed as previously described [30]. FS served as the positive control for RBV determinations, and the "no added iron" group served as the overall control (blank), as previously described [31]. Dissolution data were analyzed using one-way analysis of variance (ANOVA), and differences among HRE and RBV means tested using Tukey's multiple comparison and Duncan post hoc testing using the statistical package SAS (SAS Version 10.2, SAS Institute, Cary, NC, USA). Differences were considered significant if $p \leq 0.05$. All values are expressed as mean ± SEM. Illustration of data and results was performed using GraphPad Prism (software version 10.2; GraphPad, Boston, MA, USA).

3. Results

3.1. Elemental Iron Powder Dissolution (Solubility)

The percentage of dissolution of iron from the four elemental iron powders is shown in Table 2. Dissolution (solubility) of iron from the EL, HS, H3, and A1 EIPs at 30 min in HCl solution (pH 1.0) at 37 °C was 58.7 ± 1.1, 75.5 ± 2.7, 43.7 ± 1.4, and 72.7 ± 3.0%, respectively. Figure S1 illustrates the differences in percentage of dissolution graphically.

Table 2. Dissolution (solubility) of the four elemental iron powders [1].

Elemental Iron Powder	% Dissolution [2,3]
El-Lyte	58.7 ± 1.1 [a]
Hi-Sol	75.5 ± 2.7 [b]
H-325	43.7 ± 1.4 [c]
A-131	72.7 ± 3.0 [b]

[1] Determined at 30 min in HCl solution (pH 1.0) at 37 °C. [2] Values are mean ± SEM of triplicate assays. [3] Means in columns that do not share same letter are significantly different ($p \leq 0.05$).

3.2. Hemoglobin Change and Hemoglobin Regeneration Efficiency: No Added Iron and Ferrous Sulfate

Hemoglobin change and hemoglobin regeneration efficiency (HRE) in anemic rats fed no added iron or bakery-grade ferrous sulfate monohydrate (FS) in uncooked or cooked diets are shown in Table 3. Food intake and, thus, iron intake and weight gain were positively associated with increasing dietary iron in all treatment groups (data in Table 3). The HRE ratio calculation accounts for iron intake. The HRE of FS in uncooked and cooked diets is shown in Figure 1. Figure S2 illustrates the hemoglobin change (g/dL) in the control and FS in uncooked and cooked diets, graphically.

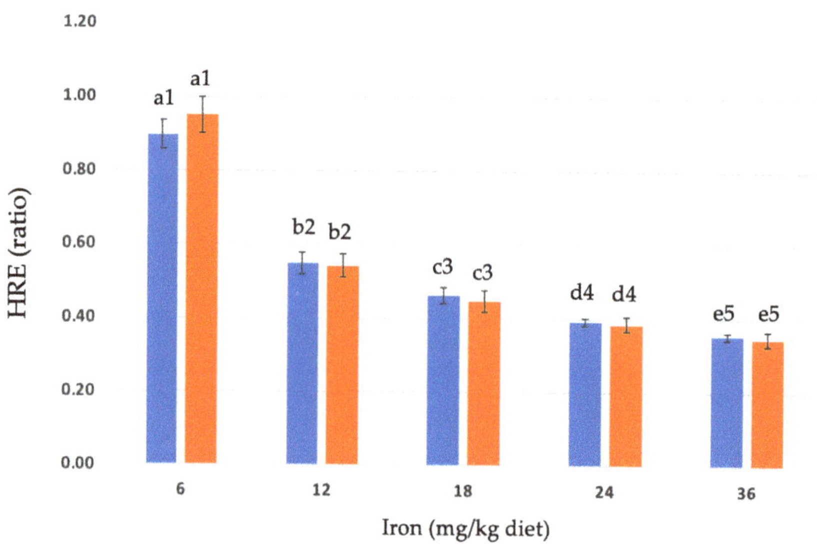

Figure 1. Hemoglobin (Hb) regeneration efficiency of ferrous sulfate monohydrate ($FeSO_4 \bullet H_2O$, FS) in uncooked and cooked diets (blue and orange bars, respectively). Values are mean ± SEM (n = 9–12/group). Different letters are used to denote significant differences ($p \leq 0.05$) between different dietary iron (Fe) concentrations (mg Fe/kg diet) of the same treatment (uncooked and cooked). Different numbers are used to designate significant differences ($p \leq 0.05$) between uncooked and cooked diets of the same Fe concentration. HRE ratio = [Final Hb Fe (mg) − Initial Hb Fe (mg)]/Fe intake (mg total consumed).

Table 3. Food and iron (Fe) intake and growth, hemoglobin change, and hemoglobin regeneration efficiency (HRE) in anemic rats fed no added iron (control) or bakery-grade ferrous sulfate monohydrate (FS) in uncooked (UC) or cooked (C) diets over a 14-day repletion period [1,2].

	Control (C) (No Added Iron)		Ferrous Sulfate (FS)									
Diet Code	C-UC	C-C	FS-1UC	FS-2UC	FS-3UC	FS-4UC	FS-5UC	FS-1C	FS-2C	FS-3C	FS-4C	FS-5C
Diet Fe (mg/kg)												
Calculated	1.6	1.6	6	12	18	24	36	6	12	18	24	36
(Analyzed)	(1.4)	(1.5)	(6.2)	(11.8)	(18.3)	(23.7)	(36.1)	(5.9)	(12.2)	(17.8)	(24.3)	(36.4)
Food intake (g/day)	11.8 ± 0.53 ^{a}	11.9 ± 0.53 ^{a}	12.1 ± 0.62 ^{a}	12.9 ± 0.61 ^{ab}	13.7 ± 0.73 ^{bc}	14.6 ± 0.77 ^{c}	14.9 ± 0.72 ^{c}	12.3 ± 0.76 ^{a}	13.1 ± 0.60 ^{ab}	13.9 ± 0.75 ^{bc}	14.5 ± 0.81 ^{bc}	14.8 ± 0.80 ^{c}
Fe intake (mg/day)	0.017 ± 7^{-04} ^{a}	0.018 ± 9^{-04} ^{a}	0.075 ± 4^{-33} ^{a}	0.152 ± 8^{-03} ^{b}	0.251 ± 0.01 ^{c}	0.346 ± 0.02 ^{d}	0.538 ± 0.02 ^{e}	0.073 ± 4^{-04} ^{a}	0.160 ± 7^{-33} ^{b}	0.247 ± 0.01 ^{c}	0.353 ± 0.02 ^{d}	0.539 ± 0.03 ^{e}
Body weight (g)												
Initial	84.1 ± 3.3 ^{a}	83.8 ± 3.2 ^{a}	82.1 ± 3.6 ^{a}	83.4 ± 3.0 ^{a}	83.9 ± 3.9 ^{a}	84.3 ± 4.1 ^{a}	84.7 ± 3.2 ^{a}	83.2 ± 3.5 ^{a}	84.1 ± 2.9 ^{a}	83.6 ± 3.5 ^{a}	84.4 ± 3.4 ^{a}	83.9 ± 3.9 ^{a}
(Gain)	(16.1 ± 0.6 ^{a})	(15.2 ± 0.5 ^{a})	(53.2 ± 2.3 ^{a})	(55.6 ± 2.7 ^{a})	(54.9 ± 2.5 ^{a})	(56.2 ± 2.3 ^{a})	(58.1 ± 2.9 ^{a})	(53.7 ± 2.3 ^{a})	(54.4 ± 2.6 ^{a})	(54.8 ± 2.9 ^{a})	(57.1 ± 2.0 ^{a})	(57.8 ± 3.2 ^{a})
Hemoglobin (g/dL)												
Initial	4.69 ± 0.2 ^{a}	4.72 ± 0.4 ^{a}	4.73 ± 0.2 ^{a}	4.75 ± 0.3 ^{a}	4.68 ± 0.2 ^{a}	4.81 ± 0.3 ^{a}	4.70 ± 0.2 ^{a}	4.77 ± 0.2 ^{a}	4.83 ± 0.3 ^{a}	4.61 ± 0.4 ^{a}	4.66 ± 0.3 ^{a}	4.65 ± 0.3 ^{a}
(Gain)	(−0.39 ± 0.02 ^{a})	(−0.41 ± 0.02 ^{a})	(1.24 ± 0.05 ^{a})	(1.83 ± 0.1 ^{b})	(3.36 ± 0.2 ^{c})	(4.10 ± 0.2 ^{d})	(6.39 ± 0.4 ^{e})	(1.29 ± 0.05 ^{a})	(2.01 ± 0.2 ^{b})	(3.14 ± 0.3 ^{c})	(4.09 ± 0.2 ^{d})	(6.33 ± 0.3 ^{e})
Hemoglobin Fe [3]												
Gain (mg)	−0.082 ± 4^{-03} ^{a}	−0.070 ± 3^{-03} ^{b}	0.941 ± 0.03 ^{a}	1.164 ± 0.04 ^{b}	1.624 ± 0.03 ^{c}	1.899 ± 0.09 ^{d}	2.662 ± 0.14 ^{e}	0.971 ± 0.03 ^{a}	1.215 ± 0.04 ^{b}	1.543 ± 0.08 ^{c}	1.903 ± 0.11 ^{d}	2.616 ± 0.15 ^{e}
HRE ratio [4]	n/a	n/a	0.896 ± 0.04 ^{a}	0.547 ± 0.03 ^{b}	0.462 ± 0.02 ^{c}	0.392 ± 0.01 ^{d}	0.353 ± 0.01 ^{e}	0.950 ± 0.05 ^{a}	0.542 ± 0.03 ^{b}	0.446 ± 0.03 ^{c}	0.386 ± 0.02 ^{d}	0.347 ± 0.02 ^{d}

[1] Values are mean ± SEM (n = 9–12/group). [2] Means in same row of same diet type (i.e., UC or C) that do not share same letter(s) are significantly different ($p \leq 0.05$). [3] Hb Fe (mg) = BW (body weight; kg) × 0.067 × Grams Hb per mL × 3.35 mg Fe. [4] HRE ratio = [Final Hb Fe (mg) − Initial Hb Fe (mg)]/Fe intake (mg total consumed).

3.3. Hemoglobin Regeneration Efficiency and Relative Iron Bioavailability of Elemental Iron Powders

Hemoglobin regeneration efficiency (HRE) and relative iron bioavailability (RBV) in anemic rats fed one of the four different elemental iron powders (EL, HS, H3, and A1) in uncooked or cooked diets are shown in Tables 4–7, which also show HRE and RBV mean ± SEM for all diet groups. Food intake and, thus, iron intake and weight gain were positively associated with increasing dietary iron in all treatment groups (see Tables 4–7). The HRE ratio and, thus, RBV calculations account for iron intake, and comparisons are diet-matched. HRE and RBV of each EIP in uncooked and cooked diets are shown in Figure 2A–H. For EL, the HRE ratios of uncooked diets at 12, 24, or 36 mg iron/kg diet (11.7, 23.5, and 36.2 mg iron/kg diet analyzed) were 0.360, 0.201, and 0.186, respectively (values ± SEM are shown in Table 4). The RBVs of iron from EL in uncooked diets at 12, 24, or 36 mg iron/kg diet were 65.8, 51.3, and 52.7, respectively. The HRE ratios of cooked diets containing EL at 12, 24, or 36 mg iron/kg diet (11.6, 24.3, and 35.8 mg iron/kg diet analyzed) were 0.369, 0.203, and 0.182, respectively. The RBVs of iron from EL in cooked diets at 12, 24, or 36 mg iron/kg diet were 68.1, 52.6, and 52.5, respectively. The overall mean RBVs (average RBV of the three dietary concentrations tested) of EL in the uncooked and cooked diets were 56.6% and 57.7%, respectively, and these did not differ significantly ($p > 0.05$). For HS, the HRE ratios of uncooked diets at 12, 24, or 36 mg iron/kg diet (12.3, 24.5, and 35.7 mg iron/kg diet analyzed) were 0.360, 0.217, and 0.220, respectively (Table 5). The RBVs of iron from HS in uncooked diets at 12, 24, or 36 mg iron/kg diet were 65.8, 55.4, and 62.3, respectively. The HRE ratios of cooked diets containing HS at 12, 24, or 36 mg iron/kg diet (11.8, 24.2, and 36.3 mg iron/kg diet analyzed) were 0.383, 0.220, and 0.219, respectively. The RBVs of iron from HS in these cooked diets at 12, 24, or 36 mg iron/kg diet were 70.7, 59.3, and 63.1, respectively. The overall mean RBVs (average of the three dietary concentrations tested) of HS in the uncooked and cooked diets were 61.2% and 64.4%, respectively, which were mildly significantly different ($p \leq 0.05$). For H3, the HRE ratios of uncooked diets at 12, 24, or 36 mg iron/kg diet (12.2, 24.6, and 36.3 mg iron/kg diet analyzed) were 0.303, 0.169, and 0.146, respectively (Table 6). The RBVs of iron from H3 in uncooked diets at 12, 24, or 36 mg iron/kg diet were 55.4, 43.1, and 41.4, respectively. The HRE ratios of cooked diets containing H3 at 12, 24, or 36 mg iron/kg diet (11.7, 23.8, and 36.2 mg iron/kg diet analyzed) were 0.327, 0.171, and 0.146, respectively. The RBVs of iron from H3 in these cooked diets at 12, 24, or 36 mg iron/kg diet were 60.3, 44.3, and 42.1, respectively. The overall mean RBVs (average of the three dietary concentrations tested) of H3 in the uncooked and cooked diets were 46.6% and 48.9%, respectively, which were not significantly different ($p > 0.05$). For A1, the HRE ratios of uncooked diets at 12, 24, or 36 mg iron/kg diet (11.5, 23.6, and 35.7 mg iron/kg diet analyzed) were 0.382, 0.216, and 0.214, respectively (Table 7). The RBVs of iron from A1 in uncooked diets at 12, 24, or 36 mg iron/kg diet were 69.8, 55.1, and 60.6, respectively. The HRE ratios of cooked diets containing A1 at 12, 24, or 36 mg iron/kg diet (11.4, 24.4, and 36.3 mg iron/kg diet analyzed) were 0.374, 0.210, and 0.205, respectively. The RBVs of iron from A1 in cooked diets at 12, 24, or 36 mg iron/kg diet were 69.0, 54.4, and 59.1, respectively. The mean RBVs (average of the three dietary concentrations tested) of A1 in the uncooked and cooked diets were 61.8% and 60.8%, respectively, and these did not significantly differ ($p > 0.05$). HRE and RBV data showing mean ± SEM for all diet groups are shown in Tables 4–7.

Table 4. Food and iron (Fe) intake and growth, hemoglobin change, hemoglobin regeneration efficiency (HRE), and relative iron bioavailability (RBV) in anemic rats fed graded quantities of the El-Lyte (EL) elemental iron powder in uncooked (UC) or cooked (C) diets over a 14-day repletion period [1,2].

	El-Lyte (EL)					
Diet Code	**EL-1UC**	**EL-2UC**	**EL-3UC**	**EL-1C**	**EL-2C**	**EL-3C**
Diet Fe (mg/kg)						
Calculated	12	24	36	12	24	36
(Analyzed)	(11.7)	(23.5)	(36.2)	(11.6)	(24.3)	(35.8)
Food intake (g/day)	13.2 ± 0.62 ^{a}	14.1 ± 0.76 ^{ab}	14.9 ± 0.86 ^{b}	12.9 ± 0.64 ^{a}	13.9 ± 0.82 ^{ab}	15.1 ± 0.91 ^{b}
Fe intake (mg/day)	0.154 ± 7$^{-03\ a}$	0.331 ± 0.02 ^{b}	0.539 ± 0.02 ^{c}	0.150 ± 7$^{-03\ a}$	0.338 ± 0.02 ^{b}	0.541 ± 0.03 ^{c}
Body weight (g)						
Initial	84.3 ± 3.4 ^{a}	83.6 ± 3.2 ^{a}	84.9 ± 3.4 ^{a}	83.8 ± 3.7 ^{a}	85.1 ± 4.4 ^{a}	84.6 ± 3.5 ^{a}
(Gain)	(53.2 ± 2.6 ^{a})	(54.7 ± 2.1 ^{a})	(54.9 ± 3.3 ^{a})	(53.6 ± 1.9 ^{a})	(54.2 ± 3.1 ^{a})	(55.3 ± 2.7 ^{a})
Hemoglobin (g/dL)						
Initial	4.85 ± 0.4 ^{a}	4.63 ± 0.2 ^{a}	4.81 ± 0.3 ^{a}	4.67 ± 0.2 ^{a}	4.79 ± 0.2 ^{a}	4.81 ± 0.3 ^{a}
(Gain)	(0.64 ± 0.03 ^{a})	(1.17 ± 0.1 ^{b})	(2.59 ± 0.2 ^{c})	(0.69 ± 0.04 ^{a})	(1.20 ± 0.1 ^{b})	(2.48 ± 0.2 ^{c})
Hemoglobin Fe [3]						
Gain (mg)	0.777 ± 0.03 ^{a}	0.932 ± 0.03 ^{b}	1.405 ± 0.05 ^{c}	0.775 ± 0.04 ^{a}	0.958 ± 0.04 ^{b}	1.376 ± 0.06 ^{c}
HRE ratio [4]	0.360 ± 0.02 ^{a}	0.201 ± 0.01 ^{b}	0.186 ± 9$^{-03\ c}$	0.369 ± 0.02 ^{a}	0.203 ± 0.01 ^{b}	0.182 ± 8$^{-03\ c}$
RBV [5]	65.8 ± 3.6 ^{a}	51.3 ± 2.6 ^{b}	52.7 ± 2.5 ^{b}	68.1 ± 3.5 ^{a}	52.6 ± 2.5 ^{b}	52.5 ± 2.3 ^{b}

[1] Values are mean ± SEM (n = 9–12/group). [2] Means in same row of same diet type (i.e., UC or C) that do not share same letter(s) are significantly different ($p \leq 0.05$). [3] Hb Fe (mg) = BW (body weight; kg) × 0.067 × Grams Hb per mL × 3.35 mg Fe. [4] HRE ratio = [Final Hb Fe (mg) − Initial Hb Fe (mg)]/Fe intake (mg total consumed). [5] RBV = Percentage (%) HRE relative to ferrous sulfate monohydrate (FS): HRE ratio of EL/HRE ratio of the FS group × 100 (diet-matched).

Table 5. Food and iron (Fe) intake and growth, hemoglobin change, hemoglobin regeneration efficiency (HRE), and relative iron bioavailability (RBV) in anemic rats fed graded quantities of the Hi-Sol (HS) elemental iron powder in uncooked (UC) or cooked (C) diets over a 14-day repletion period [1,2].

	Hi-Sol (HS)					
Diet Code	**HS-1 UC**	**HS-2UC**	**HS-3UC**	**HS-1C**	**HS-2C**	**HS-3C**
Diet Fe (mg/kg)						
Calculated	12	24	36	12	24	36
(Analyzed)	(12.3)	(24.5)	(35.7)	(11.8)	(24.2)	(36.3)
Food intake (g/day)	13.4 ± 0.53 ^{a}	14.1 ± 0.65 ^{ab}	15.2 ± 0.81 ^{b}	13.3 ± 0.67 ^{a}	13.9 ± 0.71 ^{ab}	14.7 ± 0.59 ^{b}
Fe intake (mg/day)	0.165 ± 7$^{-03\ a}$	0.345 ± 0.02 ^{b}	0.543 ± 0.03 ^{c}	0.157 ± 8$^{-03\ a}$	0.336 ± 0.02 ^{b}	0.534 ± 0.03 ^{c}
Body weight (g)						
Initial	84.7 ± 3.0 ^{a}	84.9 ± 3.6 ^{a}	85.2 ± 4.1 ^{a}	84.1 ± 3.0 ^{a}	85.3 ± 2.8 ^{a}	85.1 ± 3.3 ^{a}
(Gain)	(54.8 ± 2.6 ^{a})	(56.4 ± 2.9 ^{a})	(57.1 ± 2.2 ^{a})	(54.7 ± 2.1 ^{a})	(56.6 ± 3.4 ^{a})	(56.9 ± 2.6 ^{a})
Hemoglobin (g/dL)						
Initial	4.69 ± 0.3 ^{a}	4.71 ± 0.2 ^{a}	4.79 ± 0.3 ^{a}	4.72 ± 0.2 ^{a}	4.68 ± 0.2 ^{a}	4.78 ± 0.3 ^{a}
(Gain)	(0.81 ± 0.04 ^{a})	(1.43 ± 0.1 ^{b})	(3.31 ± 0.2 ^{c})	(0.84 ± 0.05 ^{a})	(1.51 ± 0.1 ^{b})	(3.22 ± 0.3 ^{c})
Hemoglobin Fe [3]						
Gain (mg)	0.817 ± 0.04 ^{a}	1.022 ± 0.04 ^{b}	1.638 ± 0.07 ^{c}	0.821 ± 0.03 ^{a}	1.052 ± 0.05 ^{b}	1.624 ± 0.06 ^{c}
HRE ratio [4]	0.361 ± 0.02 ^{a}	0.217 ± 0.01 ^{b}	0.220 ± 0.01 ^{b}	0.383 ± 0.02 ^{a}	0.221 ± 0.01 ^{b}	0.219 ± 0.01 ^{b}
RBV [5]	65.8 ± 3.7 ^{a}	55.4 ± 2.6 ^{b}	62.3 ± 3.9 ^{a}	70.7 ± 4.5 ^{a}	59.3 ± 3.7 ^{b}	63.1 ± 3.8 ^{b}

[1] Values are mean ± SEM (n = 9–12/group). [2] Means in same row of same diet type (i.e., UC or C) that do not share same letter(s) are significantly different ($p \leq 0.05$). [3] Hb Fe (mg) = BW (body weight; kg) × 0.067 × Grams Hb per mL × 3.35 mg Fe. [4] HRE ratio = [Final Hb Fe (mg) − Initial Hb Fe (mg)]/Fe intake (mg total consumed). [5] RBV = Percentage (%) (FS): HRE ratio of HS/HRE ratio of the FS group × 100 (diet-matched).

Table 6. Food and iron (Fe) intake and growth, hemoglobin change, hemoglobin regeneration efficiency (HRE), and relative iron bioavailability (RBV) in anemic rats fed graded quantities of the H-325 (H3) elemental iron powder in uncooked (UC) or cooked (C) diets over a 14-day repletion period [1,2].

	H-325 (H3)					
Diet Code	**H3-1UC**	**H3-2UC**	**H3-3UC**	**H3-1C**	**H3-2C**	**H3-3C**
Diet Fe (mg/kg)						
Calculated	12	24	36	12	24	36
(Analyzed)	(12.2)	(24.6)	(36.3)	(11.7)	(23.8)	(36.2)
Food intake (g/day)	13.1 ± 0.49 a	14.6 ± 0.66 b	15.1 ± 0.71 b	12.8 ± 0.67 a	14.1 ± 0.73 ab	14.7 ± 0.62 b
Fe intake (mg/day)	0.160 ± 8−03 a	0.359 ± 0.02 b	0.548 ± 0.03 c	0.150 ± 9−03 a	0.336 ± 0.05 b	0.532 ± 0.04 c
Body weight (g)						
Initial	84.8 ± 4.3 a	83.7 ± 3.6 a	84.6 ± 3.2 a	83.2 ± 3.1 a	84.1 ± 3.8 a	83.9 ± 4.1 a
(Gain)	(50.8 ± 2.5 a)	(51.2 ± 2.1 a)	(51.9 ± 3.1 a)	(51.6 ± 3.0 a)	(51.0 ± 2.1 a)	(52.6 ± 2.2 a)
Hemoglobin (g/dL)						
Initial	4.68 ± 0.2 a	4.82 ± 0.3 a	4.71 ± 0.3 a	4.73 ± 0.3 a	4.65 ± 0.5 a	4.73 ± 0.3 a
(Gain)	(0.48 ± 0.03 a)	(0.98 ± 0.1 b)	(1.86 ± 0.3 c)	(0.46 ± 0.03 a)	(0.89 ± 0.1 b)	(1.72 ± 0.2 c)
Hemoglobin Fe [3]						
Gain (mg)	0.679 ± 0.03 a	0.851 ± 0.04 b	1.119 ± 0.04 c	0.687 ± 0.03 a	0.802 ± 0.02 b	1.085 ± 0.04 c
HRE ratio [4]	0.303 ± 0.01 a	0.169 ± 7−03 b	0.146 ± 6−03 c	0.327 ± 0.02 a	0.171 ± 8−03 b	0.146 ± 7−03 c
RBV [5]	55.4 ± 2.7 a	43.1 ± 1.9 b	41.4 ± 1.7 b	60.3 ± 4.3 a	44.3 ± 2.1 b	42.1 ± 2.0 b

[1] Values are mean ± SEM (n = 9–12/group). [2] Means in same row of same diet type (i.e., UC or C) that do not share same letter(s) are significantly different ($p \leq 0.05$). [3] Hb Fe (mg) = BW (body weight; kg) X 0.067 X Grams Hb per mL X 3.35 mg Fe. [4] HRE ratio = [Final Hb Fe (mg) − Initial Hb Fe (mg)]/Fe intake (mg total consumed). [5] RBV = Percentage (%) HRE relative to ferrous sulfate monohydrate (FS): HRE ratio of H3/HRE ratio of the FS group × 100 (diet-matched).

Table 7. Food and iron (Fe) intake and growth, hemoglobin change, hemoglobin regeneration efficiency (HRE), and relative iron bioavailability (RBV) in anemic rats fed graded quantities of the A-131 (A1) elemental iron powder in uncooked (UC) or cooked (C) diets over a 14-day repletion period [1,2].

	A-131 (A1)					
Diet Code	**A1-1UC**	**A1-2UC**	**A1-3UC**	**A1-1C**	**A1-2C**	**A1-3C**
Diet Fe (mg/kg)						
Calculated	12	24	36	12	24	36
(Analyzed)	(11.5)	(23.6)	(35.7)	(11.4)	(24.4)	(36.3)
Food intake (g/day)	13.3 ± 0.70 a	13.8 ± 0.59 ab	14.8 ± 0.82 b	12.7 ± 0.58 a	13.7 ± 0.53 a	15.2 ± 0.92 b
Fe intake (mg/day)	0.153 ± 8−03 a	0.326 ± 0.02 ab	0.528 ± 0.02 b	0.145 ± 7−03 a	0.334 ± 0.02 a	0.552 ± 0.03 b
Body weight (g)						
Initial	84.2 ± 2.9 a	84.5 ± 3.0 a	84.3 ± 5.5 a	83.6 ± 3.3 a	84.6 ± 2.8 a	84.3 ± 3.9 a
(Gain)	(54.2 ± 3.1 a)	(55.2 ± 2.7 a)	(56.7 ± 2.4 a)	(53.9 ± 2.4 a)	(55.9 ± 2.6 a)	(56.7 ± 3.2 a)
Hemoglobin (g/dL)						
Initial	4.71 ± 0.3 a	4.75 ± 0.2 a	4.61 ± 0.2 a	4.69 ± 0.2 a	4.84 ± 0.3 a	4.70 ± 0.3 a
(Gain)	(0.79 ± 0.04 a)	(1.27 ± 0.05 b)	(3.06 ± 0.3 c)	(0.64 ± 0.03 a)	(1.19 ± 0.05 b)	(3.12 ± 0.2 c)
Hemoglobin Fe [3]						
Gain (mg)	0.818 ± 0.04 a	0.987 ± 0.04 b	1.581 ± 0.06 c	0.759 ± 0.03 a	0.983 ± 0.04 b	1.586 ± 0.07 c
HRE ratio [4]	0.382 ± 0.02 a	0.216 ± 0.01 b	0.214 ± 0.01 b	0.374 ± 0.02 a	0.210 ± 0.01 b	0.205 ± 0.02 b
RBV [5]	69.8 ± 3.5 a	55.1 ± 2.9 b	60.6 ± 3.3 c	69.0 ± 4.4 a	54.4 ± 3.6 b	59.1 ± 5.7 b

[1] Values are mean ± SEM (n = 9–12/group). [2] Means in same row of same diet type (i.e., UC or C) that do not share same letter(s) are significantly different ($p \leq 0.05$). [3] Hb Fe (mg) = BW (body weight; kg) X 0.067 X Grams Hb per mL X 3.35 mg Fe. [4] HRE ratio = [Final Hb Fe (mg) − Initial Hb Fe (mg)]/Fe intake (mg total consumed). [5] RBV = Percentage (%) HRE relative to ferrous sulfate monohydrate (FS): HRE ratio of A1/HRE ratio of the FS group × 100 (diet-matched).

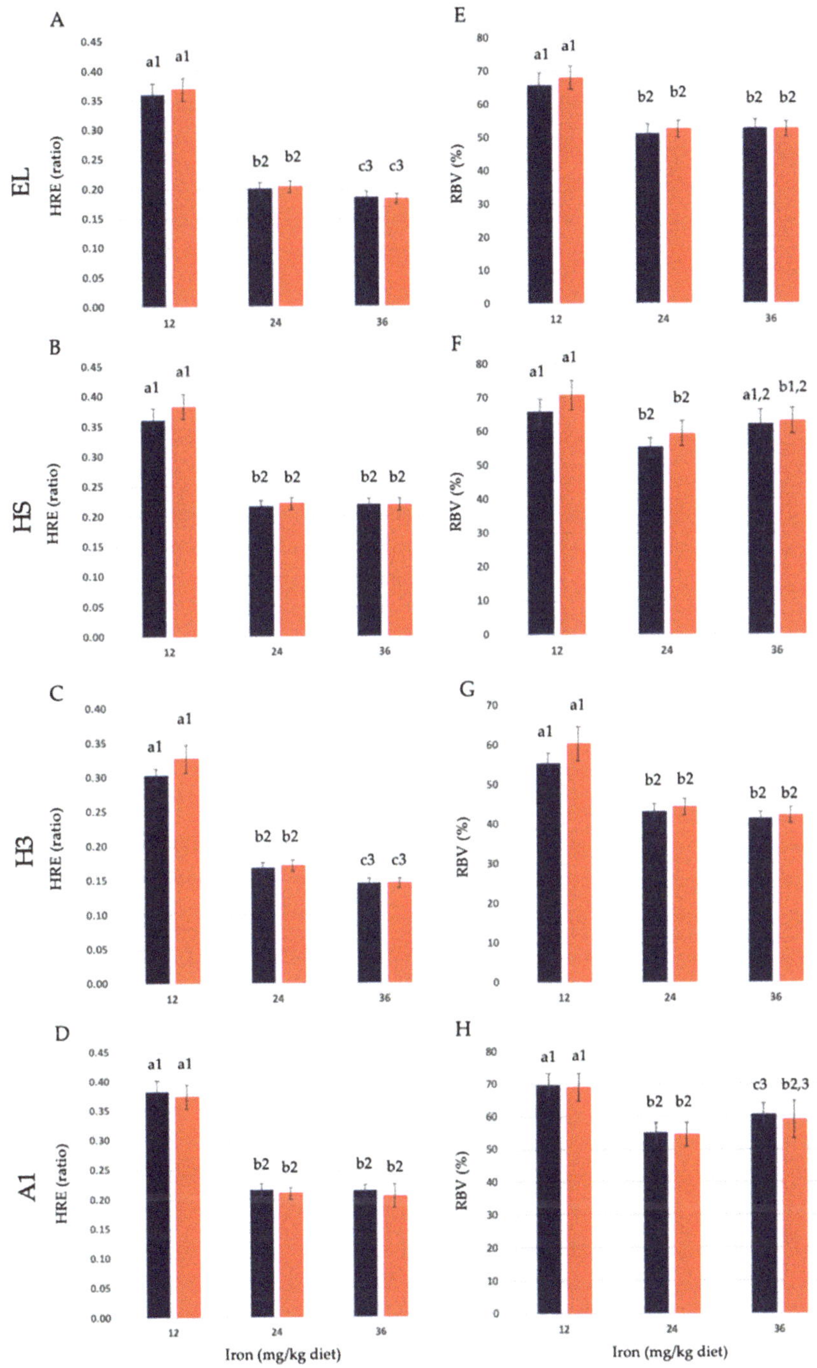

Figure 2. (**A–D**) Hemoglobin regeneration efficiency (HRE) and (**E–H**) relative iron bioavailability (RBV) of the EL (**A**,**E**), HS (**B**,**F**), H3 (**C**,**G**), and A1 (**D**,**H**) elemental iron powders (EIPs) in uncooked and cooked diets (blue and red bars, respectively). Values are mean ± SEM (n = 9–12/group). Different

letters are used to denote significant differences ($p \leq 0.05$) between different dietary iron (Fe) concentrations (mg Fe/kg diet) of the same treatment (uncooked and cooked). Different numbers are used to designate significant differences ($p \leq 0.05$) between uncooked and cooked diets of the same Fe concentration. HRE ratio = [Final Hb Fe (mg) − Initial Hb Fe (mg)]/Fe intake (mg total consumed). RBV = Percentage (%) HRE relative to ferrous sulfate monohydrate (FS): HRE ratio of EIP/HRE ratio of the FS group × 100 (diet-matched).

A combined comparison of all four EIPs' HREs and RBVs is shown in Figure 3A,B.

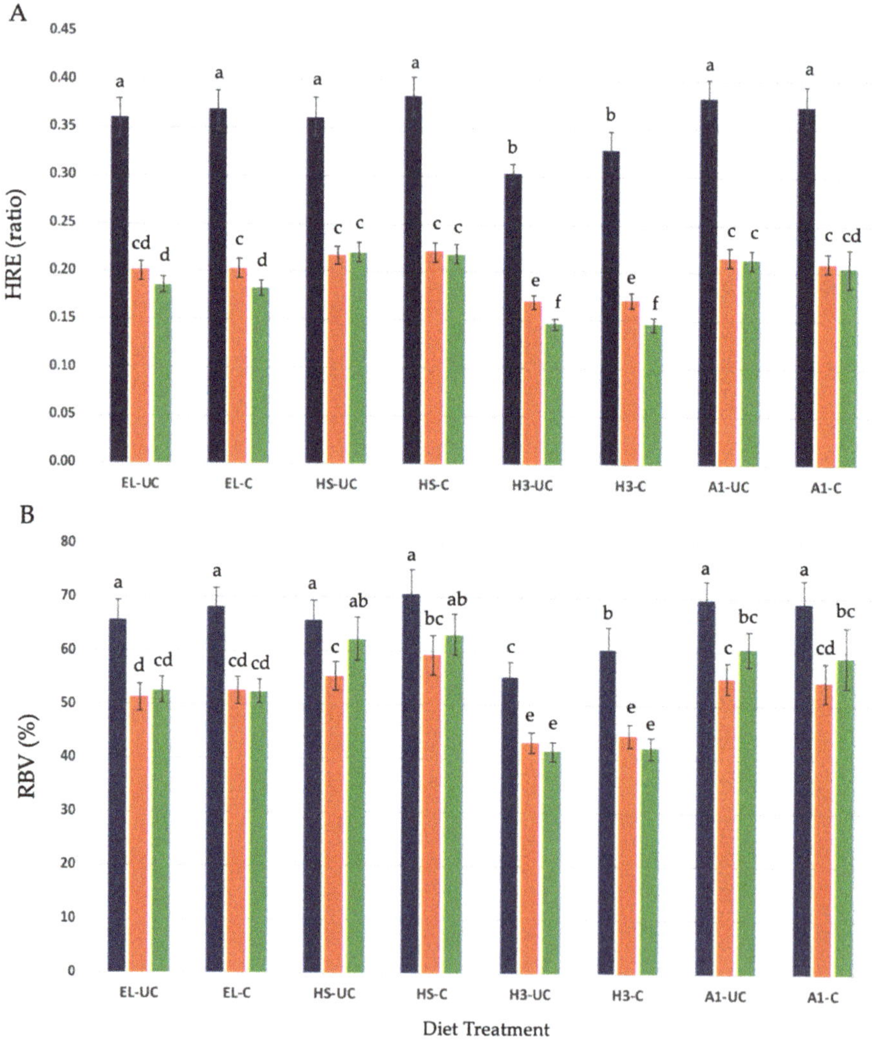

Figure 3. (**A**) Hemoglobin (Hb) regeneration efficiency (HRE) and (**B**) relative iron bioavailability (RBV) of the EL, HS, H3, and A1 elemental iron (Fe) powders (EIPs) in uncooked (UC) and cooked (C) diets, with blue, red, and green bars representing 12, 24, and 36 mg Fe/kg diet, respectively. Values are mean ± SEM (n = 9–12/group). Different letters are used to denote significant differences ($p \leq 0.05$), from higher to lower HRE and RBV. HRE ratio = [Final Hb Fe (mg) − Initial Hb Fe (mg)]/Fe intake (mg total consumed). RBV is calculated as percentage of HRE relative to ferrous sulfate monohydrate (FS) = HRE ratio of EIP/HRE ratio of the FS group × 100 (diet-matched).

3.4. Hemoglobin Maintenance Study

The ability of El-Lyte (EL) and Hi-Sol (HS) iron powders to maintain hemoglobin in non-anemic rats fed 6 mg iron/kg, uncooked and cooked diets, during a six-week period is shown in Table 8. EL and HS, at this concentration of dietary iron in uncooked and cooked diets, were both able to maintain hemoglobin levels as well as FS during the study period (mean ± SEM data for all groups are shown in Table 8).

Table 8. Ability of El-Lyte (EL) and Hi-Sol (HS) iron powders to maintain hemoglobin in non-anemic rats fed 6 mg iron/kg, uncooked (UC) and cooked (C) diets, during six-week period [1,2].

	Control (No Added Iron)		Ferrous Sulfate (FS)		EL-Lyte (EL)		Hi-Sol (HS)	
Diet Code	**C-UC**	**C-C**	**FS-UC**	**FS-C**	**EL-UC**	**EL-C**	**HS-UC**	**HS-C**
Diet Fe (mg/kg)								
Calculated	1.6	1.6	6	6	6	6	6	6
(Analyzed)	(1.4)	(1.5)	(6.2)	(6.1)	(5.8)	(6.2)	(6.1)	(5.9)
Hemoglobin (g/dL)								
Baseline (start wk 0)	13.2 ± 0.5 [a]	12.9 ± 0.6 [a]	12.8 ± 0.9 [a]	13.1 ± 0.7 [a]	13.4 ± 0.8 [a]	13.0 ± 0.7 [a]	13.3 ± 0.9 [a]	12.9 ± 0.6 [a]
Mid (end wk 3)	5.3 ± 0.3 [a]	4.9 ± 0.4 [a]	12.6 ± 0.8 [b]	12.4 ± 0.6 [b]	12.1 ± 0.5 [b]	12.0 ± 0.4 [b]	12.8 ± 0.9 [b]	12.6 ± 0.3 [b]
Post (end wk 6)	4.1 ± 0.3 [a]	4.2 ± 0.2 [a]	12.9 ± 0.8 [b]	12.8 ± 0.7 [b]	12.3 ± 0.9 [b]	12.1 ± 0.6 [b]	12.4 ± 0.4 [b]	12.7 ± 0.5 [b]

[1] Values are mean ± SEM (n = 9–12/group). [2] Means in same row with letter(s) that differ are significantly different ($p \leq 0.05$).

4. Discussion

Our findings show that at the concentrations of iron tested, the EL, HS, H3, and A1 elemental iron powders are useful fortification agents to replenish hemoglobin and correct iron deficiency anemia. Data from this study uniquely illustrate the comparative HRE ratios and RBVs of these four elemental iron powders vs. ferrous sulfate when they are tested at different concentrations in uncooked and cooked diets. In the hemoglobin maintenance arm of this study, our data show that the EL and HS (less well-known elemental iron powders) were also able to maintain hemoglobin as well as ferrous sulfate at the same level of dietary iron (6 mg/kg diet) in non-anemic rats for six weeks. Iron dissolution testing in this study revealed that overall mean RBVs of the four elemental iron powders were positively correlated with iron solubility. However, the difference observed between RBV and percentage of iron dissolution in HS and A1 was greater than the difference seen with EL and H3. Similarly, other studies have shown a relationship between elemental iron powder dissolution properties and RBV, citing a relationship between RBV, particle size [27], and structure of the iron powders tested [32–34].

Generally, HREs and RBVs of these four elemental iron powders were significantly higher ($p \leq 0.05$) at the lower concentration of dietary iron (12 mg Fe/kg diet) when compared to ferrous sulfate. Our data show that for RBV values, except for the H3 and HS samples tested at 12 mg Fe/kg diet—which had slightly higher overall mean RBVs with cooking (albeit non-significantly different; $p > 0.05$)—the effect of cooking is minimal, and cooking does not significantly ($p \leq 0.05$) increase or decrease the RBVs of these elemental iron powders when in the type of dietary mixture used in this study. When compared to ferrous sulfate, the EL, HS, and A1 samples had >50% overall mean RBV, with the following general order of RBV: HS > A1 > EL > H3. The lower overall mean RBV seen with H3 in uncooked and cooked diets (at 46.6% and 48.9%, respectively) relates to the lower RBVs seen in this study at the 24 and 36 mg Fe/kg diet concentrations, which may be due to the differences seen when compared to ferrous sulfate at the higher levels of dietary iron tested.

The rat hemoglobin repletion assay has been used previously by micronutrient researchers worldwide, especially to study the ability of different types of iron fortificants to resolve iron deficiency anemia [6,27,28,32,33,35–38]. In the field of elemental iron powder research in animals and humans, there has been an effort to identify, optimize, and standardize the amount of dietary iron used [6,27,37–39]. Some have concluded that different dietary iron concentrations do not cause differential HRE values [6,30], while others claim there are noticeable differences in HRE at different dietary iron concentrations [39]. Our data here point more toward the conclusion that dietary iron concentration itself may affect HRE calculation, especially at lower iron concentrations. In this study, bakery-grade ferrous sulfate monohydrate was used as the positive control because iron from this compound (iron salt) has consistently been shown to be highly bioavailable [38,40], and it has been used in studies comparing different types of iron fortificants [32,34,38,41,42], ranging from the assessment of iron absorption in anemic school children [35] to adult humans [38,40,42] and rats [6–8,10,27–29], with most studies testing elemental iron powders in diets that were cooked, such as in cookies [42], or, more commonly, in wheat flours [32,33,35]. Additionally, although the monohydrate form of ferrous sulfate is less likely to form a crystalline microstructure, other iron salts used as oral iron supplements clinically (such as ferrous gluconate and ferrous fumarate) may under certain conditions precipitate into "particles" that may influence the absorption of iron from these compounds.

The mean RBV found in this study for EL (a type of electrolytic iron powder) in both uncooked and cooked diets is similar (within 5–15%) to the RBVs observed for other types of electrolytic iron powders, albeit those were tested in humans consuming iron-fortified bread rolls and compared to ferrous sulfate [43]. A study in humans using an H-reduced-elemental-iron-powder-fortified flour consumed by children 5–7 years of age found an RBV of approximately 65% [32], which is within 5–10% of the overall mean RBV we observed for EL, though well above the RBV we found for the H-reduced iron powder we tested (H3), depending on the diet group. However, the RBV we found for EL in this study is 5–15% higher than values summarized in a review [2] of RBVs of many previously tested elemental iron powders using rat hemoglobin repletion assays.

A human study testing a variety of elemental iron powders found that an H-reduced iron powder resulted in an RBV of 43%, while an electrolytic iron powder yielded an RBV of 73% [41]. The RBV of the H-reduced iron powder they tested is similar to the RBVs we found for the H-reduced iron powder we tested in this study (H3) when tested at the 24 and 36 mg Fe/kg diets but lower than the RBV of H3 at 12 mg Fe/kg diet, which may illustrate the influence that dietary iron concentration has when making comparative assessments between elemental iron powders and ferrous sulfate, especially at lower concentrations. The RBV of the electrolytic iron powder they used was very similar to our findings, being within 5–10% of the RBVs we determined for the elemental iron powders we tested at 12 mg Fe/kg diet, in both uncooked and cooked diets, with our cooked diets being more closely associated for each iron powder. However, for the EL, HS, and H3 iron powders we tested at 12 mg Fe/kg diet, the RBVs were slightly higher, although the differences were not significant ($p > 0.05$). Another study using rats, testing electrolytic, H-reduced, and CO-reduced iron elemental iron powders in comparison to ferrous sulfate found RBVs of 70%, 25%, and 32%, respectively [44], again highlighting the relatively lower RBV of H-reduced iron powders in comparison to electrolytic iron powders. Overall, our data are in agreement with the findings of a previous meta-analysis [2] and an international iron fortification guidelines review task force, which studied the usefulness of elemental iron for cereal flour fortification [45].

One limitation of this study may be that other cooking methods (i.e., dry vs. moist heat) were not tested. However, the cooking method we used in this study represents a standard baking approach and how a majority of manufactured food products using fortified and enriched cereal grain flours, into which these elemental iron powders are usually added, are treated in order to produce a variety of processed foods globally. Also, in the hemoglobin maintenance arm of this study, although serum ferritin was not measured, data on the ability of the two less-well-known elemental iron powders (EL and HS) to maintain iron in the functional compartment of blood (hemoglobin) without the rats becoming anemic during this period of time may be useful when considering new food fortification policies and guidelines, as other micronutrient programs have suggested [46].

Overall, data from this study show the relative efficacy of this group of four elemental iron powders in uncooked and cooked diets. Our findings indicate that an efficacious RBV may be achieved at somewhat lower rather than higher concentrations of dietary iron; this is important because current global iron fortification guidelines regarding elemental iron powders often suggest the addition of almost double the concentration of some types of elemental iron powders based on the typical approximate 50% RBV of many iron powders. Our findings support the efficacy of the EL, HS, and A1 elemental iron powders in fortified foods and as iron supplements in clinical settings in near-equivalent dietary iron concentrations to iron salts, such as ferrous sulfate. Based on the importance of resolving iron deficiency anemia worldwide [47], it is suggested that further studies and human clinical trials be aimed at determining what an optimal level of addition of each type of elemental iron powder may be and that prospective or relatively novel elemental iron powders, such as EL and HS, be taken forward for such trials under varying food processing and cooking conditions and in the presence of both dietary inhibitors and enhancers of non-heme iron absorption in order to maximize iron bioavailability when consumed with a mixture of foods.

5. Conclusions

In sum, these four elemental iron powders are useful fortification agents to replenish hemoglobin and correct iron deficiency anemia. Taking into consideration the favorable organoleptic profiles of elemental iron powders, inclusion of these elemental iron powders as iron fortificants in a variety of foods may be both efficacious and advantageous to reduce the incidence of iron deficiency and its associated anemia.

Supplementary Materials: The following supporting information can be downloaded at: https://www.mdpi.com/article/10.3390/nu16142258/s1. Figure S1: Dissolution (% solubility) of the EL, HS, H3, and A1 elemental iron powders determined at 30 min in HCl solution (pH 1.0) at 37 °C. Figure S2: Hemoglobin change (g/dL) in the control (no added iron) and ferrous sulfate monohydrate ($FeSO_4 \bullet H_2O$) uncooked and cooked diets.

Author Contributions: Conceptualization, J.H.S.; Methodology, J.H.S.; Software, J.H.S.; Validation, J.H.S.; Formal Analysis, J.H.S.; Investigation, J.H.S.; Resources, J.H.S.; Data Curation, J.H.S., R.C.N., A.R.B., C.J.J. and E.L.Z.; Writing—Original Draft Preparation, J.H.S., R.C.N., A.R.B., C.J.J. and E.L.Z.; Writing—Review and Editing, J.H.S., R.C.N., A.R.B., C.J.J. and E.L.Z.; Visualization, J.H.S., R.C.N., A.R.B., C.J.J. and E.L.Z.; Supervision, J.H.S.; Project Administration, J.H.S.; Funding Acquisition, J.H.S. All authors have read and agreed to the published version of the manuscript.

Funding: This work was supported in part, including provision of the elemental iron powders, by North American Höganas, Hollsopple, PA USA and by Case Western Reserve University, Cleveland, OH, USA (Grant reference numbers CON110654 and HSAF-G-RB13G, respectively).

Institutional Review Board Statement: The study was conducted in accordance with the Institutional Animal Care and Use Committee (IACUC) of Case Western Reserve University, Cleveland, OH, USA (Permit number: 2011-0046. Date of approval 11 January 2011).

Informed Consent Statement: Not applicable.

Data Availability Statement: The original contributions and data based on which the findings are presented in this study are included within the article and the Supporting Material that accompany this submission.

Acknowledgments: We gratefully acknowledge the contributions of the Case Western Reserve University (CWRU) Animal Resource Center and Jayne Poyer and staff for help with animal care.

Conflicts of Interest: J.H.S. received research funding from North American Höganas, Inc., Hollsopple, PA, USA. The funder was not involved in the study design, collection, analysis, interpretation of data, the writing of this article or the decision to submit it for publication. The remaining authors declare that the research was conducted in the absence of any commercial or financial relationships that could be construed as a potential conflict of interest.

References

1. Leung, A.K.C.; Lam, J.M.; Wong, A.H.C.; Hon, K.L.; Li, X. Iron Deficiency Anemia: An Updated Review. *Curr. Pediatr. Rev.* **2024**, *20*, 339–356. [CrossRef] [PubMed]
2. Hurrell, R.F. Fortification: Overcoming technical and practical barriers. *J. Nutr.* **2002**, *132* (Suppl. S4), 806S–812S. [CrossRef] [PubMed]
3. Paim, T.C.; Wermuth, D.P.; Bertaco, I.; Zanatelli, C.; Naasani, L.I.S.; Slaviero, M.; Driemeier, D.; Schaeffer, L.; Wink, M.R. Evaluation of in vitro and in vivo biocompatibility of iron produced by powder metallurgy. *Mater. Sci. Eng. C Mater. Biol. Appl.* **2020**, *115*, 111129. [CrossRef] [PubMed]
4. Shah, B.G.; Giroux, A.; Belonje, B. Specifications for reduced iron as a food additive. *J. Agric. Food Chem.* **1977**, *25*, 592–594. [CrossRef] [PubMed]
5. Motzok, I.; Verma, R.S.; Chen, S.S.; Rasper, J.; Hancock, R.G.; Ross, H.U. Bioavailability, in vitro solubility, and physical and chemical properties of elemental iron powders. *J. Assoc. Off. Anal. Chem.* **1978**, *61*, 887–893. [CrossRef] [PubMed]
6. Forbes, A.L.; Arnaud, M.J.; Chichester, C.O.; Cook, J.D.; Harrison, B.N.; Hurrell, R.F.; Kahn, S.G.; Morris, E.R.; Tanner, J.T.; Whittaker, P.; et al. Comparison of in vitro, animal, and clinical determinations of iron bioavailability: International Nutritional Anemia Consultative Group Task Force report on iron bioavailability. *Am. J. Clin. Nutr.* **1989**, *49*, 225–238. [CrossRef] [PubMed]
7. Fritz, J.C.; Pla, G.W.; Harrison, B.N.; Clark, G.A. Estimation of the bioavailability of iron. *J. Assoc. Off. Anal. Chem.* **1975**, *58*, 902–905. [CrossRef] [PubMed]
8. Pennell, M.D.; Davies, M.I.; Rasper, J.; Motzok, I. Biological availability of iron supplements for rats, chicks and humans. *J. Nutr.* **1976**, *106*, 265–274. [CrossRef] [PubMed]
9. Coccodrilli, G.D., Jr.; Reussner, G.H.; Thiessen, R., Jr. Relative biological value of iron supplements in processed food products. *J. Agric. Food Chem.* **1976**, *24*, 351–353. [CrossRef]
10. Sacks, P.V.; Houchin, D.N. Comparative bioavailability of elemental iron powders for repair of iron deficiency anemia in rats. Studies of efficacy and toxicity of carbonyl iron. *Am. J. Clin. Nutr.* **1978**, *31*, 566–571. [CrossRef]
11. Park, Y.W.; Mahoney, A.W.; Hendricks, D.G. Bioavailability of different sources of ferrous sulfate iron fed to anemic rats. *J. Nutr.* **1983**, *113*, 2223–2228. [CrossRef] [PubMed]
12. Hurrell, R.F.; Furniss, D.E.; Burri, J.; Whittaker, P.; Lynch, S.R.; Cook, J.D. Iron fortification of infant cereals: A proposal for the use of ferrous fumarate or ferrous succinate. *Am. J. Clin. Nutr.* **1989**, *49*, 1274–1282. [CrossRef] [PubMed]
13. Elwood, P.C. A Clinical Trial of Iron-fortified Bread. *Br. Med. J.* **1963**, *1*, 224–227. [CrossRef] [PubMed]
14. Höglund, S.; Reizenstein, P. Studies in iron absorption. V. Effect of gastrointestinal factors on iron absorption. *Blood* **1969**, *34*, 486–504. [PubMed]
15. Elwood, P.C.; Waters, W.E.; Sweetnam, P. The haematinic effect of iron in flour. *Clin. Sci.* **1971**, *40*, 31–37. [CrossRef] [PubMed]
16. Cook, J.D.; Minnich, V.; Moore, C.V.; Rasmussen, A.; Bradley, W.B.; Finch, C.A. Absorption of fortification iron in bread. *Am. J. Clin. Nutr.* **1973**, *26*, 861–872. [CrossRef] [PubMed]
17. Rios, E.; Hunter, R.E.; Cook, J.D.; Smith, N.J.; Finch, C.A. The absorption of iron as supplements in infant cereal and infant formulas. *Pediatrics* **1975**, *55*, 686–693. [CrossRef] [PubMed]
18. Björn-Rasmussen, E.; Hallberg, L.; Rossander, L. Absorption of 'fortification' iron. Bioavailability in man of different samples of reduced Fe, and prediction of the effects of Fe fortification. *Br. J. Nutr.* **1977**, *37*, 375–388. [CrossRef]
19. Hallberg, L.; Brune, M.; Rossander, L. Low bioavailability of carbonyl iron in man: Studies on iron fortification of wheat flour. *Am. J. Clin. Nutr.* **1986**, *43*, 59–67. [CrossRef]
20. Walter, T.; Dallman, P.R.; Pizarro, F.; Velozo, L.; Peña, G.; Bartholmey, S.J.; Hertrampf, E.; Olivares, M.; Letelier, A.; Arredondo, M. Effectiveness of iron-fortified infant cereal in prevention of iron deficiency anemia. *Pediatrics* **1993**, *91*, 976–982. [CrossRef] [PubMed]
21. Roe, M.A.; Fairweather-Tait, S.J. High bioavailability of reduced iron added to UK flour. *Lancet* **1999**, *353*, 1938–1939. [CrossRef] [PubMed]
22. Turner, L. Monterrey Workshop summary: Evaluating the usefulness of elemental iron powders. *Nutr. Rev.* **2002**, *60* (Suppl. S7), S16–S17, discussion S43–S45. [CrossRef]
23. Whittaker, P. Iron and zinc interactions in humans. *Am. J. Clin. Nutr.* **1998**, *68* (Suppl. S2), 442S–446S. [CrossRef] [PubMed]
24. Chen, C.; Chaudhary, A.; Mathys, A. Nutrient Adequacy of Global Food Production. *Front. Nutr.* **2021**, *8*, 739755. [CrossRef] [PubMed]
25. Committee on Codex Specifications, Food and Nutrition Board, Division of Biological Sciences, Assembly of Life Sciences, National Research Council. *Food Chemicals Codex*, 13th ed.; National Academy Press: Washington, DC, USA, 2022; Available online: https://www.foodchemicalscodex.org/ (accessed on 27 June 2024).
26. Reeves, P.G.; Nielsen, F.H.; Fahey, G.C., Jr. AIN-93 purified diets for laboratory rodents: Final report of the American Institute of Nutrition ad hoc writing committee on the reformulation of the AIN-76A rodent diet. *J. Nutr.* **1993**, *123*, 1939–1951. [CrossRef] [PubMed]
27. Swain, J.H.; Newman, S.M.; Hunt, J.R. Bioavailability of elemental iron powders to rats is less than bakery-grade ferrous sulfate and predicted by iron solubility and particle surface area. *J. Nutr.* **2003**, *133*, 3546–3552. [CrossRef] [PubMed]
28. Fritz, J.C.; Pla, G.W.; Harrison, B.N.; Clark, G.A.; Smith, E.A. Measurement of the bioavailability of iron, using the rat hemoglobin repletion test. *J. Assoc. Off. Anal. Chem.* **1978**, *61*, 709–714. [CrossRef] [PubMed]

29. Whittaker, P.; Mahoney, A.W.; Hendricks, D.G. Effect of iron-deficiency anemia on percent blood volume in growing rats. *J. Nutr.* **1984**, *114*, 1137–1142. [CrossRef]
30. Lisbona, F.; Reyes-Andrada, M.D.; López-Aliaga, I.; Barrionuevo, M.; Alférez, M.J.; Campos, M.S. The importance of the proportion of heme/nonheme iron in the diet to minimize the interference with calcium, phosphorus, and magnesium metabolism on recovery from nutritional ferropenic anemia. *J. Agric. Food Chem.* **1999**, *47*, 2026–2032. [CrossRef]
31. Littell, R.C.; Henry, P.R.; Lewis, A.J.; Ammerman, C.B. Estimation of relative bioavailability of nutrients using SAS procedures. *J. Anim. Sci.* **1997**, *75*, 2672–2683. [CrossRef]
32. Walter, T.; Pizarro, F.; Abrams, S.A.; Boy, E. Bioavailability of elemental iron powder in white wheat bread. *Eur. J. Clin. Nutr.* **2004**, *58*, 555–558. [CrossRef] [PubMed]
33. Biebinger, R.; Zimmermann, M.B.; Al-Hooti, S.N.; Al-Hamed, N.; Al-Salem, E.; Zafar, T.; Kabir, Y.; Al-Obaid, I.; Petry, N.; Hurrell, R.F. Efficacy of wheat-based biscuits fortified with microcapsules containing ferrous sulfate and potassium iodate or a new hydrogen-reduced elemental iron: A randomised, double-blind, controlled trial in Kuwaiti women. *Br. J. Nutr.* **2009**, *102*, 1362–1369. [CrossRef] [PubMed]
34. Jyrwa, Y.W.; Yaduvanshi, P.S.; Sinha, G.R.; Dwarapudi, S.; Madhari, R.S.; Boiroju, N.K.; Pullakhandam, R.; Palika, R. Bioavailability of iron from novel hydrogen reduced iron powders: Studies in Caco-2 cells and rat model. *J. Food Sci.* **2021**, *86*, 3480–3491. [CrossRef] [PubMed]
35. Huang, J.; Sun, J.; Li, W.X.; Wang, L.J.; Wang, A.X.; Huo, J.S.; Chen, J.S.; Chen, C.M. Efficacy of different iron fortificants in wheat flour in controlling iron deficiency. *Biomed. Environ. Sci.* **2009**, *22*, 118–121. [CrossRef]
36. Dickmann, R.S.; Strasburg, G.M.; Romsos, D.R.; Wilson, L.A.; Lai, G.H.; Huang, H. Particle Size, Surface Area, and Amorphous Content as Predictors of Solubility and Bioavailability for Five Commercial Sources of Ferric Orthophosphate in Ready-To-Eat Cereal. *Nutrients* **2016**, *8*, 129. [CrossRef] [PubMed]
37. Zhu, Q.; Qian, Y.; Yang, Y.; Wu, W.; Xie, J.; Wei, D. Effects of carbonyl iron powder on iron deficiency anemia and its subchronic toxicity. *J. Food Drug Anal.* **2016**, *24*, 746–753. [CrossRef] [PubMed]
38. 38. da Silva Lopes, K.; Yamaji, N.; Rahman, M.O.; Suto, M.; Takemoto, Y.; Garcia-Casal, M.N.; Ota, E. Nutrition-specific interventions for preventing and controlling anaemia throughout the life cycle: An overview of systematic reviews. *Cochrane Database Syst. Rev.* **2021**, *9*, CD013092. [CrossRef]
39. Wienk, K.J.; Marx, J.J.; Beynen, A.C. The concept of iron bioavailability and its assessment. *Eur. J. Nutr.* **1999**, *38*, 51–75. [CrossRef] [PubMed]
40. Akhtar, S.; Ahmed, A.; Ahmad, A.; Ali, Z.; Riaz, M.; Ismail, T. Iron status of the Pakistani population-current issues and strategies. *Asia Pac. J. Clin. Nutr.* **2013**, *22*, 340–347. [CrossRef]
41. Hoppe, M.; Hulthén, L.; Hallberg, L. Serum iron concentration as a tool to measure relative iron absorption from elemental iron powders in man. *Scand. J. Clin. Lab. Investig.* **2003**, *63*, 489–496. [CrossRef]
42. Zimmermann, M.B.; Winichagoon, P.; Gowachirapant, S.; Hess, S.Y.; Harrington, M.; Chavasit, V.; Lynch, S.R.; Hurrell, R.F. Comparison of the efficacy of wheat-based snacks fortified with ferrous sulfate, electrolytic iron, or hydrogen-reduced elemental iron: Randomized, double-blind, controlled trial in Thai women. *Am. J. Clin. Nutr.* **2005**, *82*, 1276–1282. [CrossRef] [PubMed]
43. Hoppe, M.; Hulthén, L.; Hallberg, L. The relative bioavailability in humans of elemental iron powders for use in food fortification. *Eur. J. Nutr.* **2006**, *45*, 37–44. [CrossRef] [PubMed]
44. Hurrell, R. Linking the bioavailability of iron compounds to the efficacy of iron-fortified foods. *Int. J. Vitam. Nutr. Res.* **2007**, *77*, 166–173. [CrossRef] [PubMed]
45. Hurrell, R.; Bothwell, T.; Cook, J.D.; Dary, O.; Davidsson, L.; Fairweather-Tait, S.; Hallberg, L.; Lynch, S.; Rosado, J.; Walter, T.; et al. The usefulness of elemental iron for cereal flour fortification: A SUSTAIN Task Force report. Sharing United States Technology to Aid in the Improvement of Nutrition. *Nutr. Rev.* **2002**, *60*, 391–406. [CrossRef] [PubMed]
46. Nielsen, O.H.; Coskun, M.; Weiss, G. Iron replacement therapy: Do we need new guidelines? *Curr. Opin. Gastroenterol.* **2016**, *32*, 128–135. [CrossRef]
47. Hunt, J.R. How important is dietary iron bioavailability? *Am. J. Clin. Nutr.* **2001**, *73*, 3–4. [CrossRef]

Disclaimer/Publisher's Note: The statements, opinions and data contained in all publications are solely those of the individual author(s) and contributor(s) and not of MDPI and/or the editor(s). MDPI and/or the editor(s) disclaim responsibility for any injury to people or property resulting from any ideas, methods, instructions or products referred to in the content.

nutrients

MDPI

Review

Nutrition Facts in the Over-Eighty Population: A Narrative Review

Paolo Riccio [1,*] and Emilio Jirillo [2]

[1] Independent Researcher, 70126 Bari, Italy
[2] Interdisciplinary Department of Medicine, Medical School, University of Bari, 70100 Bari, Italy; emilio.jirillo@uniba.it
* Correspondence: paoloxriccio@gmail.com

Abstract: *Background*: For the first time, humanity is facing the worldwide challenge of global population aging over 80 years. As individuals age, energy acquisition and metabolism undergo significant changes, leading to a progressive decline in energy intake, absorption, and utilization. These changes contribute to malnutrition, loss of muscle mass, frailty, hormonal decline, mineral depletion, and impaired hydration, all of which increase the risk of morbidity and decrease quality of life. In addition, as life expectancy increases, advanced age often brings a gradual loss of autonomy, mirroring early-life dependency. *Objectives*: Addressing this age shift requires targeted interventions to support the wellness of the growing very elderly population. This review provides an overview of healthy aging through an integrated approach that includes nutritional intervention, lifestyle modifications, and targeted supplementation to support functional independence and overall well-being in older adults. The guiding principle is that longevity matters less than aging well.

Keywords: aging; sarcopenia; minerals; dietary supplements

1. Introduction

Old age has been addressed since antiquity, even at times when it barely existed. In ancient Rome, approximately 2000 years ago, famous figures such as Cicero, Seneca, and Publius Terentius Afro wrote about old age.

Due to high infant mortality, at that time, average life expectancy was approximately 25–30 years. Most people lived no more than 40–60 years and reaching 80 years was an exceptional occurrence. Despite its rarity, old age was debated in contrasting terms: Publius Terentius Afro described it as "*Senectus ipsa morbus est*" ("old age itself is a disease"), whereas Cicero valued its virtues, and Seneca argued that, when adequately supported, it could represent a positive and even essential stage for the benefit of the community.

1.1. Aging: A Natural Phase of Life

All living organisms are subject to spatial and temporal constraints that define their size range and lifespan. As in the time of Publius Terentius Afro in ancient Rome, some researchers have proposed that old age should be considered a disease [1,2]. However, aging may be more appropriately regarded as a natural phase of life [3,4], a biological process that becomes pathological only when it occurs significantly earlier than the population average, and not as a result of accidental causes.

Aging is intrinsic to biological systems because every living organism can be regarded as a form of condensed energy. Maintaining this state requires continuous energy input to preserve structural integrity and to sustain maintenance and repair mechanisms. Within our space–time framework, the ability to acquire energy is time-limited, restricting the longevity of biological systems to a finite duration. This limitation is not dictated by specific genetic programs for aging but results from the gradual loss of molecular fidelity over time [5,6]. Repair and renewal processes are inherently imperfect and never completely efficient. Consequently, indefinite survival is incompatible with the nature of life. Lifespan is ultimately determined by the balance between the frequency of structural damage and the efficiency of maintenance mechanisms at the cell, tissue, and organ levels (Table 1).

Table 1. AGING BIOLOGICAL PRINCIPLES [3,4]—redrawn and modified in part from [3].

Cause/Principle	Description
Cellular and Molecular Damage	Accumulation of oxidative stress, DNA mutations, telomere shortening, and protein misfolding, leading to functional decline and aging
Molecular Mechanistic Principle	Frequency of damage and failure of 100% maintenance/repair.
Genetic Programming	No fixed genetic "clock" for the lifespan. Apoptosis and aging-related genes exist but do not "cause" aging directly. Aging is not genetically programmed.
Species and Tissue Differences	Aging varies among species, individuals, tissues, and cell types. Different aging rates within and across organisms. Aging is heterogeneous at every biological level.
Entropy/Thermodynamic Limits	Biological systems tend toward disorder, and this makes long-term homeostasis unsustainable.

Therefore, aging results from the progressive inefficiency of maintenance and repair systems [6]. This means that although aging can be modulated, it remains fundamentally irreversible.

Maintenance is sustained by the continuous intake of exogenous molecules—primarily through feeding. Eating is not merely a matter of caloric intake; it also ensures the continuous renewal and structural preservation of the organism. The nutrients we ingest originate from other forms of life—distinct from us in their structure and in their polymeric compounds but identical to us in their monomeric fundamental units (e.g., glucose, amino acids, and fatty acids). These monomers are transformed into our own molecular constituents—thus, what we call maintenance or turnover can be interpreted as a transient propagation of life through the act of feeding ourselves.

1.2. The Challenge of Global Population Aging

In addition to the ongoing rise in the global population and the challenges posed by climate change, humanity is now facing, for the first time, the issue of the growing number of people aging over the 80 years. Currently, individuals over the age of 80 constitute approximately 1.6% of the global population, and this percentage could increase to 4% by 2050 [7–9] (Table 2).

Furthermore, according to the European Commission, the proportion of individuals aged 80 and over in the European Union increased from 3.7% in 2003 to 6.0% in 2023. Among EU member states, Italy had the highest median age, indicating a large presence of elderly individuals and a substantial aging population [10,11].

Countries like Mexico, Israel, the Philippines, and India are characterized by younger populations, in contrast to Japan, which has one of the highest proportions of elderly individuals worldwide [9] (Table 2).

Table 2. Median age and proportion of population aged 80 years and over in selected countries (latest available estimates from World Population Prospects 2022) [9].

Country/Region	Median Age (years)	Population 80+ (%)
European Union (average)	44.4	6.1
Italy	48.4	7.6
Mexico	29.2	0.9
Israel	30.4	2.0
Philippines	25.7	0.47
India	28.2	0.75
Japan	48.6	10.0
China	39.0	2.4
United States	38.5	3.8
Brazil	34.2	2.1

2. The Gastrointestinal System in Old Age

In individuals over 80 years of age, the body undergoes a marked physiological decline. While cognitive function is often mostly preserved—excluding those with neurodegenerative diseases—the gastrointestinal system, being highly susceptible to cumulative wear and tear, is among the first to experience structural and functional deterioration.

Gastrointestinal dysfunction plays a pivotal role in aging, as it is essential for nutrient absorption and energy acquisition. A decline in digestive efficiency compromises the extraction of macronutrients and micronutrients, leading to energy deficits that impair metabolism and immune defense. These deficiencies may severely compromise the individual's ability to maintain independence and quality of life.

Common digestive impairments in those over 80 years include the following:

(i) Reduced masticatory capacity [12,13], particularly when dental interventions are no longer feasible due to old age [14];
(ii) Decreased saliva production [15];
(iii) Reduced gastric acid and digestive enzyme secretion [16];
(iv) Impaired intestinal peristalsis [17];
(v) Slower gastric emptying [18], which may increase the risk of gastroesophageal reflux;
(vi) Reduced gut microbiota diversity [19].

The risk of malnutrition is further exacerbated by diminished senses of smell and taste, as well as by social isolation, dietary restrictions, and constipation, all of which may lead to reduced appetite and inadequate nutritional intake [20].

In old age, dietary adjustments should account for these physiological changes. To promote adequate intake and optimize digestive efficiency, it would be best to favor easily digestible and easily masticable foods, such as yogurt, vegetable or legume purees, creamy soups, smoothies, and juices. From a general nutritional standpoint, adherence to the principles of the Mediterranean diet, with extra-virgin olive oil as the primary source of dietary fat, is recommended [21].

Beyond overcoming masticable difficulties, the choice of soft over solid foods also has energetic implications. Solid meals require more energy for mastication, digestion, and absorption compared to equivalent soft meals. For example, a 100 kcal solid meal may require an energy expenditure of approximately 10–15 kcal, whereas a soft or liquid meal of the same caloric value may require only 5–10 kcal. This difference can be significant in preserving energy balance in the elderly.

3. Immune Aging

Aging is characterized by a progressive loss of immune function. In fact, there is evidence that hemopoietic stem cells, which differentiate into myeloid cells [granulocytes, monocytes, and dendritic cells (DCs)] and lymphoid cells [T lymphocytes, B lymphocytes, and natural killer (NK) cells)], undergo a progressive decrease in density and regeneration potential, thus contributing to age-dependent immune deficiency [22]. On the other hand, the thymus (where T lymphocytes differentiate) starts declining in early childhood, with a progressive loss of its architecture [23].

The immune system consists of two major components, innate immunity and adaptive immunity, which are both compromised with aging [24]. This implies an impaired immune responsiveness against microorganisms, cancer cells, and autoreactive clones, thus determining the condition of immune senescence [25]. "Inflammaging" is the term used for low-grade chronic inflammation, which increases with advancing age, as a result of constant antigenic exposure [26].

As far as alterations of innate immunity in human aging are concerned, a decrease in neutrophils after 65 years has been reported [27]. Furthermore, neutrophils have been shown to cause telomere damage via release of reactive oxygen species (ROS), thus inducing senescence [28]. Aged monocytes exhibit impaired phagocytosis, with increased levels of Tumor Necrosis Factor (TNF)-alpha release and decreased levels of interleukin (IL)-1 and IL-6 production [29], while dendritic cells (DCs), as antigen-presenting cells, are less tolerogenic, with reduced migration and antigen presentation [30].

Regarding adaptive immunity in aging, a decline in the production of naïve T cells and a loss of diversity in the peripheral T cell repertoire have been documented [26]. Furthermore, aged T cells lose CD28 expression, with impaired responsiveness to antigens, while CD8+ T cytotoxic cells and T regulatory (Treg) cells undergo a reduction in their anti-cancer activity and suppressive functions, respectively [31]. B cell output from the bone marrow is reduced, with a loss of class switching and somatic hypermutation for the formation of high-affinity antigen-specific antibodies [32]. In aged people, the ability of NK cells to lyse specific targets is lost, with decreased perforin release [33]. On the contrary, aging does not affect NK-mediated antibody-dependent cytotoxicity [34].

4. Gut Microbiota in Aging and Its Influence on Immune Response

The gut microbiota represents the largest microbial community in the human body, comprising bacteria, fungi, and viruses [35,36]. During aging, its composition and functionality undergo significant alterations, influenced by dietary habits, pharmacological treatments, and the age-related decline in immune competence. These changes include a reduction in microbial diversity, loss of beneficial taxa, and a relative enrichment of potentially pathogenic species [37].

In younger adults, the gut microbiota is typically dominated by *Firmicutes* and *Bacteroidetes*, together accounting for over 90% of colonic bacterial taxa. The *Firmicutes/Bacteroidetes* (F/B) ratio is often used as an ecological index; however, its variation with age is inconsistent. Some authors report a reduced F/B ratio in the elderly, reflecting a relative increase in *Bacteroidetes* [38,39], while others find the opposite trend or no significant change [40,41]. Such discrepancies may arise from differences in diet, lifestyle, genetic background, health status, and sequencing methodologies. Regardless of type of shifts, changes in these dominant *phila* are often associated with decreased production of short-chain fatty acids (SCFAs), such as acetate, propionate, and butyrate, which are crucial for intestinal barrier integrity, immune regulation, and host metabolic function.

Aging-associated dysbiosis is often linked to inflammaging, the chronic low-grade inflammatory state associated with increased intestinal permeability ("leaky gut"). This facilitates the translocation of microbial products, undigested food particles, and endotoxins into the systemic circulation, contributing to chronic inflammation, metabolic dysfunction, and age-related diseases.

Aging is specifically associated with a decline in beneficial butyrate-producing taxa, such as *Faecalibacterium prausnitzii*, *Eubacterium rectale*, *Clostridium septum*, and *Roseburia* spp., compromising epithelial barrier function and immune regulation [42,43]. Conversely, *Proteobacteria* expand, releasing immunostimulatory lipopolysaccharides (LPSs) that promote cytotoxicity, immune activation [44], and immune senescence [45]. Notably, *Bacteroides*-derived propionate has been shown to enhance resistance to Salmonella spp. in murine models [46]. More broadly, the gut microbiota confers colonization resistance by limiting pathogen overgrowth through competition for nutrients and niche exclusion [47].

Microbiota-derived metabolites also exert direct immunomodulatory effects: for example, butyrate promotes the differentiation of colonic Treg cells. In adult humans, fecal microbiota transplantation (FMT) has been shown to restore immune homeostasis following hematopoietic stem cell transplantation [48].

Reduced SCFA production and compromised barrier integrity further promote microbial translocation and systemic inflammation, which are key contributors to chronic age-related diseases [49]. Moreover, decreased *Lactobacillus*, *Akkermansia*, and certain *Bacteroidetes* spp. have been linked to weakened tight junctions and reduced mucus production [50], whereas overgrowth of pathobionts such as *Clostridium difficile* disrupts epithelial integrity and mucus layer homeostasis [51].

5. Mitochondrial Dysfunction and Cellular Senescence

Mitochondria are central regulators of cellular senescence, contributing to its initiation and progression through metabolic and redox dysregulation. During oxidative phosphorylation, mitochondria generate reactive oxygen species (ROS) as byproducts. While physiological antioxidant systems normally buffer ROS, dysfunctional mitochondria produce excessive ROS and display impaired repair responses, leading to oxidative damage of DNA, proteins, and lipids and triggering senescence [4].

Mitochondrial DNA (mtDNA) is particularly vulnerable to oxidative lesions due to its proximity to the electron transport chain and its limited repair capacity. The accumulation of mtDNA mutations further impairs respiratory chain function, reinforcing mitochondrial dysfunction in a self-amplifying cycle [4]. In parallel, aging disrupts mitochondrial dynamics, resulting in decreased ATP synthesis [52].

Aging also compromises mitophagy, the selective autophagic removal of damaged mitochondria. The accumulation of dysfunctional mitochondria exacerbates redox stress and promotes senescence [53].

Mitochondrial dysfunction acts in multiple senescence pathways [54]. Targeted strategies to restore mitochondrial integrity, such as mitochondria-directed antioxidants, mitophagy activators, and caloric restriction mimetics, have demonstrated efficacy in reducing senescence markers and increasing health span in preclinical models [55–57]. Overall, mitochondria play an active role in cellular aging, representing plausible therapeutic targets for mitigating senescence [58,59].

6. Deficiency of Mineral Salts in the Elderly

In individuals aged 80 years and over, physiological aging increases susceptibility to mineral deficiencies. The following overview summarizes the key biochemical roles of major minerals in this age group.

- Calcium, Phosphorus, and Magnesium: These minerals are vital for bone strength, neuromuscular conduction, enzyme activity, and blood coagulation. With advancing age, osteopenia and osteoporosis are common; inadequate calcium and magnesium support can accelerate bone demineralization and raise fracture risk [60].

- Potassium and Sodium are essential for fluid balance, cardiac and neuromuscular excitability, and blood pressure regulation. People aged 80 and over are most sensitive to increased sodium and may benefit from lower-salt and higher-potassium diets to prevent hypertension, and cardiovascular and renal disease [61].
- Zinc is crucial for DNA synthesis, immune competence, wound healing, and enzymatic reactions. Old people often do not meet recommended zinc intake. Lower levels of zinc exacerbate vulnerability to infections [62].
- Selenium and Copper support antioxidant defense, erythropoiesis, and collagen synthesis. Insufficiency of selenium, zinc, iodine, and copper has been observed among older adults [63].
- Iron maintains hemoglobin levels and physical performance; deficiency in the elderly impairs functional capacity [64].
- Selenium and Magnesium have been linked to reduced sarcopenia. Selenium, together with Calcium intake, is significantly associated with muscle mass, whereas together with Magnesium intake, it is significantly associated with physical performance [65].

Clinical and Nutritional Considerations

- ○ A varied diet based on DASH (Dietary Approaches to Stop Hypertension) recommendations [61] or on a Mediterranean diet with preferential use of extra-virgin olive oil generally provides adequate mineral intake. However, intestinal malabsorption may still impair mineral status.
- ○ Given the risk of deficiencies in old age, both periodic screening and, when necessary, supplementation under clinical supervision are recommended.

In summary, for individuals aged 80 years and over, ensuring adequate mineral intake is critical to maintaining systemic balance, preventing frailty and preserving physical and cognitive function. Nutritional surveillance and individualized interventions are essential to mitigate the clinical impact of deficiencies and support healthy aging and functional independence.

7. Sarcopenia in the Elderly: Prevention and Management

One of the main conditions to avoid or manage in old age is sarcopenia—the progressive loss of skeletal muscle mass, strength, and function.

Malnutrition accelerates muscle loss [66,67]. Age-related muscle decline results from hormonal changes, reduced physical activity, and chronic low-grade inflammation, but targeted nutritional strategies, exercise, and medical interventions can attenuate these effects. Loss of muscle strength leading to frailty is a key limiting factor for maintaining independence until the end of life. Nutritional interventions can not only prevent but also partially reverse sarcopenia [67–69]. A high-protein diet and regular physical activity are central components of prevention and treatment.

Exercise and Physical Activity

Regular strength and resistance training is the most effective intervention against sarcopenia. Balance training (Tai Chi, yoga, and stability exercises) helps prevent frailty [70,71]. Aerobic activities, such as walking, performed daily for 30–45 min, support cardiovascular health, mobility, and endurance. Minimizing sedentary behavior is equally important; prolonged sitting accelerates muscle atrophy, while everyday activities such as housework, gardening, or short walks promote mobility.

Sarcopenia markedly increases the risk of frailty, falls, and loss of independence. A combined approach—including adequate nutrition, resistance training, and regular medical monitoring—enhances muscle health, quality of life, and longevity. Maintaining an active lifestyle, supported by individualized supplementation, and routine check-ups is essential to preserve muscle strength and functional capacity in older adults.

The etiology of sarcopenia in very old age is multifactorial. Adequate mineral intake is associated with better physical performance and a reduced risk of sarcopenia, reinforcing the need for integrated nutritional and physical activity interventions.

8. Hormonal Therapy

As with many bioactive molecules, hormone production declines with age [72–74] This reduction is a major contributor to the structural and functional deterioration observed in aging. However, such decline should be regarded as a physiological process and, therefore, as part of normal aging. Acceptance of reduced hormone levels does not preclude intervention when clinically indicated, but in extreme old age, hormone administration should only be performed under specialist supervision and after confirming a documented deficiency, given the potential contraindications and risks.

Examples of hormone replacement strategies include the following:

- Testosterone Replacement Therapy [75].
- Estrogen and Progesterone Therapy (for postmenopausal women).
- Dehydroepiandrosterone (DHEA) supplementation [76].

DHEA—a precursor of both testosterone and estrogen—has been reported to exert immunostimulatory, anti-diabetic, anti-atherosclerotic, neuroprotective, anti-obesity, and anti-osteoporosis effects [76].

- Thyroid Hormone Replacement (Hypothyroidism); Growth Hormone (GH) Therapy; Insulin-like Growth Factor 1 (IGF-1)

As with other hormones, GH and IGF-1 levels decline with age [77,78]. IGF-1 helps prevent sarcopenia and enhances muscle repair and recovery after injury. It also influences glucose metabolism, insulin sensitivity, and fat distribution; contributes to energy balance and cellular repair mechanisms; and supports neuronal growth and protection, potentially improving memory and cognitive function.

In conclusion, hormone therapy in the elderly can provide meaningful benefits but must be implemented only after a careful, individualized risk/benefit evaluation.

9. Urinary Tract Infection and Hydration

9.1. Lower Urinary Tract Infection

Aging predisposes to recurrent urinary tract infections (UTIs), frequently caused by uropathogenic *Escherichia coli* (UPEC) migrating from the intestine to the lower urinary tract [79]. This process is facilitated by age-associated gut dysbiosis-characterized by reduced microbial diversity, *E. coli* over-representation, and impaired intestinal barrier function. Recurrent UTIs in the elderly often involve genetically identical fecal and urinary isolates, suggesting an endogenous origin [79,80].

Microbiome-based interventions, such as *Lactobacillus rhamnosus* and *L. reuteri* supplementation, have demonstrated efficacy comparable to antibiotic prophylaxis without promoting antimicrobial resistance [81].

Fecal microbiota transplantation (FMT) has also proven effective, restoring microbial diversity and reducing symptomatic episodes [82]. From this perspective, UTIs in older adults may be viewed as a "gut-mediated disease".

For the prevention of UTI, periodic urine culture testing is advisable.

9.2. Hydration Status

Hydration status is a critical yet overlooked aspect of geriatric nutrition. By the age of 80, total body water declines to 45–50% of body weight, compared to about 60% in younger adults [83], largely due to the loss of fat-free mass, particularly skeletal muscle [84].

Deydration—frequently underdiagnosed [85]—can negatively affect cognitive performance and exacerbate fatigue. Current recommendations suggest a daily fluid intake of $\geq$1.6 L for females and $\geq$2.0 L for males. Preventive strategies include ensuring water availability, consuming small amounts frequently, at regular intervals. Eventually, palatability can be enhanced with natural flavoring agents such as lemon juice, yogurt, or herbs. Moreover, adequate protein intake supports muscle mass preservation, while isotonic beverages are recommended to maintain electrolyte balance.

10. General Nutritional Strategies, Intermittent Fasting, and Supplementation in the Elderly

With advancing age, the human body undergoes significant physiological changes: our metabolism slows and energy requirements decrease. Our nutritional needs change. Maintaining a balanced eating style becomes essential for living a long, healthy life. Nutritional recommendations for older adults emphasize reducing carbohydrate intake while increasing protein and fiber. We need less energy and more bricks to repair tissues.

For those relying on liquid meals, increasing meal frequency beyond three per day may help to achieve caloric adequacy. Sugar and gluten consumption should be minimized. The same for hydrogenated fats and processed foods. Monounsaturated fats (extra-virgin olive oil) and omega-3 polyunsaturated fats (oily fish) are recommended. Whole grains and flours should replace refined products, and animal-derived foods should be consumed only occasionally. Legumes, often under-consumed, should be included daily.

A diet rich in vegetables and legumes, with moderate fruit intake, is preferable. Vegetables are preferable to fruit which, if very sugary, should be eaten in moderation. Foods should be fresh, seasonal, locally sourced, and organic.

With solid meals, each bite should be masticated for a long time. Each meal should be given the right amount of time and attention. Prolonged mastication causes the stomach to produce less ghrelin, the hormone that stimulates appetite. The intestine, on the other hand, generates more cholecystokinin, the hormone that controls the appetite center, and GLP-1, the hormone that lowers blood sugar. In other words: the more we masticate, the less we tend to eat; the less we eat, the better we feel.

Physical activity—such as walking, resistance training, dancing, or gardening—should be integrated into daily life, together with mindfulness practices to improve cognitive performance [86].

10.1. Intermittent Fasting

Throughout human evolution, intermittent food scarcity—due to hunting failures or agricultural limitations—was a recurrent condition. Consequently, intermittent fasting and episodic caloric restriction were common, in contrast to the modern Western lifestyle, which is characterized by continuous food availability [87].

Evidence suggests that intermittent fasting and/or caloric restriction may confer health benefits, including enhanced intestinal barrier integrity, increased microbial diversity, and attenuation of gut inflammation [88–93]. Periods of fasting provide physiological rest for the gastrointestinal tract, potentially supporting long-term intestinal health.

Time-restricted feeding (12–16 h fasting windows) is a practical approach: Daily caloric intake is consumed within an 8–12 h period, with only non-caloric fluids thereafter. Advancing the last meal to early evening can improve intermittent fasting.

10.2. Supplementation

In addition to dietary strategies, sustained supplementation under qualified medical supervision can play a critical role in maintaining health in the elderly. Among micronutrients, vitamin D supplementation is a priority. An intake of 5000 I.U./day is recommended to maintain serum concentrations within the optimal range of 40–60 ng/mL [94]. Co-administration with vitamin A is advised, as the receptors of both vitamins act cooperatively in regulating gene expression, upon ligand binding [95].

Additional vitamins include vitamin B12 and vitamin C. In many cases, the use of a broad-spectrum multivitamin complex may be preferable to ensure adequate coverage of essential micronutrients.

Given the high prevalence of sleep disturbances in older adults, melatonin supplementation warrants consideration. There are several similarities between melatonin and vitamin D: both act as hormones, possess immunomodulatory and anti-inflammatory properties, are present in the skin, and respond to environmental light cycles—vitamin D to sunlight, melatonin to darkness [96,97]. Thus, the two molecules seem to have overlapping functions and complementary activities related to light exposure. Their combined use may provide synergistic benefits. Endogenous levels of both decline with age [98,99].

Other recommended supplements include the following:

(1) Essential amino acids, to support protein synthesis;
(2) Complex carbohydrates such as inulin, for prebiotic effects;
(3) Probiotics, to maintain microbiota balance [100];
(4) Butyrate, to preserve or restore intestinal barrier integrity.

For neuroprotection, phospholipids such as L-alpha-glycerophosphorylethanolamine may be beneficial, particularly in formulations combined with plant extracts, coenzyme Q10, and a mixture of antioxidants [101]. Among antioxidants, the modified polyphenol *polydatin* (resveratrol glucoside) is of particular interest for its high bioavailability.

To prevent or counteract sarcopenia, supplementation with L-leucine, omega-3 fatty acids, and *Lactobacillus paracasei* (OLEP) has been classified as food for special medical purposes, for the dietary management of sarcopenia, and for the maintenance and/or recovery of muscle strength and mass [102].

11. Conclusions

This review focuses on individuals over 80 years of age who are free from disease. Given the heterogeneity of this population, a multidimensional and individualized approach is essential to preserve their quality of life. The primary objective should be to ensure that people over 80 enjoy a dignified and as independent as possible old age.

Actions to Be Taken

To achieve this goal, a dedicated task force on aging should be established, and special geriatric care units should be created to address the specific needs of this population. Preventive strategies to reduce frailty—including regular physical activity, adequate nutritional support, and vitamin D supplementation—are strongly recommended to maintain functional capacity and prevent disability. Routine cognitive screening should enable early detection of cognitive decline and facilitate timely, targeted interventions.

Access to and training in telemedicine should be promoted to support remote monitoring and continuity of care. These tools should also be made available to caregivers, thereby enhancing their capacity to assist and coordinate care effectively. Furthermore, when requested, access to social networks; email communication; and more broadly, the use of digital devices should be supported to help reduce both social and cognitive isolation.

Ultimately, the question is, above all, an ethical one: Should our attention be directed primarily toward the very elderly or the young? This dilemma recalls the debate in ancient Rome between the positions of Publius Terentius Afro, on the one hand, and the reflections of Cicero and Seneca, on the other, concerning the value of preserving the role of the elderly within society.

Author Contributions: This work was equally conceived and written by the two authors. Both authors approved the final version. The preparation of Table 2 was assisted by ChatGPT-5 (OpenAI, San Francisco, CA, USA). Both authors reviewed, verified, and approved all AI-assisted content. All authors have read and agreed to the published version of the manuscript.

Funding: This research received no external funding.

Acknowledgments: The authors thank Salvatore Passarella for careful reading of the manuscript.

Conflicts of Interest: The authors declare no conflicts of interest.

References

1. De Winter, G. Aging as disease. *Med. Health Care Philos.* **2015**, *18*, 237–243. [CrossRef] [PubMed]
2. Khaltourina, D.; Matveyev, Y.; Alekseev, A.; Cortese, F.; Ioviţă, A. Aging Fits the Disease Criteria of the International Classification of Diseases. *Mech. Ageing Dev.* **2020**, *189*, 111230. [CrossRef] [PubMed]
3. Rattan, S.I.S. Molecular and cellular basis of aging. In *Molecular Basis of Nutrition and Aging*; Malavolta, M., Mocchegiani, E., Eds.; Academic Press: London, UK, 2016; pp. 3–9.
4. López-Otín, C.; Blasco, M.A.; Partridge, L.; Serrano, M.; Kroemer, G. Hallmarks of aging: An expanding universe. *Cell* **2023**, *186*, 243–278. [CrossRef] [PubMed]
5. Hayflick, L. Biological aging is no longer an unsolved problem. *Ann. N.Y. Acad. Sci.* **2007**, *1100*, 1–13. [CrossRef] [PubMed]
6. Rattan, S.I.S. Biogerontology: Research status, challenges and opportunities. *Acta Biomed.* **2018**, *89*, 291–301. [CrossRef] [PubMed] [PubMed Central]
7. SilverEco. (n.d.). The Ageing Population: Numbers and Statistics. SilverEco.org. Available online: https://www.silvereco.org/en/statistics/ (accessed on 12 August 2025).
8. World Health Organization. Ageing and Health. Available online: https://www.who.int/news-room/fact-sheets/detail/ageing-and-health (accessed on 12 August 2025).
9. United Nations, Department of Economic and Social Affairs, Population Division. World Population Prospects 2022. 2022. Available online: https://population.un.org/wpp/ (accessed on 12 August 2025).
10. European Commission. On 1 January 2023, Across EU Members, the Median Age Ranged Between 38.4 Years in Cyprus and 48.4 Years in Italy. Eurostat News Release [Internet]. 15 February 2024. Available online: https://ec.europa.eu/eurostat/web/products-eurostat-news/w/ddn-20240215-1 (accessed on 12 August 2025).
11. European Union. Demographic Crisis: Aging EU Population Relies on Immigration, with Italy Leading the Decline. EU News [Internet]. 20 May 2025. Available online: https://www.eunews.it/en/2025/05/20/demographic-crisis-aging-eu-population-relies-on-immigration-with-italy-leading-the-decline/ (accessed on 12 August 2025).
12. Watanabe, E.; Maeno, M.; Kayashita, J.; Miyamoto, K.I.; Kogirima, M. Cooking methods for a soft diet using chicken based on food texture analysis. *J. Nutr. Sci. Vitaminol.* **2017**, *63*, 256–262. [CrossRef] [PubMed]
13. Lamster, I.B.; Asadourian, L.; Del Carmen, T.; Friedman, P.K. The aging mouth: Differentiating normal aging from disease. *Periodontol 2000* **2016**, *72*, 96–107. [CrossRef] [PubMed]
14. Sessle, B.J. Can you be too old for oral implants? An update on ageing and plasticity in the oro-facial sensorimotor system. *J. Oral Rehabil.* **2019**, *46*, 936–951. [CrossRef] [PubMed]
15. Xu, F.; Laguna, L.; Sarkar, A. Aging-related changes in quantity and quality of saliva: Where do we stand in our understanding? *J. Texture Stud.* **2019**, *50*, 27–35. [CrossRef] [PubMed]
16. Feldman, M.; Cryer, B.; McArthur, K.E.; Huet, B.A.; Lee, E. Effects of aging and gastritis on gastric acid and pepsin secretion in humans: A prospective study. *Gastroenterology* **1996**, *110*, 1043–1052. [CrossRef] [PubMed]
17. Szymaszkiewicz, A.; Szymaszkiewicz, K.; Fichna, J.; Zielińska, M. Age-related alterations in the enteric nervous system and their impact on peristalsis. *Postepy Biochem.* **2021**, *67*, 34–43. [CrossRef] [PubMed]
18. O'Donovan, D.; Hausken, T.; Lei, Y.; Russo, A.; Keogh, J.; Horowitz, M.; Jones, K.L. Effect of aging on transpyloric flow, gastric emptying, and intragastric distribution in healthy humans--impact on glycemia. *Dig. Dis. Sci.* **2005**, *50*, 671–676. [CrossRef] [PubMed]

19. Badal, V.D.; Vaccariello, E.D.; Murray, E.R.; Yu, K.E.; Knight, R.; Jeste, D.V.; Nguyen, T.T. The gut microbiome, aging, and longevity: A systematic review. *Nutrients* **2020**, *12*, 3759. [CrossRef] [PubMed] [PubMed Central]
20. Arganini, C.; Sinesio, F. Chemosensory impairment does not diminish eating pleasure and appetite in independently living older adults. *Maturitas* **2015**, *82*, 241–244. [CrossRef] [PubMed]
21. Mazza, E.; Ferro, Y.; Pujia, R.; Mare, R.; Maurotti, S.; Montalcini, T.; Pujia, A. Mediterranean Diet In Healthy Aging. *J. Nutr. Health Aging.* **2021**, *25*, 1076–1083. [CrossRef] [PubMed] [PubMed Central]
22. Florian, M.C.; Klose, M.; Sacma, M.; Jablanovic, J.; Knudson, L.; Nattamai, K.J.; Marka, G.; Vollmer, A.; Soller, K.; Sakk, V.; et al Aging alters the epigenetic asymmetry of HSC division. *PLoS Biol.* **2018**, *16*, e2003389. [CrossRef] [PubMed] [PubMed Central]
23. Mittelbrunn, M.; Kroemer, G. Hallmarks of T cell aging. *Nat. Immunol.* **2021**, *22*, 687–698. [CrossRef] [PubMed]
24. Liu, Z.; Liang, Q.; Ren, Y.; Guo, C.; Ge, X.; Wang, L.; Cheng, Q.; Luo, P.; Zhang, Y.; Han, X. Immunosenescence: Molecular mechanisms and diseases. *Signal Transduct. Target Ther.* **2023**, *8*, 200. [CrossRef] [PubMed] [PubMed Central]
25. Ajoolabady, A.; Pratico, D.; Tang, D.; Zhou, S.; Franceschi, C.; Ren, J. Immunosenescence and inflammaging: Mechanisms and role in diseases. *Ageing Res. Rev.* **2024**, *101*, 102540. [CrossRef] [PubMed]
26. Saavedra, D.; Añé-Kourí, A.L.; Barzilai, N.; Caruso, C.; Cho, K.H.; Fontana, L.; Franceschi, C.; Frasca, D.; Ledón, N.; Niedernhofer, L.J.; et al. Aging and chronic inflammation: Highlights from a multidisciplinary workshop. *Immun. Ageing* **2023**, *20*, 25. [CrossRef] [PubMed] [PubMed Central]
27. Sato, Y. Immune Aging and its Implication for Age-Related Disease Progression. *Physiology* **2025**, *40*, 363–373. [CrossRef]
28. Lagnado, A.; Leslie, J.; Ruchaud-Sparagano, M.H.; Victorelli, S.; Hirsova, P.; Ogrodnik, M.; Collins, A.L.; Vizioli, M.G.; Habiballa, L.; Saretzki, G.; et al. Neutrophils induce paracrine telomere dysfunction and senescence in ROS-dependent manner. *EMBO J.* **2021**, *40*, e106048. [CrossRef] [PubMed] [PubMed Central]
29. Basu, S.; Ulbricht, Y.; Rosol, M. Healthy and Premature Aging of Monocytes and Macrophages. *Front. Immunol.* **2025**, *16*, 1506165. [CrossRef] [PubMed]
30. Malik, J.A.; Zafar, M.A.; Lamba, T.; Nanda, S.; Khan, M.A.; Agrewala, J.N. The impact of aging-induced gut microbiome dysbiosis on dendritic cells and lung diseases. *Gut Microbes* **2023**, *15*, 2290643. [CrossRef] [PubMed] [PubMed Central]
31. Vallejo, A.N. CD28 extinction in human T cells: Altered functions and the program of T-cell senescence. *Immunol. Rev.* **2005**, *205*, 158–169. [CrossRef] [PubMed]
32. Frasca, D.; Diaz, A.; Romero, M.; Garcia, D.; Blomberg, B.B. B cell immunosenescence. *Annu. Rev. Cell Dev. Biol.* **2020**, *36*, 551–574. [CrossRef] [PubMed] [PubMed Central]
33. Hazeldine, J.; Hampson, P.; Lord, J.M. Reduced release and binding of perforin at the immunological synapse underlies the age-related decline in natural killer cell cytotoxicity. *Aging Cell* **2012**, *11*, 751–759. [CrossRef] [PubMed]
34. Fernandes, G.; Gupta, S. Natural killing and antibody-dependent cytotoxicity by lymphocyte subpopulations in young and aging humans. *J. Clin. Immunol.* **1981**, *1*, 141–148. [CrossRef] [PubMed]
35. Sender, R.; Fuchs, S.; Milo, R. Revised estimates for the number of human and bacterial cells in the body. *PLoS Biol.* **2016**, *14*, e1002533. [CrossRef] [PubMed] [PubMed Central]
36. Riccio, P.; Rossano, R. The human gut microbiota is neither an organ nor a commensal. *FEBS Lett.* **2020**, *594*, 3262–3271. [CrossRef] [PubMed]
37. Kim, S.; Jazwinski, S.M. The Gut Microbiota and Healthy Aging: A Mini-Review. *Gerontology* **2018**, *64*, 513–520. [CrossRef] [PubMed] [PubMed Central]
38. Arike, L.; Hansson, G.C. The densely O-glycosylated MUC2 mucin protects the intestine and provides food for the commensal bacteria. *J. Mol. Biol.* **2016**, *428*, 3221–3229. [CrossRef] [PubMed] [PubMed Central]
39. Magrone, T.; Jirillo, E. The interplay between the gut immune system and microbiota in health and disease: Nutraceutical intervention for restoring intestinal homeostasis. *Curr. Pharm. Des.* **2013**, *19*, 1329–1342. [CrossRef] [PubMed]
40. Odamaki, T.; Kato, K.; Sugahara, H.; Hashikura, N.; Takahashi, S.; Xiao, J.Z.; Abe, F.; Osawa, R. Age-related changes in gut microbiota composition from newborn to centenarian: A cross-sectional study. *BMC Microbiol.* **2016**, *16*, 90. [CrossRef] [PubMed]
41. Rampelli, S.; Candela, M.; Turroni, S.; Biagi, E.; Collino, S.; Franceschi, C.; O'Toole, P.W.; Brigidi, P. Functional metagenomic profiling of intestinal microbiome in extreme ageing. *Aging* **2013**, *5*, 902–912. [CrossRef]
42. Biagi, E.; Franceschi, C.; Rampelli, S.; Severgnini, M.; Ostan, R.; Turroni, S.; Consolandi, C.; Quercia, S.; Scurti, M.; Monti, D.; et al. Gut microbiota and extreme longevity. *Curr. Biol.* **2016**, *26*, 1480–1485. [CrossRef] [PubMed]
43. Magrone, T.; Jirillo, E. The interaction between gut microbiota and age-related changes in immune function and inflammation. *Immun. Ageing* **2013**, *10*, 31. [CrossRef] [PubMed] [PubMed Central]
44. Shin, N.R.; Whon, T.W.; Bae, J.W. Proteobacteria: Microbial signature of dysbiosis in gut microbiota. *Trends Biotechnol.* **2015**, *33*, 496–503. [CrossRef] [PubMed]
45. Franceschi, C.; Garagnani, P.; Vitale, G.; Capri, M.; Salvioli, S. Inflammaging and 'Garb-aging'. *Trends Endocrinol. Metab.* **2017**, *28*, 199–212. [CrossRef] [PubMed]

46. Jacobson, A.; Lam, L.; Rajendram, M.; Tamburini, F.; Honeycutt, J.; Pham, T.; Van Treuren, W.; Pruss, K.; Stabler, S.R.; Lugo, K.; et al. A gut commensal-produced metabolite mediates colonization resistance to *Salmonella* infection. *Cell Host Microbe* **2018**, *24*, 296–307.e7. [CrossRef] [PubMed] [PubMed Central]
47. Litvak, Y.; Mon, K.K.Z.; Nguyen, H.; Chanthavixay, G.; Liou, M.; Velazquez, E.M.; Kutter, L.; Alcantara, M.A.; Byndloss, M.X.; Tiffany, C.R.; et al. Commensal Enterobacteriaceae protect against *Salmonella* via oxygen competition. *Cell Host Microbe* **2019**, *25*, 128–139.e5. [CrossRef] [PubMed] [PubMed Central]
48. Schluter, J.; Peled, J.U.; Taylor, B.P.; Markey, K.A.; Smith, M.; Taur, Y.; Niehus, R.; Staffas, A.; Dai, A.; Fontana, E.; et al. Gut microbiota is associated with immune cell dynamics in humans. *Nature* **2020**, *588*, 303–307. [CrossRef] [PubMed] [PubMed Central]
49. Martens, E.C.; Neumann, M.; Desai, M.S. Interactions of commensal and pathogenic microorganisms with the intestinal mucosal barrier. *Nat. Rev. Microbiol.* **2018**, *16*, 457–470. [CrossRef] [PubMed]
50. Lewis, K.; Lutgendorff, F.; Phan, V.; Söderholm, J.D.; Sherman, P.M.; McKay, D.M. Enhanced translocation of bacteria across metabolically stressed epithelia is reduced by butyrate. *Inflamm. Bowel Dis.* **2010**, *16*, 1138–1148. [CrossRef] [PubMed]
51. Desai, M.S.; Seekatz, A.M.; Koropatkin, N.M.; Kamada, N.; Hickey, C.A.; Wolter, M.; Pudlo, N.A.; Kitamoto, S.; Terrapon, N.; Muller, A.; et al. A dietary fiber-deprived gut microbiota degrades the colonic mucus barrier and enhances pathogen susceptibility. *Cell* **2016**, *167*, 1339–1353.e21. [CrossRef] [PubMed] [PubMed Central]
52. Sun, N.; Youle, R.J.; Finkel, T. The mitochondrial basis of aging. *Mol. Cell* **2016**, *61*, 654–666. [CrossRef]
53. Wai, T.; Langer, T. Mitochondrial dynamics and metabolic regulation. *Trends Endocrinol. Metab.* **2016**, *27*, 105–117. [CrossRef]
54. Palikaras, K.; Lionaki, E.; Tavernarakis, N. Mechanisms of mitophagy in cellular homeostasis, physiology and pathology. *Nat. Cell Biol.* **2018**, *20*, 1013–1022. [CrossRef] [PubMed]
55. López-Otín, C.; Galluzzi, L.; Freije, J.M.P.; Madeo, F.; Kroemer, G. Metabolic control of longevity. *Cell* **2016**, *166*, 802–821. [CrossRef]
56. Ryu, D.; Mouchiroud, L.; Andreux, P.A.; Katsyuba, E.; Moullan, N.; Nicolet-Dit-Félix, A.A.; Williams, E.G.; Jha, P.; Lo Sasso, G.; Huzard, D.; et al. Urolithin A induces mitophagy and prolongs lifespan in *C. elegans* and increases muscle function in rodents. *Nat. Med.* **2016**, *22*, 879–888. [CrossRef] [PubMed]
57. Smith, R.A.J.; Murphy, M.P. Mitochondria-targeted antioxidants as therapies. *Discov. Med.* **2011**, *11*, 106–114.
58. Andreux, P.A.; van Diemen, M.P.J.; Heezen, M.R.; Auwerx, J.; Rinsch, C.; Groeneveld, G.J.; Singh, A. Mitochondrial function is impaired in the skeletal muscle of pre-frail elderly. *Sci. Rep.* **2018**, *8*, 8548. [CrossRef] [PubMed] [PubMed Central]
59. Miwa, S.; Kashyap, S.; Chini, E.; von Zglinicki, T. Mitochondrial dysfunction in cell senescence and aging. *J. Clin. Investig.* **2022**, *132*, e158447. [CrossRef] [PubMed] [PubMed Central]
60. Barbagallo, M.; Dominguez, L.J. Magnesium and aging. *Curr. Pharm. Des.* **2010**, *16*, 832–839. [CrossRef]
61. Sacks, F.M.; Svetkey, L.P.; Vollmer, W.M.; Appel, L.J.; Bray, G.A.; Harsha, D.; Obarzanek, E.; Conlin, P.R.; Miller, E.R.; Simons-Morton, D.G.; et al. Effects on blood pressure of reduced dietary sodium and the Dietary Approaches to Stop Hypertension (DASH) diet. *N. Engl. J. Med.* **2001**, *344*, 3–10. [CrossRef] [PubMed]
62. Prasad, A.S. Zinc deficiency. *BMJ* **2003**, *326*, 409–410. [CrossRef]
63. Uriu-Adams, J.Y.; Keen, C.L. Copper, oxidative stress, and human health. *Mol. Asp. Med.* **2005**, *26*, 268–298. [CrossRef]
64. Murata, S.; Ebeling, M.; Meyer, A.C.; Schmidt-Mende, K.; Hammar, N.; Modig, K. Blood biomarker profiles and exceptional longevity: Comparison of centenarians and non-centenarians in a 35-year follow-up of the Swedish AMORIS cohort. *Geroscience* **2024**, *46*, 1693–1702, Erratum in: *Geroscience* **2024**, *46*, 2793–2794. https://doi.org/10.1007/s11357-023-00996-y. [CrossRef] [PubMed] [PubMed Central]
65. van Dronkelaar, C.; van Velzen, A.; Abdelrazek, M.; van der Steen, A.; Weijs, P.J.M.; Tieland, M. Minerals and Sarcopenia; The Role of Cal-cium, Iron, Magnesium, Phosphorus, Potassium, Selenium, Sodium, and Zinc on Muscle Mass, Muscle Strength, and Physical Per-formance in Older Adults: A Systematic Review. *J. Am. Med. Dir. Assoc.* **2018**, *19*, 6–11.e3. [CrossRef] [PubMed]
66. Tomasiewicz, A.; Polański, J.; Tański, W. Advancing the Understanding of Malnutrition in the Elderly Population: Current Insights and Future Directions. *Nutrients* **2024**, *16*, 2502. [CrossRef] [PubMed] [PubMed Central]
67. Sieber, C.C. Malnutrition and sarcopenia. *Aging Clin. Exp. Res.* **2019**, *31*, 793–798. [CrossRef]
68. Verlaan, S.; Aspray, T.J.; Bauer, J.M.; Cederholm, T.; Hemsworth, J.; Hill, T.R.; McPhee, J.S.; Piasecki, M.; Seal, C.; Sieber, C.C.; et al. Nutritional status, body composition, and quality of life in community-dwelling sarcopenic and non-sarcopenic older adults: A case-control study. *Clin. Nutr.* **2017**, *36*, 267–274. [CrossRef]
69. Cochet, C.; Belloni, G.; Buondonno, I.; Chiara, F.; D'Amelio, P. The Role of Nutrition in the Treatment of Sarcopenia in Old Patients: From Restoration of Mitochondrial Activity to Improvement of Muscle Performance, a Systematic Review. *Nutrients* **2023**, *15*, 3703. [CrossRef] [PubMed] [PubMed Central]
70. Horiuchi, M.; Okita, K. Blood flow restricted exercise and vascular function. *Int. J. Vasc. Med.* **2012**, *2012*, 543218. [CrossRef] [PubMed] [PubMed Central]

71. Huang, C.Y.; Mayer, P.K.; Wu, M.Y.; Liu, D.H.; Wu, P.C.; Yen, H.R. The effect of Tai Chi in elderly individuals with sarcopenia and frailty: A systematic review and meta-analysis of randomized controlled trials. *Ageing Res. Rev.* **2022**, *82*, 101747. [CrossRef] [PubMed]
72. Lamberts, S.W.; van den Beld, A.W.; van der Lely, A.J. The endocrinology of aging. *Science* **1997**, *278*, 419–424. [CrossRef] [PubMed]
73. Arlt, W.; Hewison, M. Hormones and immune function: Implications of aging. *Aging Cell* **2004**, *3*, 209–216. [CrossRef]
74. Villareal, D.T.; Morley, J.E. Trophic factors in aging. Should older people receive hormonal replacement therapy? *Drugs Aging* **1994**, *4*, 492–509. [CrossRef] [PubMed]
75. Zitzmann, M. Testosteron –„Benzin" für den alten Mann? [Testosterone-"Fuel" for old men?]. *Inn. Med.* **2025**, *66*, 114–123. [CrossRef] [PubMed]
76. Nawata, H.; Yanase, T.; Goto, K.; Okabe, T.; Ashida, K. Mechanism of action of anti-aging DHEA-S and the replacement of DHEA-S. *Mech. Ageing Dev.* **2002**, *123*, 1101–1106. [CrossRef]
77. Vanitallie, T.B. Frailty in the elderly: Contributions of sarcopenia and visceral protein depletion. *Metabolism* **2003**, *52* (Suppl. S2), 22–26. [CrossRef] [PubMed]
78. Toth, L.; Czigler, A.; Hegedus, E.; Komaromy, H.; Amrein, K.; Czeiter, E.; Yabluchanskiy, A.; Koller, A.; Orsi, G.; Perlaki, G.; et al. Age-related decline in circulating IGF-1 associates with impaired neurovascular coupling responses in older adults. *Geroscience* **2022**, *44*, 2771–2783. [CrossRef] [PubMed] [PubMed Central]
79. Stepanova, N. How advanced is our understanding of the role of intestinal barrier dysfunction in the pathogenesis of recurrent urinary tract infections. *Front. Pharmacol.* **2022**, *13*, 780122. [CrossRef] [PubMed]
80. Brigida, M.; Saviano, A.; Petruzziello, C.; Manetti, L.L.; Migneco, A.; Ojetti, V. Gut microbiome implication and modulation in the management of recurrent urinary tract infection. *Pathogens* **2024**, *13*, 1028. [CrossRef]
81. Sturov, N.V.; Popov, S.V.; Zhukov, V.A.; Lyapunova, T.V.; Rusanova, E.I.; Kobylyanu, G.N. Intestinal microbiota correction in the treatment and prevention of urinary tract infection. *Turk. J. Urol.* **2022**, *48*, 406–414. [CrossRef]
82. Jeney, S.E.S.; Lane, F.; Oliver, A.; Whiteson, K.; Dutta, S. Fecal microbiota transplantation for the treatment of refractory recurrent urinary tract infection. *Obstet. Gynecol.* **2020**, *136*, 771–773. [CrossRef]
83. Virgili, F.; D'Amicis, A.; Ferro-Luzzi, A. Body composition and body hydration in old age estimated by means of skinfold thickness and deuterium dilution. *Ann. Hum. Biol.* **1992**, *19*, 57–66. [CrossRef] [PubMed]
84. Lorenzo, I.; Serra-Prat, M.; Yébenes, J.C. The Role of Water Homeostasis in Muscle Function and Frailty: A Review. *Nutrients* **2019**, *11*, 1857. [CrossRef] [PubMed]
85. Hooper, L.; Bunn, D.K.; Downing, A.; Jimoh, F.O.; Groves, J.; Free, C.; Cowap, V.; Potter, J.F.; Hunter, P.R.; Shepstone, L. Which frail older people are dehydrated? The UK DRIE study. *J. Gerontol. A Biol. Sci. Med. Sci.* **2016**, *71*, 1341–1347. [CrossRef]
86. Whitfield, T.; Barnhofer, T.; Acabchuk, R.; Cohen, A.; Lee, M.; Schlosser, M.; Arenaza-Urquijo, E.M.; Böttcher, A.; Britton, W.; Coll-Padros, N.; et al. The effect of mindfulness-based programs on cognitive function in adults: A systematic review and meta-analysis. *Neuropsychol. Rev.* **2022**, *32*, 677–702. [CrossRef]
87. Riccio, P.; Rossano, R. Undigested food and gut microbiota may cooperate in the pathogenesis of neuroinflammatory diseases: A matter of barriers and a proposal on the origin of organ specificity. *Nutrients* **2019**, *11*, 2714. [CrossRef]
88. Choi, I.Y.; Piccio, L.; Childress, P.; Bollman, B.; Ghosh, A.; Brandhorst, S.; Suarez, J.; Michalsen, A.; Cross, A.H.; Morgan, T.E.; et al. A diet mimicking fasting promotes regeneration and reduces autoimmunity and multiple sclerosis symptoms. *Cell Rep.* **2016**, *15*, 2136–2146. [CrossRef] [PubMed]
89. Paoli, A.; Tinsley, G.; Bianco, A.; Moro, T. The influence of meal frequency and timing on health in humans: The role of fasting. *Nutrients* **2019**, *11*, 719. [CrossRef]
90. Cignarella, F.; Cantoni, C.; Ghezzi, L.; Salter, A.; Dorsett, Y.; Chen, L.; Phillips, D.; Weinstock, G.M.; Fontana, L.; Cross, A.H.; et al. Intermittent fasting confers protection in CNS autoimmunity by altering the gut microbiota. *Cell Metab.* **2018**, *27*, 1222–1235.e6. [CrossRef]
91. Rangan, P.; Choi, I.; Wei, M.; Navarrete, G.; Guen, E.; Brandhorst, S.; Enyati, N.; Pasia, G.; Maesincee, D.; Ocon, V.; et al. Fasting-Mimicking Diet Modulates Microbiota and Promotes Intestinal Regeneration to Reduce Inflammatory Bowel Disease Pathology. *Cell Rep.* **2019**, *26*, 2704–2719.e6. [CrossRef] [PubMed] [PubMed Central]
92. Catterson, J.H.; Khericha, M.; Dyson, M.C.; Vincent, A.J.; Callard, R.; Haveron, S.M.; Rajasingam, A.; Ahmad, M.; Partridge, L. Short-term, intermittent fasting induces long-lasting gut health and TOR-independent lifespan extension. *Curr. Biol.* **2018**, *28*, 1714–1724.e4. [CrossRef]
93. Surugiu, R.; Iancu, M.A.; Vintilescu, Ș.B.; Stepan, M.D.; Burdusel, D.; Genunche-Dumitrescu, A.V.; Dogaru, C.-A.; Dumitra, G.G. Molecular mechanisms of healthy aging: The role of caloric restriction, intermittent fasting, Mediterranean diet, and ketogenic diet—A scoping review. *Nutrients* **2024**, *16*, 2878. [CrossRef]
94. Riccio, P. Vitamin D, the sunshine molecule that makes us strong: What does its current global deficiency imply? *Nutrients* **2024**, *16*, 2015. [CrossRef]

95. Riccio, P.; Rossano, R. Diet, gut microbiota, and vitamins D + A in multiple sclerosis. *Neurotherapeutics* **2018**, *15*, 75–91. [CrossRef] [PubMed]
96. Minich, D.M.; Henning, M.; Darley, C.; Fahoum, M.; Schuler, C.B.; Frame, J. Is melatonin the "next vitamin D"?: A review of emerging science, clinical uses, safety, and dietary supplements. *Nutrients* **2022**, *14*, 3934. [CrossRef] [PubMed]
97. Ahmad, S.B.; Ali, A.; Bilal, M.; Rashid, S.M.; Wani, A.B.; Bhat, R.R.; Rehman, M.U. Melatonin and health: Insights of melatonin action, biological functions, and associated disorders. *Cell. Mol. Neurobiol.* **2023**, *43*, 2437–2458. [CrossRef] [PubMed] [PubMed Central]
98. Gallagher, J.C. Vitamin D and aging. *Endocrinol. Metab. Clin. N. Am.* **2013**, *42*, 319–332. [CrossRef] [PubMed]
99. Grivas, T.B.; Savvidou, O.D. Melatonin the "light of night" in human biology and adolescent idiopathic scoliosis. *Scoliosis* **2007**, *2*, 6. [CrossRef] [PubMed]
100. Magrone, T.; Jirillo, E. Prebiotics and probiotics in aging population: Effects on the immune-gut microbiota axis. In *Molecular Basis of Nutrition and Aging*; Malavolta, M., Mocchegiani, E., Eds.; Academic Press: London, UK, 2016; pp. 693–705.
101. Amara, F.; Berbenni, M.; Fragni, M.; Leoni, G.; Viggiani, S.; Ippolito, V.M.; Larocca, M.; Rossano, R.; Alberghina, L.; Riccio, P.; et al. Neuroprotection by Cocktails of Dietary Antioxidants under Conditions of Nerve Growth Factor Deprivation. *Oxidative Med. Cell. Longevity.* **2015**, *2015*, 217258. [CrossRef] [PubMed]
102. Rondanelli, M.; Gasparri, C.; Barrile, G.C.; Battaglia, S.; Cavioni, A.; Giusti, R.; Mansueto, F.; Moroni, A.; Nannipieri, F.; Patelli, Z.; et al. Effectiveness of a novel food composed of leucine, omega-3 fatty acids and probiotic Lactobacillus paracasei PS23 for the treatment of sarcopenia in elderly subjects: A 2-month randomized double-blind placebo-controlled trial. *Nutrients* **2022**, *14*, 4566. [CrossRef]

Disclaimer/Publisher's Note: The statements, opinions and data contained in all publications are solely those of the individual author(s) and contributor(s) and not of MDPI and/or the editor(s). MDPI and/or the editor(s) disclaim responsibility for any injury to people or property resulting from any ideas, methods, instructions or products referred to in the content.

nutrients

MDPI

Review

Minerals and Human Health: From Deficiency to Toxicity

Mohammed S. Razzaque [1] and Sunil J. Wimalawansa [2,*]

[1] Department of Medical Education, University of Texas, Rio Grande Valley, Edinburg, TX 78520, USA; msr.boston@gmail.com
[2] Cardiometabolic Institute, Brunswick, NJ 08902, USA
* Correspondence: suniljw@hotmail.com

Abstract: Minerals are essential nutrients that play critical roles in human health by regulating various physiological functions. Examples include bone development, enzyme function, nerve signaling, and the immune response. Both the deficiencies and toxicities of minerals can have significant health implications. Deficiencies in macrominerals such as calcium, magnesium, and phosphate can lead to osteoporosis (associated with falls and fractures), cardiovascular events, and neuromuscular dysfunction. Trace mineral deficiencies, such as iron and zinc. Selenium deficiency impairs oxygen transport, immune function, and antioxidant defenses, contributing to anemia, delaying wound healing, and increasing susceptibility to infectious diseases. Conversely, excessive intake of minerals can have severe health consequences. Hypercalcemia can cause kidney stones and cardiac arrhythmias as well as soft-tissue calcification, whereas excessive iron deposition can lead to oxidative stress and organ/tissue damage. Maintaining adequate mineral levels through a balanced diet, guided supplementation, and monitoring at-risk populations is essential for good health and preventing disorders related to deficiencies and toxicities. Public health interventions and education about dietary sources of minerals are critical for minimizing health risks and ensuring optimal well-being across populations. While a comprehensive analysis of all macro and micronutrients is beyond the scope of this article, we have chosen to focus on calcium, magnesium, and phosphate. We summarize the consequences of deficiency and the adverse events associated with the overconsumption of other minerals.

Keywords: biological functions; calcium; magnesium; micronutrients; public health; nutrients; physiology

1. Introduction

Adequate mineral balance is crucial for maintaining normal human health and preventing disease. Minerals, which are essential for sustaining various physiological functions, are classified into two main groups: (1) macrominerals, which primarily include calcium (Ca), phosphorus, magnesium (Mg), sodium, and potassium; and (2) microminerals (trace minerals), such as iron, zinc, copper, iodine, selenium, and manganese [1–3]. These minerals are vital for the growth and development of bones and teeth and the physiological functions of muscles and nerves. Additionally, minerals help maintain fluid balance, regulate pH levels, and serve as cofactors for numerous enzymes and coenzymes.

Specific mineral ions are involved in selective functions, ranging from exerting antioxidant effects (selenium, zinc, and copper) to hormone production (iodine for thyroid hormones). Iron deficiency is one of the most prevalent nutritional deficiencies worldwide; it affects more than 25% of the global population and is a major cause of anemia [4]. Vitamin D deficiency is also widespread, with approximately 23% of children in Africa estimated to have vitamin D deficiency [5]. Iodine deficiency can lead to an enlarged thyroid gland (goiter), fatigue, weight gain, dry skin, and impaired cognitive development in children.

An adequate amount of zinc is essential for physiological cellular functions. It is a cofactor for more than 600 enzymes, facilitating DNA and RNA synthesis and playing critical roles in immune function and wound healing. Zinc is also involved in protein synthesis and cell division [1]. Zinc deficiency can slow wound healing, impair oral health [6], and lead to symptoms such as loss of taste or smell, hair loss, skin lesions, and diarrhea [7]. Selenium is an antioxidant in physiological amounts that protects cells from oxidative damage caused by reactive oxygen species (ROS) and reactive nitrogen species [8]. It also supports thyroid hormone metabolism and has immunoregulatory functions [9,10]. Figure 1 illustrates several essential minerals and their fundamental functions in humans.

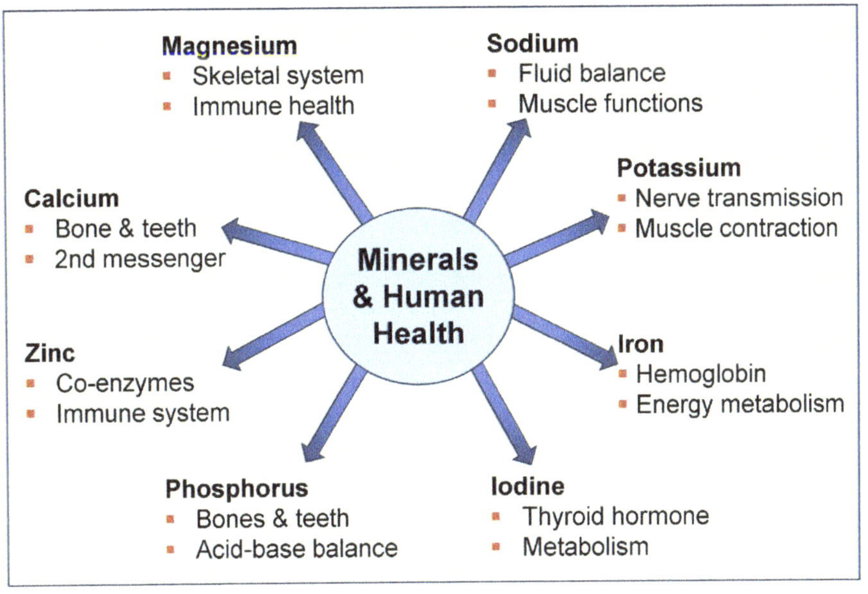

Figure 1. The figure depicts essential minerals with primary functions.

This review focused on essential minerals Ca, Mg, and phosphate because of their physiological importance; they account for around 98% of the body's mineral content by weight. More importantly, Ca, Mg, and phosphate have agonist–antagonist and interrelated metabolic pathways (see Section 2). For instance, parathyroid hormones regulate the levels of all three minerals, making their study as a group particularly relevant. The focused approach presented here comprehensively describes the most abundant and interconnected minerals while acknowledging the importance of other minerals in health and disease.

1.1. The Importance of Micro-Minerals in Physiological Functions

The physiological functions of iron vary from being an essential component of hemoglobin and myoglobin to being crucial for oxygen transport in the blood and energy metabolism (Table 1). Zinc, selenium, and iron inadequacy can impair immune responses and increase infection susceptibility [11]. Potassium and Mg help regulate blood pressure and heart rhythm, and the inadequacy of these minerals may increase the risk of hypertension and cardiovascular diseases. Below, we explain the clinical aspects of Ca, Mg, and phosphate and summarize the importance of other essential minerals for human health.

Table 1. Common minerals, their sources, physiological functions, and potential toxicity in the human body.

Minerals	Sources	Main Physiological Functions	Potential Human Toxicity
Boron	Fruits (avocados, raisins, peaches, apples, grapes, oranges, bananas); vegetables and legumes (broccoli, potatoes, carrots, celery); nuts and seeds (peanuts, almonds, Brazil nuts and hazelnuts)	It affects the metabolism of steroid hormones (estrogen, testosterone, and vitamin D), improves cognitive performance in older adults, and improves antioxidant activity and wound healing.	Skin reactions (rashes and dermatitis) and neurological symptoms (headaches, restlessness, and convulsions) affect fertility and male reproductive organs
Calcium	Milk and milk products; canned fish with bones (salmon and sardines); fortified tofu and soy beverage; greens (broccoli and mustard greens); legumes	Important for bone and teeth health; second messenger.	Kidney stones and nephrocalcinosis, abnormal heart rhythms, and vascular calcification
Chloride	Table salt, soy sauce, and processed foods	Maintaining pH levels in the body (particularly in the blood) is essential for producing hydrochloric acid (in the stomach), facilitating the transmission of nerve impulses, and maintaining proper fluid balance and blood pressure.	Impair sodium or potassium metabolism
Chromium	Whole grains (bread and cereals, oatmeal, and barley); seafood (mussels, oysters, and shrimp); fruits (apples, bananas, and grapes); lean meats (beef, turkey, and chicken); vegetables (broccoli, green beans, potatoes, and asparagus)	Blood sugar regulation is achieved by improving insulin sensitivity; protein metabolism is achieved through the breakdown and absorption of proteins; and fatty acid and cholesterol synthesis is stimulated.	Skin issues (contact dermatitis, skin ulcers, and sensitization), liver and kidney damage; increased risk of lung cancer
Cobalt	Fish and shellfish (oysters, clams, and mussels), meat (mainly liver and kidney), milk and dairy products, legumes, beans and nuts	Essential components of vitamin B12 (cobalamin), erythropoiesis, cofactor for methyl malonyl-CoA mutase, and methionine synthase.	Chronic exposure may lead to asthma-like symptoms, toxic cardiomyopathy, cognitive decline, and polycythemia
Copper	Legumes, nuts and seeds, whole grains, organ meats, and drinking water	Many enzymes are needed for iron metabolism.	Liver damage, kidney failure, neurological effects (mood changes, depression, anxiety, irritability, and difficulty focusing), hematological effects (hemolytic anemia)
Fluoride	Beverages (black tea and coffee, grape juice, chocolate, and almond milk); fruits (grapes and raisins, apples, strawberries, bananas, peaches, watermelon, and cherries); vegetables (spinach, potatoes, carrots, and asparagus); seafood (shrimp, crab, and oysters)	Preventing and reversing dental caries by strengthening tooth enamel, contributing to the mineralization and strength of skeletal tissues, increases the stability and crystallinity of bone apatite structures.	Dental fluorosis (white chalky opacities on tooth enamel, brownish discoloration or pigmentation and pitting of tooth); skeletal fluorosis (increased bone density and decreased elasticity, joint pain and decreased mobility, and increased risk of fractures)

Table 1. *Cont.*

Minerals	Sources	Main Physiological Functions	Potential Human Toxicity
Iodine	Seafood, foods grown in iodine-rich soil, iodized salt, bread, and dairy products	Iodine is present in the thyroid hormone, which helps regulate growth, development, and metabolism.	Thyroid dysfunction (hypothyroidism, hyperthyroidism, thyroiditis, or increased risk of thyroid cancer); neurological effects (delirium, seizures, and stupor)
Iron	Organ meats, red meats, fish, poultry, shellfish (especially clams), egg yolks, legumes, dried fruits, dark leafy greens, and iron-enriched breads and cereals	Part of a molecule (hemoglobin) found in red blood cells that carries oxygen in the body needed for energy metabolism.	Liver damage (chronic liver disease, cirrhosis, and increased risk of hepatocellular carcinoma); cardiac effects (heart failure and arrhythmias); neurological effects (potential acceleration of neurodegenerative diseases)
Magnesium	Nuts and seeds, legumes, leafy green vegetables, seafood, and chocolate	Maintain skeletal system and immune system health.	Gastrointestinal effects (diarrhea and abdominal discomfort); cardiovascular effects (hypotension, bradycardia, and heart blocks); neuromuscular effects (muscle weakness and paralysis)
Manganese	Nuts and seeds (hazelnuts, pecans, and pine nuts); legumes (chickpeas, soybeans, and lentils); shellfish (mussels, oysters, and clams)	A key component of the antioxidant enzyme superoxide dismutase (SOD); it plays a role in blood clotting and hemostasis; acts as a cofactor for various enzymes.	Cognitive impairment, increased susceptibility to respiratory tract infections, and slurred speech
Molybdenum	Legumes (black-eyed peas, lima beans, lentils, and pinto beans); whole grains (oats, barley, and brown rice); dairy products (milk, yogurt, and cheese); vegetables (spinach, potatoes, and asparagus)	The enzyme cofactor for xanthine oxidase and aldehyde oxidase plays a role in the liver's phase I and II detoxification pathways.	Joint pain and gout-like symptoms, anemia, and neurological effects (seizures and hallucinations)
Phosphorus	Meat, fish, poultry, eggs, and milk	Important for healthy bones and teeth; maintains acid-base balance.	Cardiovascular calcification, impaired renal functions, and dysregulation in bone metabolism
Potassium	Meats, milk, fresh fruits and vegetables, whole grains, and legumes	Needed for proper fluid balance, nerve transmission, and muscle contraction.	Cardiovascular effects (palpitations, arrhythmias, and potential heart attack); neurological effects (fatigue, headache, delirium, or seizures); muscle-related effects (weakness, pain, and in severe cases, and paralysis)

Table 1. *Cont.*

Minerals	Sources	Main Physiological Functions	Potential Human Toxicity
Selenium	Meat, seafood, and grains	A key component of antioxidant enzymes, particularly glutathione peroxidases, is important for male fertility and spermatogenesis in regulating thyroid hormones. Selenium-containing enzymes help make DNA and protect against cell damage.	Dermatitis, alopecia, nail discoloration, peripheral neuropathy, decreased cognitive function, and cardiovascular issues (tachycardia and palpitations)
Silicon	Legumes (soybeans, tofu, and red lentils); nuts and seeds (almonds, peanuts, and sunflower seeds); whole grains (oats, barley, and brown rice)	It plays a role in maintaining the structural integrity and elasticity of skin, hair, and nails, crucial for synthesizing and stabilizing collagen.	Excessive exposure can cause silicosis (a progressive and irreversible lung disease); exposure to silica dust can also increase the risk of lung cancer, chronic obstructive pulmonary disease (COPD), and tuberculosis; long-term exposure increases the risk of autoimmune diseases.
Sodium	Table salt, soy sauce, and processed foods	Plays a key role in regulating blood pressure by influencing blood volume and vascular tone; critical for the conduction of nerve impulses, allowing proper communication between nerve cells; essential for normal muscle contraction and relaxation; involved in the transport of various nutrients across cell membranes, including glucose, amino acids, and phosphate.	Neurological effects (confusion, seizures, coma, potential cerebrovascular damage, thirst and dehydration, muscle weakness, and pain)
Zinc	Meats, fish, poultry, whole grains, and vegetables	They are needed for making protein and genetic material and have a role in taste perception, wound healing, normal fetal development, sperm production, normal growth and sexual maturation, and immune system health.	Copper deficiency (resulting in anemia and neutropenia); impaired immune function; and neurological effects (lethargy, dizziness, and, in severe cases, convulsions)

1.2. Broader Physiological Functions of Macro-Minerals

Physiological Mg balance is crucial for muscle contraction and nerve function. Mg also activates enzymes and supports intracellular signaling; it plays an essential role in DNA and RNA synthesis, stabilizing membrane structures and potential [1]. Mg deficiency can induce muscle cramps, twitches, fatigue, irregular heartbeats, osteoporosis, hypertension, and migraines [12,13]. In addition, it may contribute to steroidogenesis through its effects on enzyme activity and cellular signaling pathways [14,15].

Vitamin D is an essential micronutrient critical for regulating the metabolism of mineral ions, including Ca and phosphorus. It is also necessary for normal muscle function, bone health, and immune system regulation. Vitamin D deficiency can impair intestinal mineral ion absorption, renal mineral ion reabsorption, and bone resorption. Insufficient levels of vitamin D may lead to bone pain, muscle weakness, an increased risk of fractures, fatigue, and depression. These symptoms are mediated partially by Ca inadequacy. Ma and zinc are also required to properly process and activate vitamin D and its receptors [16,17].

Both mineral deficiency and excess can have adverse health consequences. For instance, the increased consumption of processed foods containing phosphate additives can lead to chronic phosphate toxicity, which may cause bone deformities, cardiovascular calcification, metabolic disorders, and accelerated aging [18–22]. While a balanced diet typically provides sufficient minerals, understanding their specific roles and ensuring adequate intakes are essential for overall well-being and disease prevention.

Notably, certain groups may be at risk of mineral deficiencies, including pregnant women, elderly individuals, vegetarians/vegans, people with certain medical conditions, and those following restrictive diets. In such cases, targeted supplementation under medical supervision may be necessary. Nutrient deficiencies manifest nonspecific symptoms so that blood levels may offer valuable insights into the underlying pathophysiology. Table 1 lists common minerals, their sources, and their physiological functions.

Most of the information in the Table was compiled from the Dietary Supplement of the Institute of Medicine and Fact Sheets and recommended daily allowances from the National Institute of Health (NIH), Office of Dietary Supplements (ODS) (https://ods.od.nih.gov/, Accessed on 3 January 2025).

2. Calcium—From Physiology to Pathology

Calcium (Ca) is essential for numerous physiological processes, including bone health, muscle contraction, nerve signaling, and blood clotting. Approximately 99% of the body's Ca is stored in bones and teeth, providing structural support and serving as a reservoir to maintain extracellular Ca levels. Adequate Ca intake prevents osteoporosis and maintains bone density, particularly in aging populations and postmenopausal women [23]. Beyond skeletal health, Ca is a key regulator of cellular functions such as enzyme activity and signal transduction, emphasizing its importance in overall metabolic processes [24].

Normal Ca balance is essential for neuronal functions, cardiac activities, and blood coagulation [25,26]. Ca deficiency is common and can cause various symptoms, including numbness and tingling in the fingers and toes, muscle cramps, arrhythmia, bone deformities (osteoporosis), and fractures. C deficiency can also lead to serious health issues, such as cardiac arrhythmias, which may require immediate medical intervention [27]. In pregnant women, there is an increased risk of preeclampsia and related complications associated with insufficient Ca (and vitamin D) intake [28].

2.1. The Impact of Global Calcium Deficiency

An estimated 3.5 billion people worldwide are at risk of Ca deficiency due to inadequate dietary intake. Populations in low- and middle-income countries, particularly in parts of Asia, Africa, and South America, are at risk of low Ca intake. Conversely, people in Western countries have an excessive Ca intake. Consequently, they tend to have higher Ca:Mg increase risks for hypercalcemia, potentially leading to kidney stones, vascular calcification, and impaired renal function [29]. Balancing Ca levels through diet and supplementation, as needed, is essential for supporting long-term health and preventing complications. Foods such as dairy products, leafy greens, and fortified alternatives remain primary sources of dietary Ca, underscoring their accessibility and importance in public health initiatives. The primary physiological regulators of Ca metabolism and related mechanisms are illustrated in Figure 2.

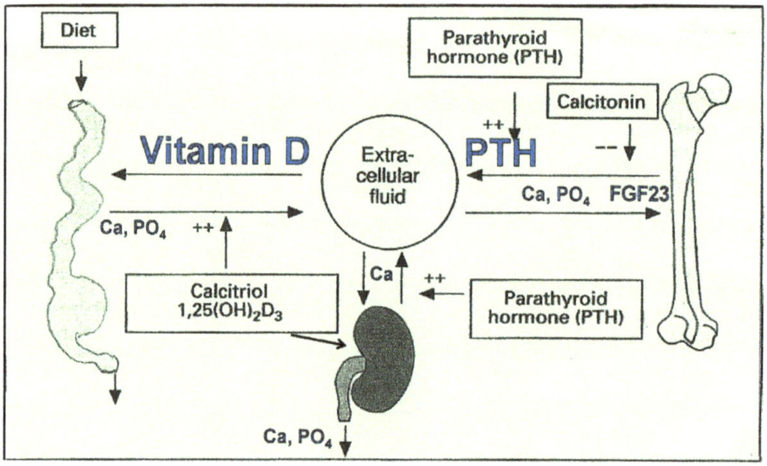

Figure 2. Major physiological and biological regulators and controls are illustrated [Fibroblast growth factor 23 (FGF23)].

2.2. Causes of Hypocalcemia

Causes of hypocalcemia are multi-factorial and include dietary insufficiencies, vitamin D deficiency, hypoparathyroidism (due to any case), and renal dysfunction. Acute symptomatic hypocalcemia presents with tetany, irritability, mental issues, or cardiac arrhythmias; these may require immediate attention with intravenous administration of Ca gluconate or Ca chloride to restore serum Ca levels [27]. Monitoring serum Ca, Mg, and phosphate levels ensures effective correction and prevents recurrence.

2.3. Treatment of Hypocalcemia

Hypocalcemia is managed by addressing the underlying causes, such as vitamin D deficiency, hypoparathyroidism, or renal dysfunction. Acute symptomatic hypocalcemia, which presents with tetany or cardiac arrhythmias, requires immediate intravenous administration of Ca gluconate or Ca chloride to restore serum Ca levels [27]. To increase intestinal Ca absorption, chronic hypocalcemia is typically treated with oral Ca supplements and active vitamin D, calcitriol or its synthetic analogs [27]. In addition, it can be treated with newer agents, like Yorvipath (palopegteriparatide) [30], to increase intestinal Ca absorption [27]. Monitoring serum Ca, Mg, and phosphate levels ensures effective correction and prevents recurrence [31].

2.4. Causes of Hypercalcemia

Hypercalcemia, characterized by elevated serum Ca levels, is most commonly caused by primary hyperparathyroidism and malignancies [32,33] that lead to increased bone resorption or ectopic production of parathyroid hormone (PTH)-related proteins [29,31]. Other causes include excessive Ca intake and/or vitamin D supplements (milk-alkali syndrome) and granulomatous diseases such as sarcoidosis, where activated macrophages produce excessive calcitriol [31,34]. Medications such as thiazide diuretics and lithium can also contribute to hypercalcemia by reducing Ca excretion or increasing parathyroid hormone (PTH) secretion.

2.5. Treatment of Hypercalcemia

Elevated serum Ca levels define hypercalcemia. Its treatment is based on the underlying etiology, development (acute vs. chronic), and severity, such as malignancy, hyperparathyroidism, or excessive Ca intake. Mild cases often require hydration and dietary modification to reduce Ca levels. More severe cases necessitate aggressive rehydration with intravenous saline to promote renal Ca excretion, followed by loop diuretics such as furosemide if needed. Bisphosphonates, such as zoledronic acid, effectively reduce bone resorption in hypercalcemia of malignancy [32], while calcitonin provides a rapid, albeit temporary, reduction in serum Ca [35]. In refractory or life-threatening cases, dialysis may be required to remove excess Ca, particularly in patients with renal impairment [32].

3. Magnesium—From Physiology to Pathology

Magnesium (Mg) is essential for many physiological functions, including enzymatic activity, hormone synthesis and release, neuromuscular function, cellular energy balance, and receptor activation [36,37]. Consequently, hypomagnesemia can have adverse effects on human health. Individuals may experience muscle weakness, cramps, tremors, and fatigue when Mg levels fall below the normal range [38]. Hypomagnesemia can also lead to cardiac arrhythmias, as Mg is crucial for maintaining proper heart rhythm and regulating potassium and Ca ions in cardiac tissues [39].

Chronic hypomagnesemia has been linked to a range of long-term health problems, including an increased risk of falls, osteoporosis, hypertension, and insulin resistance. Circulatory Mg concentrations do not accurately reflect tissue Mg levels [13]. Lower tissue Mg can lead to cardiac arrhythmia, nerve conduction defects, and muscle weakness [37]. It may also impair parathyroid gland function and exacerbate conditions such as diabetes and metabolic syndrome [39]. In severe cases, hypomagnesemia can lead to life-threatening complications, such as respiratory distress and seizures [40]. Early diagnosis and management, through dietary adjustments or supplementation, are crucial to prevent adverse outcomes and restore Mg balance in the body [31,34].

Hypomagnesemia also disrupts hormone release and vitamin D activity [41]. It is associated with decreased synthesis and activation of vitamin D and its receptor (CTR)interactions, increased oxidative stress, and exacerbated cytotoxic activity in T lymphocytes, promoting the cytokine storm [42,43]. In addition, hypomagnesemia causes abnormal platelet aggregation and coagulation abnormalities [44,45], damage to the myocardium [46], and endothelial dysfunction [47,48]. These compounds are present in severely ill COVID-19 patients and are associated with adverse effects related to COVID-19 vaccines [49,50]. In addition to vitamin D and Mg, there are other essential micro-nutrients. These include iodine, vitamins C and A, B_2, B_6, and vitamin B_{12}, omega-3 fatty acids, zinc, selenium, copper, and folate. Most of the world's population is deficient in several of these nutrients, thus increasing their vulnerability to diseases.

3.1. *Effects of Hypermagnesemia*

Hypermagnesemia can cause neuromuscular dysfunction, manifesting as lethargy, muscle weakness, diminished deep tendon reflexes, hypotension, nausea, facial flushing, ileus, and flaccid muscle paralysis, which can progress to paralysis in severe cases [38,51]. It can also lead to cardiovascular complications, including hypotension, bradycardia, and, in extreme cases, cardiac arrest due to the inhibitory effect of Mg on Ca -mediated cardiac conduction pathways [52].

Additionally, elevated Mg levels can depress central nervous system activity, leading to respiratory depression or coma [53]. Severe hypermagnesemia (>6.0 mg/dL) is a medical emergency that often requires interventions such as intravenous Ca gluconate and renal replacement therapy to reduce Mg levels rapidly [51]. Early recognition and prompt management are critical to prevent life-threatening outcomes [38].

3.2. Treatment of Hypo- and Hypermagnesemia

In clinical practice, the management of hypomagnesemia involves identifying and addressing the underlying causes, such as poor dietary intake, gastrointestinal losses, or renal waste [38]. Mild cases can typically be treated with oral Mg supplements, such as Mg oxide or citrate [41,48]. For more severe cases, particularly when symptoms like arrhythmias or seizures are present, intravenous Mg sulfate is administered to restore levels rapidly [38]. Monitoring serum Mg levels during treatment is essential to prevent overcorrection and avoid complications such as hypermagnesemia [37,48].

4. Effects of the Ca–Mg Balance in Physiology

In addition to bones and teeth, Ca is found mainly outside the cells, while Mg is primarily intracellular. Ca is involved in several pathophysiological aspects of the interaction between SARS-CoV-2 and human host cells [54,55]. Similar interactions between hypocalcemia and the severity and mortality of SARS-CoV-2 infection have also been reported [56,57]. In addition, low serum Ca (and low Mg) levels are considered prognostic factors for determining the severity of the disease [58] and eventual clinical outcomes [56,59].

4.1. Healthy Balance of Ca:Mg Ratio

The balance between Ca and Mg is crucial for maintaining proper physiological functions, as both minerals play key roles in cellular signaling, muscle contraction, nerve transmission, and cardiovascular health [48]. However, this is neglected in clinical practice. Most guidelines and recommendations do not even mention such. Ca and Mg interact to ensure proper neuromuscular function and cellular stability and play a role in aging [60]. Mg acts as a natural Ca blocker, helping to prevent excessive Ca influx into cells. Disruption of this balance can lead to a variety of health issues. For example, low Mg levels can increase Ca entry into cells, leading to muscle spasms, arrhythmias, and hypertension.

Conversely, elevated circulatory Ca levels can inhibit Mg actions, exacerbating symptoms of Mg deficiency [61]. The calculated Ca:Mg ratios from the diet and dietary supplements (total intake) would benefit from using Mg supplements to establish a favorable ratio to counter adverse effects. The data confirmed that a proper Ca:Mg ratio (another biomarker of health) could mitigate certain chronic diseases in the long term. The optimal intake of Ca:Mg and circulatory ratios is approximately 2 [62]. Either a high dietary Ca:Mg (>2.60) ratio, or a low intake ratio (<1.70) are unphysiological and increase health risks and disorders [63]. Therefore, maintaining the correct Ca to Mg ratio around 2.0 is necessary for optimal health [63].

Cellular influx/effluxes of Mg involve the same transporters as those for Ca; Mg acts physiologically to counter Ca [58,64–67]. Consequently, one needs to focus on the relationships of the serum Mg-to-Ca ratio with disease conditions and severity, including mortality from severe COVID-19 [58]. Mg is critical for inhibiting ROMK potassium channels in the principal cells of collecting tubules and ducts. When Mg levels decrease, ROMK channels become hyperactive. Consequently, hypomagnesemia can lead to hypokalemia [64,68].

4.2. Ca-to-Mg Ratio for Physiological Functions

Mg is essential for many enzymatic and B-vitamin functions, acting as a critical cofactor in synthesizing all CYP450 enzymes and hormones, vitamin D, and melatonin, as well as in the methylation of proteins and DNA [69,70]. Additionally, Ca and Mg are ligands compete for the same Ca-sensing receptor (CaSR). The Western diet typically has a high Ca-to-Mg ratio. Supplemental Ca was popular, especially among women, but this changed significantly after the Women's Health Initiative study [71] and similar studies published in subsequent years. As a result, the intake ratio increased from less than 2.5 to over 3.0, which is considered unphysiological [72].

A higher Ca-to-Mg ratio impairs vitamin D synthesis and displaces binding to the VDR, reducing its potential to interact with other essential elements [73]. The Ca:Mg ratio should be considered a "messenger" from the circulation to their target cells—especially those with CaSR—such as parathyroid cells, renal tubular cells, and the brain [74]. CaSR is a G-protein coupled receptor that detects extracellular Ca levels to maintain Ca homeostasis. The activation of CaSR in parathyroid cells reduces the secretion of PTH, while activating the renal CaSR promotes the urinary excretion of Ca [75].

The Ca-to-Mg (Ca:Mg) ratio is equally vital in maintaining systemic equilibrium [76]. A high Ca intake without adequate Mg can suppress PTH levels, negatively impacting bone remodeling and mineralization. An imbalanced Ca:Mg ratio can also exacerbate chronic conditions, including cardiovascular diseases, due to improper Ca deposition in arterial walls [16]. Ensuring a balanced ratio, ideally around 2:1 (Ca:Mg), is critical for optimizing the synergistic effects of these minerals on vitamin D metabolism, bone health, and overall physiological well-being [77]. Ca/Mg ratio (or higher blood Ca or lower Mg levels) worsens metabolic disorders. For example, persons with poorly controlled type 2 diabetes have a higher Ca:Mg ratio than those with better control [78].

4.3. Ca-to-Mg Ratio also Affects Non-Mineral Functions

The data suggest that Ca:Mg ratio (and Mg) are important variables that indicate glycemic control and complications. Other studies [75] have reported a high Ca:Mg ratio is associated with higher mortality in those with severe SARS-CoV-2 infections [79]. Others have reported that a high Ca:Mg ratio can be used as a biomarker of clinical outcomes for chronic disease, and its correction is beneficial [62]. High and low Ca:Mg ratios increase cardiovascular and all-cause mortality [80]. Ca/Mg ratios above 3.5 and below 1.70 are independently associated with an increased risk of chronic conditions, like cardiovascular disease, cancer, metabolic syndrome, type 2 diabetes, as well as all-cause mortality [81].

Evaluations and conclusions cannot be made on the efficacy of D_3 supplementation without concomitant knowledge and attention to the Ca:Mg ratio in the circulation. In addition, patients consume D_3 and Mg for many illnesses, which significantly impacts Ca:Mg ratio and associated therapies. Although it is not routine to measure Ca^{2+} and Mg^{2+}, these can be relevant in certain circumstances and disorders. Furthermore, low adequate D_3 is linked to gut dysbiosis, compromising the absorption of all these elements.

Mg is crucial for many biological activities, including hormone synthesis and release, as well as facilitating calcitriol-VDR receptor interactions [16,82]. Adequate Mg is essential for CTR interactions and can help reduce complications and mortality from post-COVID syndrome [83]. However, increasing Mg intake without addressing Ca overload may decrease PTH levels and impair Mg absorption. While lower circulating Mg levels typically stimulate PTH synthesis, Mg is required for both PTH synthesis and its release, so this may paradoxically reduce PTH levels [16]. The goal is to maintain the Ca:Mg ratio closer to 2.0, significantly enhancing enzymatic functions and improving vitamin D efficacy [16,77,84].

4.4. Importance of the Ca-to-Mg Ratio for Vitamin D and CTR Functions

Vitamin D is crucial for Ca and phosphorus metabolism and maintaining bone health and various physiological processes [85]. However, its effectiveness is significantly influenced by the balance between Ca and Mg (their ratio) (Ca:Mg) in the body [72,73]. Mg is a cofactor for enzymes that activate vitamin D, converting it into calcitriol [16,77]. Vitamin D and its receptor VDR function optimally at a Ca-to-Mg ratio of around 2.0 [84,86], supporting Ca absorption in the intestines and preventing calcification in soft tissues.

As with vitamin D_3 and PTH [87,88], responses have many non-linear U- or J-shaped activities/curves, including Mg [61]. In addition, there are localized overlapping or built redundant feedback systems, such as 24-hydroxylase enzymes in the skin, to prevent excess vitamin D3 from entering the circulation [89]. This is not represented in circulatory or whole-body concentrations. Consequently, cause-and-effect relationships are not necessarily apparent: they are subtle but interconnected and complex, making and distinguishing them challenges. These interactions behave differently when the ratios, like Ca:Mg or high T4/T3 (thyroiditis) to low T4/T3 (Graves), make interactions even more complicated.

A sufficient amount of vitamin K_2 in the circulation can mitigate this imbalance [90,91]. The disruption of this balance, particularly Mg deficiency, can impair vitamin D activation [73,77], leading to suboptimal Ca regulation and increased risks of bone disorders like osteoporosis [16]. Deficiencies in either mineral may lead to pathological conditions such as skeletal deformities, metabolic syndrome/obesity, cardiovascular diseases, and other inflammation-related disorders [16,72,91]. Individuals with a higher Ca:Mg ratio may also require circulating 25(OH)D concentrations above the recommended standard levels to obtain its biological activities, which may mimic pseudo vitamin D resistance. Therefore, it is advisable to supplement with Mg after assessing the Ca:Mg ratio before providing higher doses of vitamin D from Mg.

Increasing Ca intake in individuals with lower Ca:Mg ratios does not necessarily resolve accompanying vitamin D insufficiency. Addressing this requires D_3 supplementation or exposure to UVB rays to stimulate endogenous production. As highlighted, tissue sensitivity can vary based on the Ca:Mg ratio (e.g., high vs. low Ca:Mg). However, there is no evidence that the physiological cut-off concentrations of circulating 25(OH)D differ by ethnicity, age, sex, or sun exposure [87,92,93]. Emerging evidence, nevertheless, suggests that optimal 25(OH)D levels vary, indicating that different tissue sensitivities and disorders require distinct (higher) serum 25(OH)D concentrations, as depicted in Figure 3.

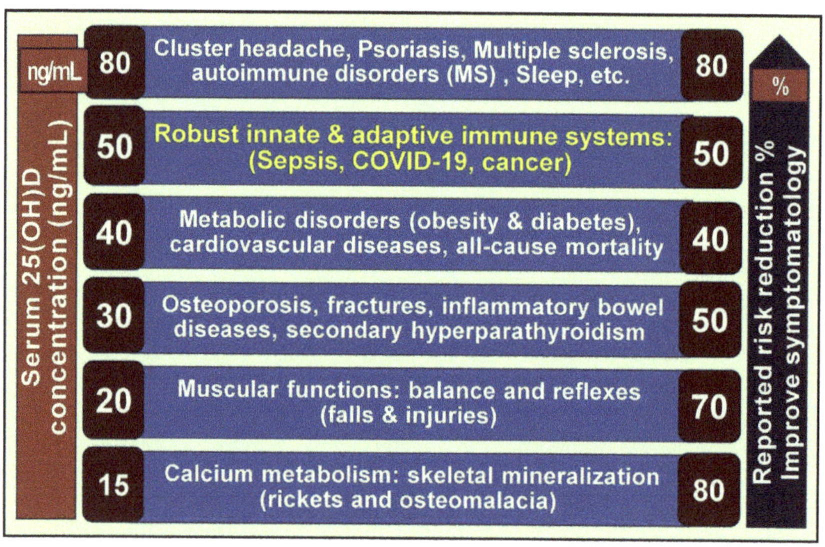

Figure 3. Different body systems, tissues, and diseases require varying steady-state serum 25(OH)D concentrations to prevent initiation and progression. The left side of the figure presents the minimum (average) serum 25(OH)D concentrations needed to obtain optimum clinical outcomes. In contrast, the right side depicts the percentage risk reduction for each disease entity [25(OH)D;25-hydroxyvitamin D] (modified from Wimalawansa) [94].

The dose–response curves following vitamin D supplementation are curvilinear [95–97]. This response can become exaggerated in individuals with imbalanced Ca:Mg ratios—at high or low extremes [88]. As a result, the efficacy of D_3 supplementation—and consequently clinical outcomes—can vary significantly unless the unbalanced Ca:Mg ratio is addressed beforehand.

4.5. Regulation of Ca and Phosphate Through Parathyroid Hormone

Parathyroid hormone (PTH) regulates Ca and Mg; however, Ca is the primary determinant of PTH levels, with PTH profoundly affecting ionized Ca [98]. Ca and phosphate physiology and metabolism are regulated by the hormones PTH, vitamin D, fibroblast growth factor 23 (FGF23), and calcitonin [99]. Mg is also a Ca antagonist in various ways, including functioning as a Ca channel blocker [73,100]. Understanding these interactions is crucial for comprehending the physiology of Ca regulation.

Low endogenous vitamin D_3 levels could arise from low Mg and excess Ca, which could overwhelm Mg, leading to the downregulation of CaSRs [100]. Mg also mitigates vascular calcification and osteogenic differentiation in the presence of high Ca and phosphate in the circulation [85,92]. Commonly, low D_3 levels result from reduced D_3 synthesis in the skin and/or inadequate dietary intake of D_3. In states of a high Ca:Mg ratio, excess Ca (due to high intake or increased absorption, consequently, elevated Ca in the circulation) suppresses PTH and calcitriol synthesis [73]. This suppression leads to decreased Ca and Mg absorption and reduced renal tubular resorption. Additionally, insufficient Mg intake exacerbates the synthesis of calcifediol and calcitriol.

Conversely, insufficient Ca stimulates PTH secretion at a low Ca:Mg ratio. Increasing Ca intake under such conditions further suppresses PTH and calcitriol synthesis [98]. It also inhibits Mg absorption and resorption, particularly in individuals already deficient in D_3 due to inadequate sun exposure and/or low dietary or supplemental D_3 intake [100]. Furthermore, elevated 25(OH)D levels (e.g., >40 ng/mL) may still be associated with a higher Ca:Mg ratio. Increasing Mg intake could lower 25(OH)D concentrations while improving the Ca:Mg balance [73]. Therefore, clinicians and researchers must remain aware of these delicate balances and interactions.

Most human hormone synthesis and release processes require physiological Mg concentrations in tissue [99]. While tissue Mg concentrations are important, no routine laboratory tests measure such levels. However, these processes can be impaired when the Ca:Mg ratio is unphysiological [93]. Therefore, the influence of Ca:Mg should be considered in specific vitamin D-resistant syndromes before initiating pharmacological vitamin D therapies [92,101]. Additionally, hypovitaminosis D disrupts the intestinal microbiome, negatively affecting mineral absorption and maintaining mineral balance. Ca and phosphate are essential for human physiology, including skeletal mineralization and neuromuscular function [98]. Some of these are discussed in the next section.

4.6. Importance of the Ca-to-Mg Ratio in Non-Mineral Disorders

At lower serum 25(OH)D concentrations, particularly below 20 ng/mL, circulatory PTH concentrations exhibit an exponential inverse relationship with vitamin D status [102]. A similar relationship exists between the Ca:Mg ratio and serum 25(OH)D levels, indirectly reflecting bioavailable PTH in circulation [103]. For instance, low blood Ca levels (hypocalcemia with low Ca:Mg ratio) contribute to secondary hyperparathyroidism, whereas Mg deficiency (hypomagnesemia with higher Ca:Mg ratio) can manifest as idiopathic hypoparathyroidism [81,104]. A thorough understanding of these relationships enables clinicians to manage Ca metabolism disorders and parathyroid dysfunctions better, potentially preventing unnecessary parathyroidectomy surgeries.

Ca and Mg generally compete and have opposed actions [104]. The Western diets are rich in Ca and low in Mg. Consequently, many have higher ratios than the optimal [81]. Asian and South American diets are generally lower in Ca than Western diets [103], thus generally having a lower Ca:Mg ratio, except for those who regularly consume hard water (primarily groundwater) containing high Ca salts [105,106]. In the latter group, due to very high Ca: Mg ratios (I.e., above 4.0), those who have sustained dehydration and also indulged in daily alcohol for several years develop Ca carbonate crystallization in renal tubules and tissues [107,108], causing unusual chronic renal failure affecting in tropical regions. In the presence of fluoride, these nanocrystals become stable and gradually grow [108,109]. This used to be called CKD of unknown etiology but is now remaned as CKD of crystalloid-tubular nephropathy (CKD-CTN) [107,108].

Before stage IIIB, the disease is reversible by drinking plenty of portable water. REF This is another illustration where investigating Ca: Mg ratios can help diagnose and manage such patients without drastic and expensive therapies [110]. In those with low Mg-derived high Ca:Mg ratios, Mg supplementation is the treatment of choice [110]. Most affected are farmers; thus, they are exposed to plenty of sunlight in the tropics. As a result, there is no reason to supplement them with vitamin D. In some, the 25(OH)D defects can be corrected with Mg supplements. In these situations, a more accurate picture reflecting actual ratios is obtained by expressing Ca and Mg in mmol/L (mM concentrations) than mg/L.

5. Phosphate and Human Health

Phosphate is another essential mineral found in both natural and processed foods. It is absorbed in the intestines, with any excess filtered by the kidneys and excreted in the urine [111]. Phosphate is crucial for various cellular processes, including ATP production, intracellular signal transduction, bone mineralization, cell membrane formation, and DNA and RNA synthesis [112,113]. Phosphate homeostasis is regulated by a complex interplay of several organs and hormones, including PTH, FGF23, klotho, and vitamin D [111]. High dietary phosphate intake can precipitate diseases such as type 2 diabetes [114].

Ca and phosphorus are involved in multiple physiological activities that affect most body systems [85]. Consequently, abnormalities in Ca and phosphorus metabolism lead to several pathological conditions, including skeletal- and cardiovascular-related disorders and premature mortality, such as renal failure [92,98]. The Recommended Dietary Allowance (RDA) for phosphorus is approximately 700 mg/day [115]. The European Food Safety Authority (EFSA) set an Adequate Intake of 550 mg/day for healthy adults, including those during pregnancy and lactation [115].

The Nordic Nutrition Recommendations (NNRs) from 2012 endorsed a slightly higher recommended intake (RI) of 600 mg/day for adults [116]. Most healthy individuals in North America consume nearly twice as much phosphate as the RDA [113]. Dietary phosphate overload is a health risk, especially for low-income populations, who often consume more inexpensive processed foods and phosphate-containing fizzy drinks containing inorganic phosphate additives [113,117]. These inorganic additives are absorbed more readily than organic phosphates in natural foods [115], leading to disorders [114].

5.1. *Phosphate Interactions with Other Minerals*

Pre-clinical and clinical research has shown that chronic dietary phosphorus excess has toxic effects, leading to adverse health outcomes [85,118]. Improved food labeling regulations and nutritional education are necessary to help individuals make informed dietary choices and reduce the risks associated with high phosphate intake [112,113]. Elevated phosphate levels have been linked to an increased risk of metabolic syndrome [98,118]. Mg is known to counteract the harmful effects of excess phosphate [98,119]. Sustained high phosphate levels trigger inflammatory responses that contribute to vascular stiffness (as in hypertension) [112], impair insulin sensitivity, and disrupt lipid metabolism (e.g., raising low-density lipoprotein levels) [117,118].

High phosphate levels decrease pancreatic insulin production, increasing the risk of type 2 diabetes and cardiometabolic diseases [112,117]. These findings emphasize the importance of proper phosphorus management in maintaining overall metabolic health [118]. In the absence of interventions, those with high phosphate diets and/or renal impairment are likely to develop phosphate toxicity [98,113], which can affect metabolic, cardiovascular, and renal health, leading to premature death and increasing the risk of tumor formation [117,119].

5.2. Causes of Hyperphosphatemia

Hyperphosphatemia is characterized by elevated levels of phosphate in the blood, typically exceeding 4.5 mg/dL in adults [120]. It is often associated with impaired renal dysfunction—impaired renal filtration leads to phosphate accumulation [113]. Other causes include excessive dietary phosphate intake, hypoparathyroidism, and using phosphate-containing medications [113,117]. Elevated phosphate levels disrupt the delicate balance of mineral metabolism, leading to complications such as vascular calcification, secondary hyperparathyroidism, and bone disorders [121]. Chronic hyperphosphatemia has been implicated in increased cardiovascular morbidity and mortality, particularly in individuals with chronic kidney disease (CKD), and increased mortality [120].

5.3. Treatment of Hyperphosphatemia

Treating hyperphosphatemia involves maintaining good hydration, combining dietary modifications and pharmacological interventions, and addressing underlying conditions [122]. Dietary phosphate restriction focuses on reducing the intake of phosphate-rich foods, mainly processed foods with inorganic phosphate additives. Phosphate binders, such as Ca acetate, sevelamer, or lanthanum carbonate, are commonly prescribed to bind dietary phosphate in the gastrointestinal tract, preventing absorption. In severe cases, particularly in individuals with advanced CKD [107,120], dialysis may be required to remove excess phosphate from the bloodstream [113,123]. Monitoring serum phosphate levels and Ca and parathyroid hormone levels is essential to guide treatment and prevent complications [121].

5.4. Causes of Hypophosphatemia

Hypophosphatemia is characterized by abnormally low levels of phosphate in the blood, typically below 2.5 mg/dL in adults [124]. Insufficient phosphate can disrupt physiological processes, including energy production, cell signaling, and bone mineralization [122]. The causes of hypophosphatemia include malnutrition, chronic alcohol use, vitamin D deficiency, prolonged use of antacids, refeeding syndrome, and certain medical conditions like hyperparathyroidism or Fanconi syndrome [113,121]. Symptoms can range from mild fatigue and muscle weakness to severe complications, such as respiratory failure, hemolysis, rhabdomyolysis, and impaired neurological function, depending on the severity and duration of the phosphate deficiency [124].

5.5. Treatment of Hypophosphatemia

The treatment of hypophosphatemia focuses on addressing the underlying cause and restoring normal phosphate levels [124]. Mild cases can often be managed with dietary modifications, such as increasing the intake of phosphate-rich foods like dairy products, nuts, and meats. Oral phosphate supplements, such as potassium phosphate or sodium phosphate, are commonly prescribed for moderate to severe cases or when dietary intake is insufficient [121]. In cases of severe hypophosphatemia with symptomatic manifestations, intravenous phosphate administration may be necessary [113]. Monitoring serum phosphate, Ca, and renal function during treatment is crucial to prevent tissue complications such as hyperphosphatemia or Ca–phosphate precipitation [121,124].

6. Additional Functions of Common Minerals

Minerals are crucial for maintaining various physiological functions in the human body. Essential minerals such as Ca, Mg, and phosphorus contribute to bone health, muscle function, and nerve transmission [85]. Ca, for example, is vital for bone strength and proper muscle contraction, whereas Mg supports enzymatic reactions and energy production [125]. Phosphorus is a key component of DNA, RNA, and ATP and plays a fundamental role in cellular energy metabolism.

Trace minerals, including iron, zinc, and selenium, are equally important for overall health. Iron is essential for oxygen transport in the blood through hemoglobin [126,127], while zinc supports immune function, wound healing [128], and DNA synthesis [85]. Selenium is a powerful antioxidant that protects cells from oxidative stress and supports thyroid function. A balanced intake of these minerals is critical to prevent deficiencies and ensure optimal health and well-being [125].

6.1. Effects of Untreated Mineral Deficiencies

Untreated mineral deficiencies can have serious long-term effects on human health. Mineral deficiencies during childhood and adolescence can significantly impact growth and development [129]. Clinically, Ca and vitamin D deficiencies can lead to poor bone mineralization, resulting in rickets in children and osteoporosis later in life. Similarly, iron deficiency can cause cognitive impairments and developmental delays in children [130,131].

Iodine deficiency during pregnancy and early childhood can result in irreversible brain damage and intellectual disabilities [129]. Numerous meta-analyses have revealed a strong relationship between iodine status during pregnancy and neonatal and maternal outcomes [132]. Similarly, a meta-analysis of 2190 pregnant women revealed a pooled prevalence of iodine deficiency during pregnancy in Ethiopia [133]. Zinc deficiency has been associated with depression and cognitive impairment in children and adults [134]. Zinc importers and exporters control the cellular zinc balance. The zinc importer family (ZIP) allows zinc to accumulate in the cytosol, while the zinc exporter family (ZnT) transports zinc out of the cytosol [135]. When required, excess zinc is eliminated through the kidneys, skin, and intestines [128].

6.2. Effects of Other Trace Minerals on Human Health

Several other minerals play crucial roles in maintaining a functional immune system: zinc deficiency can impair immune cell development and function, increasing susceptibility to infections. In contrast, selenium deficiency may reduce the body's ability to fight viral infections and increase the risk of certain tumors. A prospective, randomized, double-blind, placebo-controlled multicenter trial revealed that, compared with placebo, oral zinc supplementation decreased 30-day mortality and ICU admission rates in COVID-19 patients [136].

The antimicrobial and immune-boosting effects of zinc are thought to be partly responsible for reducing infection and inflammation-related burdens [137,138]. Iron deficiency can compromise immune responses, increasing the susceptibility of individuals to infections [139]. Iron absorption in the intestine is a coordinated process in which ascorbic acid (vitamin C) creates an acidic environment, and enteric brush border Na+/H+ exchangers generate an H+ gradient for apical Fe^{2+} (ferrous iron) absorption. This process is facilitated by divalent metal transporter 1 (DMT1). Iron is then transported across the basolateral surface of enterocytes by the iron exporter protein, ferroportin, and eventually enters the bloodstream for distribution throughout the body [140].

Long-term mineral deficiencies are associated with an increased risk of various chronic diseases. For example, Mg deficiency has been linked to cardiovascular diseases, type 2 diabetes, and osteoporosis [13,141]. Ca deficiency can lead to osteoporosis and increased fracture risk, especially in older adults. Iron deficiency anemia can cause fatigue, weakness, and, in severe cases, heart problems. Minerals play essential roles in various metabolic processes; chromium deficiency can impair glucose tolerance and insulin resistance, potentially increasing the risk of type 2 diabetes [142–144].

Copper deficiency may cause anemia and neutropenia, affecting oxygen transport and immune function. Both zinc and Mg inadequacy are also linked to metabolic disorders [145]. Mineral deficiencies can also impact reproductive health and pregnancy outcomes; for example, iron deficiency during pregnancy increases the risk of preterm birth and low birth weight, and zinc deficiency may contribute to complications during pregnancy and childbirth [146].

Essential trace elements, including molybdenum, cobalt, and chromium, are also important for human health. Molybdenum acts as a cofactor for enzymes involved in the metabolism of sulfur-containing amino acids [147]. It helps in the production of uric acid and involves glucose metabolism. Although its deficiency is rare in humans, it may cause headaches, seizures, visual impairments, and neurological disorders. A low risk of toxicity in humans and high consumption (10–15 mg/day) could induce gout-like symptoms and joint pain [147].

Cobalt is an essential vitamin B_{12} (cobalamin) component (Figure 1), and its deficiency can lead to anemia [148]. Chromium, however, is crucial for protein, fat, and carbohydrate metabolism, and its deficiency can cause impaired glucose tolerance. While chromium is necessary for these metabolic processes, excessive intake can result in skin diseases and liver and kidney damage and may even promote various tumors. These elements are vital for human health, thus, maintaining a healthy physiological balance, and avoiding deficiency and toxicity is crucial but can be delicate, especially for micro minerals like chromium and molybdenum [149].

6.3. *Importance of Minerals During Pregnancy and Lactation*

Minerals play crucial roles in pregnancy and lactation, and their inadequacy could significantly impact the health of both mothers and fetuses. For example, zinc is important for fetal development and milk secretion [150]. Maintaining Ca is essential for fetal growth of the musculoskeletal system and teeth [150]. Adequate Ca balance is also needed to reduce the risk of maternal hypertension and preeclampsia. Similarly, iodine is necessary for healthy brain development [151,152]. Mineral balance is important during breastfeeding, and mineral requirements increase during lactation [153,154].

Higher Mg intake is recommended for fetal growth and development, especially during early lactation [37]. Adequate mineral intake during pregnancy is essential for maternal health, fetal growth and development, and pregnancy and lactation. Although increased mineral intake through diet and supplements can meet maternal nutritional needs, avoiding excessive intake is also important [152]. Personalized dietary counseling and supplemental recommendations can help pregnant women meet the minimum mineral requirements to ensure adequate and prevent excessive intake (Figure 4).

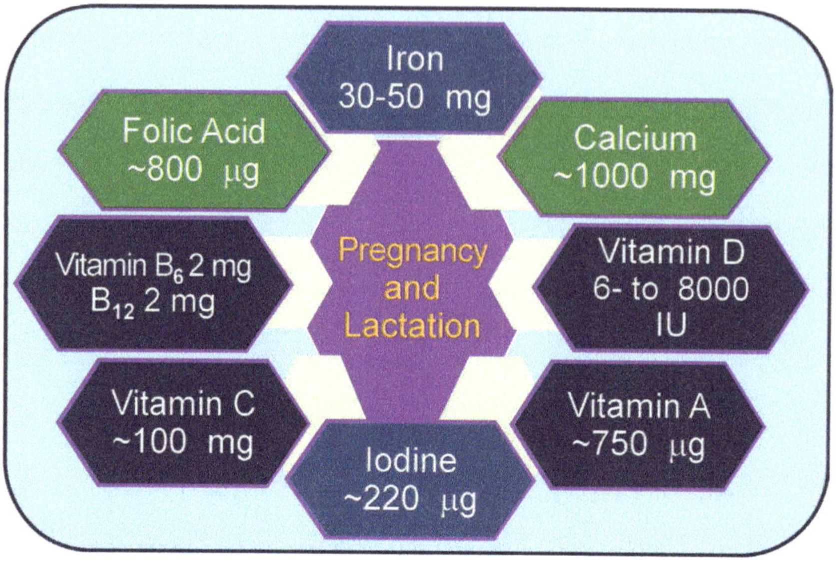

Figure 4. A few important nutrients with approximate total (diet plus supplements) daily requirements during pregnancy and lactation.

6.4. Abnormal Mineral Metabolism Secondary to Genetic Abnormalities

Genetic mutations can lead to mineral metabolism disorders in several ways. Many genetic mutations affect enzymes involved in mineral metabolism pathways. For example, mutations in the CYP27B1 gene, which encodes the 1α-hydroxylase enzyme, can cause vitamin D-dependent rickets type 1A by impairing the synthesis of calcitriol (the active component of vitamin D) [155]. Mutations in the CYP24A1 gene, which encodes a 24-hydroxylase gene, can lead to idiopathic infantile hypercalcemia by disrupting vitamin D degradation. Mutations can also affect receptors crucial for mineral homeostasis.

VDR gene mutations can cause vitamin D-dependent type 2 rickets, impairing the body's response to vitamin D. CaSR gene mutations can lead to familial hypocalciuric hypercalcemia or autosomal dominant hypocalcemia [156]. Similarly, mutations in genes encoding mineral transporters can also disrupt mineral absorption or excretion, as mutations affecting sodium-phosphate co-transporters can lead to renal phosphate waste and induce skeletal deformities [157]. Genetic defects can impact hormones regulating mineral metabolism, and mutations in the PTH gene can cause hypoparathyroidism. Similarly, mutations affecting FGF23 can lead to disorders of phosphate homeostasis [158,159].

Defects in genes encoding transcription factors that regulate mineral metabolism genes can have widespread effects. For example, GATA3 mutations can cause hypoparathyroidism, deafness, and renal dysplasia syndrome [160]. Some genetic disorders affect multiple systems, including mineral metabolism. Examples include multiple endocrine neoplasia type 1 (MEN1) and autoimmune poly-endocrinopathy-candidiasis-ectodermal dystrophy (APECED) [161]. These genetic defects can result in a wide range of mineral metabolism disorders, from relatively mild imbalances to severe, life-threatening conditions. The specific effects depend on the gene involved, the nature of the mutation, and environmental factors. Understanding these genetic mechanisms is crucial for diagnosing, treating, and potentially developing gene therapies for mineral metabolism disorders [162].

6.5. Strategies to Reduce the Burden of Mineral Deficiencies

The safest and most cost-effective way to reduce mineral deficiency is to increase the consumption of micronutrient-rich foods [163]. There should be a combined approach addressing supply and demand by increasing the production of micronutrient-rich foods and promoting increased consumption through education and behavior change efforts. Food-based fortification strategies should be implemented by promoting small-scale vegetable and fruit gardens (home gardens), encouraging the production of small animals, poultry, and fish as sources of bioavailable micronutrients [164], and supporting efficient large-scale commercial production of fruits and vegetables [163,165]. In addition, food fortification by adding synthetic nutrients to commonly consumed staple foods and biofortification by breeding crops to increase their nutritional value should be considered [166], including food literacy [167].

Moreover, education and increasing awareness of communication techniques to promote changes in eating practices would provide long-term benefits [167]. Furthermore, community involvement by ensuring community participation, especially of women, in program design and implementation with the participation of local people in assessment, analysis, and action to increase acceptance would be beneficial. The most effective approach is likely to be a combined strategy that addresses increased production and consumption while meeting the needs of particular groups, such as children and women of childbearing age [168].

7. Mineral Toxicity and Human Health

Mineral toxicity occurs when the concentration of certain minerals in the body becomes abnormally high, leading to adverse health consequences. Excess sodium in the bloodstream, known as hypernatremia, can be highly harmful; the normal sodium level in plasma is 136–145 mM, and >152 mM is considered toxic. Hypernatremia can lead to confusion, seizures, coma, paralysis of the lung muscles, and death. Hypernatremia causes brain cells to shrink, leading to severe neurological effects [169].

Hyperkalemia can also be life-threatening. The normal concentrations range from 3.5 to 5.0 mM. When the concentration is high (6.3–8.0 mM), severe toxicity, including cardiac arrhythmias, cardiac arrest, and death, occurs [170]. The body usually minimizes potassium toxicity by inducing a vomiting reflex and increasing kidney function [123]. Hyperkalemia often occurs in individuals with various kidney diseases (unable to remove excess potassium) or individuals taking certain medications (potassium-sparing diuretics, ACE inhibitors) [170,171].

Iron poisoning is particularly threatening for children. Mineral toxicity is most common in U.S. children under 6 years of age [172]. Approximately 20,000 cases of accidental iron ingestion in children occur annually, with symptoms of nausea, vomiting, abdominal pain, shock, liver failure, and death in severe cases [173,174]. Iron poisoning can be acute (from a single large dose) or chronic (from long-term accumulation). The severity of iron poisoning depends on the amount of elemental iron ingested, with children being particularly vulnerable due to their smaller size [175].

Although not conclusively proven, aluminum exposure may be a risk factor for Alzheimer's disease [176]. Other effects include nausea, skin ulcers and rashes, vomiting, diarrhea, arthritic pain, and nervous system issues (memory loss and balance problems). Cadmium exposure is often occupational but can occur through contaminated water or food [177]. Acute toxicity can induce abdominal pain, vomiting, and diarrhea, whereas chronic toxicity can induce kidney disease, lung damage, and bone weakness [178].

Several genetic disorders can increase the risk of mineral toxicity by affecting the body's ability to regulate certain minerals properly [179]. Hemochromatosis is one of the most common genetic disorders associated with mineral toxicity. It affects approximately 1 in 10 people in the United States who carry genetic mutations, causing excessive absorption and iron accumulation in the body, leading to iron overload and toxicity if left untreated [162]. Approximately 1 million people in the U.S. have hemochromatosis [180,181].

Wilson's disease is a rare inherited disorder of copper metabolism caused by a mutation in the ATP7B gene encoding a canalicular copper-transporting ATPase [182]. It affects approximately 1 in 30,000 people, causing copper to accumulate in the liver, brain, and other organs [183]. If not managed, it can lead to copper toxicity [182]. Copper is an essential mineral that provides antioxidant protection to the body, and its inadequacy reduces the body's ability to cope with oxidative stress [183,184].

Menke's disease is a rare X-linked recessive disorder affecting copper metabolism that primarily affects males. The incidence ranges from 1 in 50,000 to 1 in 250,000 individuals, causing copper deficiency in the brain and other tissues [185], but can paradoxically lead to copper accumulation in some organs. Importantly, having a genetic predisposition does not always ensure that mineral toxicity will occur. Environmental factors, diet, and overall health also play significant roles.

8. Discussion

Minerals are essential in food and vital for the development and physiological functions of the body. Minerals play important roles, including structural functions such as building and maintaining bones, teeth, and soft tissues (e.g., Ca, phosphorus, and Mg), as cofactors in enzymatic reactions, neural functions, and even receptor interactions (as with vitamin D) and membrane stability (Figure 1). Minerals regulate many biological processes, such as electrolytes and fluid balance (e.g., sodium, potassium, and chloride), maintaining heart rhythm, nerve responses, and signal conduction (e.g., Ca, potassium, and sodium), as well as blood clotting (e.g., Ca).

Iron is crucial for forming hemoglobin, which transports oxygen to the blood. Minerals also support immune function (e.g., zinc, copper, and selenium) and aid in metabolism and energy regulation, such as thyroid hormone production (e.g., iodine). Additionally, they are required to synthesize DNA and the function of antioxidants, such as glutathione (e.g., Mg and zinc), and breakdown proteins, carbohydrates, and cholesterol (e.g., manganese). Vital dietary sources of minerals include fruits, vegetables, whole grains, cereals, lentils, lean meats, and dairy products.

Untreated mineral deficiencies can significantly impact overall health, affecting physical growth, cognitive function, and disease susceptibility. Maintaining adequate mineral intake through a balanced diet or supervised supplementation is crucial for long-term well-being. Proper diagnosis and management of genetic conditions related to mineral metabolism can help prevent toxicity risks. Balancing mineral intake is essential for optimal health outcomes.

9. Conclusions

Mineral dysregulation contributes to a substantial economic burden on societies, particularly in developing countries with high prevalence rates. These inadequacies reduce work productivity and increase healthcare costs, leading to significant economic losses associated with chronic diseases. Much of this economic toll could be avoided through proper prevention and treatment of mineral deficiencies or toxicity. Excessive supplementation should be avoided, and occupational exposure should be safe to prevent mineral toxicity. While minerals are vital for health, maintaining proper balance is essential; consultation with a healthcare professional is recommended if mineral toxicity is suspected. In summary, minerals play essential roles in maintaining the functions of most body systems, and maintaining an adequate balance is vital for health and well-being.

Funding: This research received no funding or aid from funding agencies or commercial or non-profit sectors.

Data Availability Statement: Data included in the article are referenced in the article.

Acknowledgments: Some of the information has been collected from online sources, inlcuding dietary guilines.

Conflicts of Interest: The authors have no competing interests to declare.

References

1. Weyh, C.; Krüger, K.; Peeling, P.; Castell, L. The Role of Minerals in the Optimal Functioning of the Immune System. *Nutrients* **2022**, *14*, 644. [CrossRef] [PubMed]
2. Faba, L.; Gasa, J.; Tokach, M.D.; Varella, E.; Sola-Oriol, D. Effects of supplementing organic microminerals and methionine during the rearing phase of replacement gilts on lameness, growth, and body composition. *J. Anim. Sci.* **2018**, *96*, 3274–3287. [CrossRef] [PubMed]
3. Islam, M.R.; Islam, M.R.; Shalahuddin Qusar, M.M.A.; Islam, M.S.; Kabir, M.H.; Mustafizur Rahman, G.K.M.; Islam, M.S.; Hasnat, A. Alterations of serum macro-minerals and trace elements are associated with major depressive disorder: A case-control study. *BMC Psychiatry* **2018**, *18*, 94. [CrossRef] [PubMed]
4. Safiri, S.; Kolahi, A.A.; Noori, M.; Nejadghaderi, S.A.; Karamzad, N.; Bragazzi, N.L.; Sullman, M.J.M.; Abdollahi, M.; Collins, G.S.; Kaufman, J.S.; et al. Burden of anemia and its underlying causes in 204 countries and territories, 1990-2019: Results from the Global Burden of Disease Study 2019. *J. Hematol. Oncol.* **2021**, *14*, 185. [CrossRef]
5. Mogire, R.M.; Muriuki, J.M.; Morovat, A.; Mentzer, A.J.; Webb, E.L.; Kimita, W.; Ndungu, F.M.; Macharia, A.W.; Cutland, C.L.; Sirima, S.B.; et al. Vitamin D Deficiency and Its Association with Iron Deficiency in African Children. *Nutrients* **2022**, *14*, 1372. [CrossRef]
6. Uwitonze, A.M.; Ojeh, N.; Murererehe, J.; Atfi, A.; Razzaque, M.S. Zinc Adequacy Is Essential for the Maintenance of Optimal Oral Health. *Nutrients* **2020**, *12*, 949. [CrossRef]
7. Chao, H.C. Zinc Deficiency and Therapeutic Value of Zinc Supplementation in Pediatric Gastrointestinal Diseases. *Nutrients* **2023**, *15*, 4093. [CrossRef]
8. Zoidis, E.; Seremelis, I.; Kontopoulos, N.; Danezis, G.P. Selenium-Dependent Antioxidant Enzymes: Actions and Properties of Selenoproteins. *Antioxidants* **2018**, *7*, 66. [CrossRef]
9. Huwiler, V.V.; Maissen-Abgottspon, S.; Stanga, Z.; Mühlebach, S.; Trepp, R.; Bally, L.; Bano, A. Selenium Supplementation in Patients with Hashimoto Thyroiditis: A Systematic Review and Meta-Analysis of Randomized Clinical Trials. *Thyroid* **2024**, *34*, 295–313. [CrossRef]
10. Sadler, R.A.; Mallard, B.A.; Shandilya, U.K.; Hachemi, M.A.; Karrow, N.A. The Immunomodulatory Effects of Selenium: A Journey from the Environment to the Human Immune System. *Nutrients* **2024**, *16*, 3324. [CrossRef]
11. Gombart, A.F.; Pierre, A.; Maggini, S. A review of micronutrients and the immune system-working in harmony to reduce the risk of infection. *Nutrients* **2020**, *12*, 236. [CrossRef] [PubMed]
12. Erem, S.; Atfi, A.; Razzaque, M.S. Anabolic effects of vitamin D and magnesium in aging bone. *J. Steroid Biochem. Mol. Biol.* **2019**, *193*, 105400. [CrossRef]
13. Razzaque, M.S. Magnesium: Are We Consuming Enough? *Nutrients* **2018**, *10*, 1863. [CrossRef]
14. Schutten, J.C.; Joris, P.J.; Minović, I.; Post, A.; van Beek, A.P.; de Borst, M.H.; Mensink, R.P.; Bakker, S.J.L. Long-term magnesium supplementation improves glucocorticoid metabolism: A post-hoc analysis of an intervention trial. *Clin. Endocrinol.* **2021**, *94*, 150–157. [CrossRef]
15. Stocco, D.M.; Wang, X.; Jo, Y.; Manna, P.R. Multiple signaling pathways regulating steroidogenesis and steroidogenic acute regulatory protein expression: More complicated than we thought. *Mol. Endocrinol.* **2005**, *19*, 2647–2659. [CrossRef]
16. Uwitonze, A.M.; Razzaque, M.S. Role of magnesium in vitamin D activation and function. *J. Am. Osteopath. Assoc.* **2018**, *118*, 181–189. [CrossRef]
17. Amos, A.; Razzaque, M.S. Zinc and its role in vitamin D function. *Curr. Res. Physiol.* **2022**, *5*, 203–207. [CrossRef]
18. Michigami, T.; Yamazaki, M.; Razzaque, M.S. Extracellular Phosphate, Inflammation and Cytotoxicity. *Adv. Exp. Med. Biol.* **2022**, *1362*, 15–25. [CrossRef]
19. Razzaque, M.S.; Sitara, D.; Taguchi, T.; St-Arnaud, R.; Lanske, B. Premature aging-like phenotype in fibroblast growth factor 23 null mice is a vitamin D-mediated process. *FASEB J.* **2006**, *20*, 720–722. [CrossRef]

20. Razzaque, M.S.; Lanske, B. Hypervitaminosis D and premature aging: Lessons learned from Fgf23 and Klotho mutant mice *Trends Mol. Med.* **2006**, *12*, 298–305. [CrossRef]
21. Vervloet, M. Modifying Phosphate Toxicity in Chronic Kidney Disease. *Toxins* **2019**, *11*, 522. [CrossRef] [PubMed]
22. Lanske, B.; Razzaque, M.S. Mineral metabolism and aging: The fibroblast growth factor 23 enigma. *Curr. Opin. Nephrol. Hypertens.* **2007**, *16*, 311–318. [CrossRef]
23. Beto, J.A. The role of calcium in human aging. *Clin. Nutr. Res.* **2015**, *4*, 1–8. [CrossRef] [PubMed]
24. Berridge, M.J.; Bootman, M.D.; Roderick, H.L. Calcium signalling: Dynamics, homeostasis and remodelling. *Nat. Rev. Mol. Cell Biol.* **2003**, *4*, 517–529. [CrossRef]
25. Brini, M.; Calì, T.; Ottolini, D.; Carafoli, E. Neuronal calcium signaling: Function and dysfunction. *Cell Mol. Life Sci.* **2014**, *71*, 2787–2814. [CrossRef]
26. LeBrasseur, N. Calcium for strong clotting. *J. Cell Biol.* **2003**, *160*, 980. [CrossRef]
27. Pepe, J.; Colangelo, L.; Biamonte, F.; Sonato, C.; Danese, V.C.; Cecchetti, V.; Occhiuto, M.; Piazzolla, V.; De Martino, V.; Ferrone, F.; et al. Diagnosis and management of hypocalcemia. *Endocrine* **2020**, *69*, 485–495. [CrossRef]
28. Mahadevan, S.; Kumaravel, V.; Bharath, R. Calcium and bone disorders in pregnancy. *Indian J. Endocrinol. Metab.* **2012**, *16*, 358–363. [CrossRef]
29. Walker, M.D.; Shane, E. Hypercalcemia: A Review. *Jama* **2022**, *328*, 1624–1636. [CrossRef]
30. Rejnmark, L.; Gosmanova, E.O.; Khan, A.A.; Makita, N.; Imanishi, Y.; Takeuchi, Y.; Sprague, S.; Shoback, D.M.; Kohlmeier, L.; Rubin, M.R.; et al. Palopegteriparatide treatment Improves renal function in adults with chronic hypoparathyroidism: 1-year results from the phase 3 PaTHway trial. *Adv. Ther.* **2024**, *41*, 2500–2518. [CrossRef]
31. Wimalawansa, S.J. Significance of plasma PTH-rp in patients with hypercalcemia of malignancy treated with bisphosphonate. *Cancer* **1994**, *73*, 2223–2230. [CrossRef] [PubMed]
32. Rouf, R.; Bhuiyan, A.; Alam, A.; Chowdhury, M.K. Treatment utcomes of hypercalcemia of malignancy among advanced cancer patients attending palliative care Unit of a tertiary care hospital in Bangladesh: An observational study. *Health Sci. Rep.* **2024**, *7*, e70247. [CrossRef] [PubMed]
33. Sternlicht, H.; Glezerman, I.G. Hypercalcemia of malignancy and new treatment options. *Ther. Clin. Risk Manag.* **2015**, *11*, 1779–1788. [CrossRef] [PubMed]
34. Cooper, C.; Li, H.; Wimalawansa, S.J. Cancer-associated hypercalcemia and parathyroid hormone-related peptide: A new peptide with diverse roles. *Reg. Pept. Lett.* **1997**, *7*, 39–42.
35. El-Hajj Fuleihan, G.; Clines, G.A.; Hu, M.I.; Marcocci, C.; Murad, M.H.; Piggott, T.; Van Poznak, C.; Wu, J.Y.; Drake, M.T. Treatment of Hypercalcemia of Malignancy in Adults: An Endocrine Society Clinical Practice Guideline. *J. Clin. Endocrinol. Metab.* **2023**, *108*, 507–528. [CrossRef]
36. Fatima, G.; Dzupina, A.; Alhmadi, H.B.; Magomedova, A.; Siddiqui, Z.; Mehdi, A.; Hadi, N. Magnesium Matters: A Comprehensive Review of Its Vital Role in Health and Diseases. *Cureus* **2024**, *16*, e71392. [CrossRef]
37. Fanni, D.; Gerosa, C.; Nurchi, V.M.; Manchia, M.; Saba, L.; Coghe, F.; Crisponi, G.; Gibo, Y.; Van Eyken, P.; Fanos, V.; et al. The role of magnesium in pregnancy and in fetal programming of adult diseases. *Biol. Trace Elem. Res.* **2021**, *199*, 3647–3657. [CrossRef]
38. Ayuk, J.; Gittoes, N.J. Contemporary view of the clinical relevance of magnesium homeostasis. *Ann. Clin. Biochem.* **2014**, *51*, 179–188. [CrossRef]
39. Oost, L.J.; Tack, C.J.; de Baaij, J.H.F. Hypomagnesemia and Cardiovascular Risk in Type 2 Diabetes. *Endocr. Rev.* **2023**, *44*, 357–378. [CrossRef]
40. Mawri, S.; Gildeh, E.; Joseph, N.; Rabbani, B.; Zweig, B. Cardiac Dysrhythmias and Neurological Dysregulation: Manifestations of Profound Hypomagnesemia. *Case Rep. Cardiol.* **2017**, *2017*, 6250312. [CrossRef]
41. Hansen, B.A.; Bruserud, O. Hypomagnesemia in critically ill patients. *J. Intensive Care* **2018**, *6*, 21. [CrossRef] [PubMed]
42. Suhail, S.; Zajac, J.; Fossum, C.; Lowater, H.; McCracken, C.; Severson, N.; Laatsch, B.; Narkiewicz-Jodko, A.; Johnson, B.; Liebau, J.; et al. Role of oxidative stress on SARS-CoV (SARS) and SARS-CoV-2 (COVID-19) infection: A review. *Protein J.* **2020**, *39*, 644–656. [CrossRef] [PubMed]
43. Vabret, N.; Britton, G.J.; Gruber, C.; Hegde, S.; Kim, J.; Kuksin, M.; Levantovsky, R.; Malle, L.; Moreira, A.; Park, M.D.; et al. Immunology of COVID-19: Current state of the science. *Immunity* **2020**, *52*, 910–941. [CrossRef] [PubMed]
44. Gomez-Mesa, J.E.; Galindo-Coral, S.; Montes, M.C.; Munoz Martin, A.J. Thrombosis and Coagulopathy in COVID-19. *Curr. Probl. Cardiol.* **2021**, *46*, 100742. [CrossRef] [PubMed]
45. Ali, M.A.M.; Spinler, S.A. COVID-19 and thrombosis: From bench to bedside. *Trends Cardiovasc. Med.* **2021**, *31*, 143–160. [CrossRef]
46. Babapoor-Farrokhran, S.; Gill, D.; Walker, J.; Rasekhi, R.T.; Bozorgnia, B.; Amanullah, A. Myocardial injury and COVID-19: Possible mechanisms. *Life Sci.* **2020**, *253*, 117723. [CrossRef]
47. Nagele, M.P.; Haubner, B.; Tanner, F.C.; Ruschitzka, F.; Flammer, A.J. Endothelial dysfunction in COVID-19: Current findings and therapeutic implications. *Atherosclerosis* **2020**, *314*, 58–62. [CrossRef]

48. Di Mario, F.; Regolisti, G.; Greco, P.; Maccari, C.; Superchi, E.; Morabito, S.; Pistolesi, V.; Fiaccadori, E. Prevention of hypomagnesemia in critically ill patients with acute kidney injury on continuous kidney replacement therapy: The role of early supplementation and close monitoring. *J. Nephrol.* **2021**, *34*, 1271–1279. [CrossRef]
49. Dominguez, L.J.; Veronese, N.; Guerrero-Romero, F.; Barbagallo, M. Magnesium in infectious diseases in older people. *Nutrients* **2021**, *13*, 180. [CrossRef]
50. Gunay, S.; Caliskan, S.; Sigirli, D. Relationship of magnesemia with myocardial damage and mortality in patients with COVID-19. *Magnes. Res.* **2021**, *34*, 93–102. [CrossRef]
51. Aal-Hamad, A.H.; Al-Alawi, A.M.; Kashoub, M.S.; Falhammar, H. Hypermagnesemia in clinical practice. *Medicina* **2023**, *59*, 1190. [CrossRef] [PubMed]
52. Nishikawa, M.; Shimada, N.; Kanzaki, M.; Ikegami, T.; Fukuoka, T.; Fukushima, M.; Asano, K. The characteristics of patients with hypermagnesemia who underwent emergency hemodialysis. *Acute Med. Surg.* **2018**, *5*, 222–229. [CrossRef]
53. Tosto, F.; Magro, G.; Laterza, V.; Romozzi, M. Neurological manifestations of hypermagnesemia: A narrative review. *Acta Neurol. Belg.* **2024**. [CrossRef]
54. Wan, Y.; Shang, J.; Graham, R.; Baric, R.S.; Li, F. Receptor recognition by the novel coronavirus from Wuhan: An analysis based on decade-long structural studies of SARS coronavirus. *J. Virol.* **2020**, *94*, e00127-20. [CrossRef]
55. Dakal, T.C. SARS-CoV-2 attachment to host cells is possibly mediated via RGD-integrin interaction in a calcium-dependent manner and suggests pulmonary EDTA chelation therapy as a novel treatment for COVID 19. *Immunobiology* **2021**, *226*, 152021. [CrossRef]
56. Alemzadeh, E.; Alemzadeh, E.; Ziaee, M.; Abedi, A.; Salehiniya, H. The effect of low serum calcium level on the severity and mortality of COVID patients: A systematic review and meta-analysis. *Immun. Inflamm. Dis.* **2021**, *9*, 1219–1228. [CrossRef]
57. Raesi, A.; Saedi Dezaki, E.; Moosapour, H.; Saeidifard, F.; Habibi, Z.; Rahmani, F.; Kheiri, S.; Taheri, E. Hypocalcemia in COVID-19: A prognostic marker for severe disease. *Iran. J. Pathol.* **2021**, *16*, 144–153. [CrossRef]
58. Yogi, A.; Callera, G.E.; Antunes, T.T.; Tostes, R.C.; Touyz, R.M. Vascular biology of magnesium and its transporters in hypertension. *Magnes. Res.* **2010**, *23*, S207–S215. [CrossRef]
59. Sun, J.K.; Zhang, W.H.; Zou, L.; Liu, Y.; Li, J.J.; Kan, X.H.; Dai, L.; Shi, Q.K.; Yuan, S.T.; Yu, W.K.; et al. Serum calcium as a biomarker of clinical severity and prognosis in patients with coronavirus disease 2019. *Aging* **2020**, *12*, 11287–11295. [CrossRef]
60. Souza, A.C.R.; Vasconcelos, A.R.; Dias, D.D.; Komoni, G.; Name, J.J. The integral role of magnesium in muscle integrity and aging: A comprehensive review. *Nutrients* **2023**, *15*, 5127. [CrossRef]
61. Wu, Z.; Yang, X.; Ruan, Z.; Li, L.; Wu, J.; Wang, B. Nonlinear relationship between dietary calcium and magnesium intake and peripheral neuropathy in the general population of the United States. *Front. Nutr.* **2023**, *10*, 1217465. [CrossRef] [PubMed]
62. Costello, R.B.; Rosanoff, A.; Dai, Q.; Saldanha, L.G.; Potischman, N.A. Perspective: Characterization of dietary supplements containing calcium and magnesium and their respective ratio-Is a rising ratio a cause for concern? *Adv. Nutr.* **2021**, *12*, 291–297. [CrossRef] [PubMed]
63. Dai, Q.; Shu, X.O.; Deng, X.; Xiang, Y.B.; Li, H.; Yang, G.; Shrubsole, M.J.; Ji, B.; Cai, H.; Chow, W.H.; et al. Modifying effect of calcium/magnesium intake ratio and mortality: A population-based cohort study. *BMJ Open* **2013**, *3*, e002111. [CrossRef]
64. Yang, L.; Frindt, G.; Palmer, L.G. Magnesium modulates ROMK channel-mediated potassium secretion. *J. Am. Soc. Nephrol.* **2010**, *21*, 2109–2116. [CrossRef]
65. Sontia, B.; Touyz, R.M. Magnesium transport in hypertension. *Pathophysiology* **2007**, *14*, 205–211. [CrossRef]
66. Hoorn, E.J.; Zietse, R. Disorders of calcium and magnesium balance: A physiology-based approach. *Pediatr. Nephrol.* **2013**, *28*, 1195–1206. [CrossRef]
67. Suzuki, Y.; Bürzle, M.; Hediger, M.A. 369Physiology and pathology of calcium and magnesium transport. In *The Spectrum of Mineral and Bone Disorders in Chronic Kidney Disease, Olgaard, K., Salusky, I.B., Silver, J., Eds.*; Oxford University Press: Oxford, UK, 2010; pp. 369–377.
68. Adomako, E.A.; Yu, A.S.L. Magnesium Disorders: Core Curriculum 2024. *Am. J. Kidney Dis.* **2024**, *83*, 803–815. [CrossRef]
69. Hosseini, S.M.; Panahi-Azar, A.; Sheybani-Arani, M.; Morovatshoar, R.; Mirzadeh, M.; Salimi Asl, A.; Naghdipour Mirsadeghi, M.; Khajavi-Mayvan, F. Vitamins, minerals and their maternal levels' role in brain development: An updated literature-review. *Clin. Nutr. ESPEN* **2024**, *63*, 31–45. [CrossRef]
70. Fiorentini, D.; Cappadone, C.; Farruggia, G.; Prata, C. Magnesium: Biochemistry, nutrition, detection, and social impact of diseases linked to Its deficiency. *Nutrients* **2021**, *13*, 1136. [CrossRef]
71. Cauley, J.A.; Chlebowski, R.T.; Wactawski-Wende, J.; Robbins, J.A.; Rodabough, R.J.; Chen, Z.; Johnson, K.C.; O'Sullivan, M.J.; Jackson, R.D.; Manson, J.E. Calcium plus vitamin D supplementation and health outcomes five years after active intervention ended: The Women's Health Initiative. *J. Womens Health* **2013**, *22*, 915–929. [CrossRef]
72. Rosanoff, A.; Weaver, C.M.; Rude, R.K. Suboptimal magnesium status in the United States: Are the health consequences underestimated? *Nutr. Rev.* **2012**, *70*, 153–164. [CrossRef] [PubMed]

73. Dai, Q.; Zhu, X.; Manson, J.E.; Song, Y.; Li, X.; Franke, A.A.; Costello, R.B.; Rosanoff, A.; Nian, H.; Fan, L.; et al. Magnesium status and supplementation influence vitamin D status and metabolism: Results from a randomized trial. *Am. J. Clin. Nutr.* **2018**, *108*, 1249–1258. [CrossRef] [PubMed]
74. Dimke, H. New insights into renal calcium-sensing receptor activation. *Curr. Opin. Nephrol. Hypertens.* **2024**, *33*, 433–440. [CrossRef]
75. Tian, L.; Andrews, C.; Yan, Q.; Yang, J.J. Molecular regulation of calcium-sensing receptor (CaSR)-mediated signaling. *Chronic Dis. Transl. Med.* **2024**, *10*, 167–194. [CrossRef]
76. Fouhy, L.E.; Mangano, K.M.; Zhang, X.; Hughes, B.D.; Tucker, K.L.; Noel, S.E. Association between a Calcium-to-Magnesium Ratio and Osteoporosis among Puerto Rican Adults. *J. Nutr.* **2023**, *153*, 2642–2650. [CrossRef]
77. Rosanoff, A.; Dai, Q.; Shapses, S.A. Essential nutrient Interactions: Does low or suboptimal magnesium status interact with vitamin D and/or calcium status? *Adv. Nutr.* **2016**, *7*, 25–43. [CrossRef]
78. Yücel, K.; Gürbüz, A.F. Evaluation of calcium/magnesium ratio in patients with type 2 diabetes mellitus. *Turk J. Biochem.* **2023**, *48*, 327–334. [CrossRef]
79. Guerrero-Romero, F.; Mercado, M.; Rodriguez-Moran, M.; Ramirez-Renteria, C.; Martinez-Aguilar, G.; Marrero-Rodriguez, D.; Ferreira-Hermosillo, A.; Simental-Mendia, L.E.; Remba-Shapiro, I.; Gamboa-Gomez, C.I.; et al. Magnesium-to-calcium ratio and mortality from COVID-19. *Nutrients* **2022**, *14*, 1686. [CrossRef]
80. Li, Q.; Chen, Q.; Zhang, H.; Xu, Z.; Wang, X.; Pang, J.; Ma, J.; Ling, W.; Li, D. Associations of serum magnesium levels and calcium-magnesium ratios with mortality in patients with coronary artery disease. *Diabetes Metab.* **2020**, *46*, 384–391. [CrossRef]
81. Escobedo-Monge, M.F.; Barrado, E.; Parodi-Roman, J.; Escobedo-Monge, M.A.; Torres-Hinojal, M.C.; Marugan-Miguelsanz, J.M. Magnesium status and Ca/Mg ratios in a series of children and adolescents with chronic diseases. *Nutrients* **2022**, *14*, 2941. [CrossRef]
82. Azem, R.; Daou, R.; Bassil, E.; Anvari, E.M.; Taliercio, J.J.; Arrigain, S.; Schold, J.D.; Vachharajani, T.; Nally, J.; Na Khoul, G.N. Serum magnesium, mortality and disease progression in chronic kidney disease. *BMC Nephrol.* **2020**, *21*, 49. [CrossRef]
83. La Carrubba, A.; Veronese, N.; Di Bella, G.; Cusumano, C.; Di Prazza, A.; Ciriminna, S.; Ganci, A.; Naro, L.; Dominguez, L.J.; Barbagallo, M.; et al. Prognostic Value of Magnesium in COVID-19: Findings from the COMEPA Study. *Nutrients* **2023**, *15*, 830. [CrossRef] [PubMed]
84. DeLuccia, R.; Cheung, M.; Ng, T.; Ramadoss, R.; Altasan, A.; Sukumar, D. Calcium to magnesium Ratio higher than optimal across age groups. *Curr. Dev. Nutr.* **2019**, *3*, nzz034-P10. [CrossRef]
85. Ciosek, Z.; Kot, K.; Kosik-Bogacka, D.; Lanocha-Arendarczyk, N.; Rotter, I. The effects of calcium, magnesium, phosphorus, fluoride, and lead on bone tissue. *Biomolecules* **2021**, *11*, 506. [CrossRef]
86. Steck, S.E.; Omofuma, O.O.; Su, L.J.; Maise, A.A.; Woloszynska-Read, A.; Johnson, C.S.; Zhang, H.; Bensen, J.T.; Fontham, E.T.H.; Mohler, J.L.; et al. Calcium, magnesium, and whole-milk intakes and high-aggressive prostate cancer in the North Carolina-Louisiana Prostate Cancer Project (PCaP). *Am. J. Clin. Nutr.* **2018**, *107*, 799–807. [CrossRef]
87. Wimalawansa, S.J.; Weiss, S.T.; Hollis, B.W. Integrating Endocrine, Genomic, and Extra-Skeletal Benefits of Vitamin D into National and Regional Clinical Guidelines. *Nutrients* **2024**, *16*, 3969. [CrossRef]
88. Wimalawansa, S.J. Controlling chronic diseases and acute Infections with vitamin D sufficiency. *Nutrients* **2023**, *15*, 3623. [CrossRef]
89. Bikle, D.D. Vitamin D metabolism and function in the skin. *Mol. Cell Endocrinol.* **2011**, *347*, 80–89. [CrossRef]
90. Yan, Q.; Zhang, T.; O'Connor, C.; Barlow, J.W.; Walsh, J.; Scalabrino, G.; Xu, F.; Sheridan, H. The biological responses of vitamin K2: A comprehensive review. *Food Sci. Nutr.* **2023**, *11*, 1634–1656. [CrossRef]
91. Hariri, E.; Kassis, N.; Iskandar, J.P.; Schurgers, L.J.; Saad, A.; Abdelfattah, O.; Bansal, A.; Isogai, T.; Harb, S.C.; Kapadia, S. Vitamin K(2)-a neglected player in cardiovascular health: A narrative review. *Open Heart* **2021**, *8*, e001715. [CrossRef]
92. Alharazy, S.; Robertson, M.D.; Lanham-New, S.; Naseer, M.I.; Chaudhary, A.G.; Alissa, E. Directly measured free and total 25-hydroxyvitamin D levels in relation to metabolic health in multi-ethnic postmenopausal females in Saudi Arabia. *Endocr. Connect.* **2021**, *10*, 1594–1606. [CrossRef] [PubMed]
93. Wimalawansa, S.J. Physiology of Vitamin D-Focusing on Disease Prevention. *Nutrients* **2024**, *16*, 1666. [CrossRef] [PubMed]
94. Wimalawansa, S.J. Non-musculoskeletal benefits of vitamin D. *J. Steroid Biochem. Mol. Biol.* **2018**, *175*, 60–81. [CrossRef] [PubMed]
95. Wimalawansa, S. Enhancing the design of nutrient clinical trials for disease prevention: A focus on vitamin D: A systematic review. In *Nutrition Reviews*; Oxford University Press: Oxford, UK, 2025; pp. 1–41. [CrossRef]
96. Yao, P.; Lu, L.; Hu, Y.; Liu, G.; Chen, X.; Sun, L.; Ye, X.; Zheng, H.; Chen, Y.; Hu, F.B.; et al. A dose-response study of vitamin D3 supplementation in healthy Chinese: A 5-arm randomized, placebo-controlled trial. *Eur. J. Nutr.* **2016**, *55*, 383–392. [CrossRef]
97. Grant, W.; Wimalawansa, S.J.; Pludowski, P.; Cheng, R. Vitamin D: Evidence-based health benefits and recommendations for population guidelines. *Nutrients* **2025**, *17*, 277. [CrossRef]
98. Sun, M.; Wu, X.; Yu, Y.; Wang, L.; Xie, D.; Zhang, Z.; Chen, L.; Lu, A.; Zhang, G.; Li, F. Disorders of calcium and phosphorus metabolism and the proteomics/metabolomics-based research. *Front. Cell Dev. Biol.* **2020**, *8*, 576110. [CrossRef]

99. Ducy, P. A central regulation of PTH secretion and function. *Neuron* **2023**, *111*, 1847–1849. [CrossRef]
100. Quinn, S.J.; Thomsen, A.R.; Egbuna, O.; Pang, J.; Baxi, K.; Goltzman, D.; Pollak, M.; Brown, E.M. CaSR-mediated interactions between calcium and magnesium homeostasis in mice. *Am. J. Physiol. Endocrinol. Metab.* **2013**, *304*, E724–E733. [CrossRef]
101. Wimalawansa, S.J. Physiological Basis for Using Vitamin D to Improve health. *Biomedicines* **2023**, *11*, 1542. [CrossRef]
102. Wimalawansa, S.J. Biology of vitamin D. *J. Steroids Horm. Sci.* **2019**, *10*, 1–8. [CrossRef]
103. Guo, J.H.; Zhang, X.S.; Vogt, R.D.; Xiao, J.S.; Zhao, D.W.; Xiang, R.J.; Luo, J.H. Evaluating controlling factors to Al(i)/(Ca + Mg) molar ratio in acidic soil water, southern and southwestern China: Multivariate approach. *Environ. Monit. Assess.* **2007**, *129*, 321–329. [CrossRef] [PubMed]
104. Wimalawansa, S.J. Vitamin D: An essential component for skeletal health. *Ann. N. Y. Acad. Sci.* **2011**, *1240*, E1–E12. [CrossRef] [PubMed]
105. Dunuweera, R.; Shimomura, R.M.G.; Priyankarage, J.V.; Jayasingha, P.; Wimalawansa, S.J. Chronic kidney disease of multifunctional origin (CKDmfo) prevailing in Sri Lanka: Re-evaluated. *World J. Pharma. Res.* **2017**, *6*, 33–66.
106. Wimalawansa, S.J. Effect of Water Hardness on Non-Communicable Diseases, Including Chronic Kidney Disease of Multifactorial Origin (CKDmfo/CKDuo). *J. Environ. Health Sci.* **2016**, *2*, 1–11. [CrossRef]
107. Wimalawansa, S.J.; Dissanayake, C.B. Nanocrystal-induced chronic tubular-nephropathy in tropical countries: Diagnosis, mitigation, and eradication. *Eur. J. Med. Res.* **2023**, *28*, 221. [CrossRef]
108. Wimalawansa, S.J.; Dissanayake, C.B. Factors Affecting the Environmentally Induced, Chronic Kidney Disease of Unknown Aetiology in Dry Zonal Regions in Tropical Countries—Novel Findings. *Environments* **2019**, *7*, 2. [CrossRef]
109. Wimalawansa, S.J. Molecular and cellular toxicity of fluoride in mystery, tubulointerstitial chronic kidney disease: A systematic review. *Rev. Environ. Sci. Bio/Technol.* **2019**, *19*, 117–147. [CrossRef]
110. Wimalawansa, S.J. Public health interventions for chronic diseases: Cost-benefit modelizations for eradicating chronic kidney disease of multifactorial origin (CKDmfo/CKDu) from tropical countries. *Heliyon* **2019**, *5*, e02309. [CrossRef]
111. Razzaque, M.S. Interactions between FGF23 and vitamin D. *Endocr. Connect.* **2022**, *11*, e220239. [CrossRef]
112. Hong, S.H.; Park, S.J.; Lee, S.; Kim, S.; Cho, M.H. Biological effects of inorganic phosphate: Potential signal of toxicity. *J. Toxicol. Sci.* **2015**, *40*, 55–69. [CrossRef]
113. Razzaque, M.S. Phosphate metabolism: From physiology to toxicity. *Adv. Exp. Med. Biol.* **2022**, *1362*, 1–6. [CrossRef] [PubMed]
114. Mancini, F.R.; Affret, A.; Dow, C.; Balkau, B.; Clavel-Chapelon, F.; Bonnet, F.; Boutron-Ruault, M.C.; Fagherazzi, G. High dietary phosphorus intake is associated with an increased risk of type 2 diabetes in the large prospective E3N cohort study. *Clin. Nutr.* **2018**, *37*, 1625–1630. [CrossRef] [PubMed]
115. Miyamoto, K.I.; Oh, J.; Razzaque, M.S. Common dietary sources of natural and artificial phosphate in food. *Adv. Exp. Med. Biol.* **2022**, *1362*, 99–105. [CrossRef]
116. Itkonen, S.T.; Lamberg-Allardt, C. Phosphorus—A scoping review for Nordic Nutrition Recommendations 2023. *Food Nutr. Res.* **2023**, *67*. [CrossRef]
117. Gross, P.; Six, I.; Kamel, S.; Massy, Z.A. Vascular toxicity of phosphate in chronic kidney disease: Beyond vascular calcification. *Circ. J.* **2014**, *78*, 2339–2346. [CrossRef]
118. Mironov, N.; Haque, M.; Atfi, A.; Razzaque, M.S. Phosphate dysregulation and metabolic syndrome. *Nutrients* **2022**, *14*, 4477. [CrossRef]
119. Sakaguchi, Y.; Hamano, T.; Isaka, Y. Effects of magnesium on the phosphate Toxicity in chronic kidney disease: Time for intervention studies. *Nutrients* **2017**, *9*, 112. [CrossRef]
120. Hruska, K.A.; Mathew, S.; Lund, R.; Qiu, P.; Pratt, R. Hyperphosphatemia of chronic kidney disease. *Kidney Int.* **2008**, *74*, 148–157. [CrossRef]
121. Geerse, D.A.; Bindels, A.J.; Kuiper, M.A.; Roos, A.N.; Spronk, P.E.; Schultz, M.J. Treatment of hypophosphatemia in the intensive care unit: A review. *Crit. Care* **2010**, *14*, R147. [CrossRef]
122. Kilpatrick, R.D.; Danese, M.D.; Belozeroff, V.; Smirnakis, K.; Goodman, W.G.; Rothman, K.J. The association of vitamin D use with hypercalcemia and hyperphosphatemia in hemodialysis patients: A case-crossover study. *Pharmacoepidemiol. Drug Saf.* **2011**, *20*, 914–921. [CrossRef]
123. Bianchi, S.; Aucella, F.; De Nicola, L.; Genovesi, S.; Paoletti, E.; Regolisti, G. Management of hyperkalemia in patients with kidney disease: A position paper endorsed by the Italian Society of Nephrology. *J. Nephrol.* **2019**, *32*, 499–516. [CrossRef]
124. Amanzadeh, J.; Reilly, R.F., Jr. Hypophosphatemia: An evidence-based approach to its clinical consequences and management. *Nat. Clin. Pr. Nephrol.* **2006**, *2*, 136–148. [CrossRef]
125. Ali, A.A.H. Overview of the vital roles of macro minerals in the human body. *J. Trace Elem. Miner.* **2023**, *4*, 100076. [CrossRef]
126. Lakhal-Littleton, S.; Robbins, P.A. The interplay between iron and oxygen homeostasis with a particular focus on the heart. *J. Appl. Physiol. (1985)* **2017**, *123*, 967–973. [CrossRef]
127. Hirota, K. An intimate crosstalk between iron homeostasis and oxygen metabolism regulated by the hypoxia-inducible factors (HIFs). *Free Radic. Biol. Med.* **2019**, *133*, 118–129. [CrossRef]

128. Lin, P.H.; Sermersheim, M.; Li, H.; Lee, P.H.U.; Steinberg, S.M.; Ma, J. Zinc in wound healing modulation. *Nutrients* **2017**, *10*, 16. [CrossRef]
129. Soliman, A.T.; Alaaraj, N.; Noor, H.; Alyafei, F.; Ahmed, S.; Shaat, M.; Itani, M.; Elalaily, R.; Soliman, N. Nutritional interventions during adolescence and their possible effects. *Acta Biomed.* **2022**, *93*, e2022087. [CrossRef]
130. Mireku, M.O.; Davidson, L.L.; Boivin, M.J.; Zoumenou, R.; Massougbodji, A.; Cot, M.; Bodeau-Livinec, F. Prenatal Iron Deficiency, Neonatal Ferritin, and Infant Cognitive Function. *Pediatrics* **2016**, *138*, e20161319. [CrossRef]
131. Jauregui-Lobera, I. Iron deficiency and cognitive functions. *Neuropsychiatr. Dis. Treat.* **2014**, *10*, 2087–2095. [CrossRef]
132. Croce, L.; Chiovato, L.; Tonacchera, M.; Petrosino, E.; Tanda, M.L.; Moleti, M.; Magri, F.; Olivieri, A.; Pearce, E.N.; Rotondi, M. Iodine status and supplementation in pregnancy: An overview of the evidence provided by meta-analyses. *Rev. Endocr. Metab. Disord.* **2023**, *24*, 241–250. [CrossRef]
133. Kabthymer, R.H.; Shaka, M.F.; Ayele, G.M.; Malako, B.G. Systematic review and meta-analysis of iodine deficiency and its associated factors among pregnant women in Ethiopia. *BMC Pregnancy Childbirth* **2021**, *21*, 106. [CrossRef] [PubMed]
134. Black, M.M. The evidence linking zinc deficiency with children's cognitive and motor functioning. *J. Nutr.* **2003**, *133*, 1473S–1476S. [CrossRef]
135. Kambe, T.; Tsuji, T.; Hashimoto, A.; Itsumura, N. The Physiological, Biochemical, and Molecular Roles of Zinc Transporters in Zinc Homeostasis and Metabolism. *Physiol. Rev.* **2015**, *95*, 749–784. [CrossRef]
136. Ben Abdallah, S.; Mhalla, Y.; Trabelsi, I.; Sekma, A.; Youssef, R.; Bel Haj Ali, K.; Ben Soltane, H.; Yacoubi, H.; Msolli, M.A.; Stambouli, N.; et al. Twice-Daily Oral Zinc in the Treatment of Patients With Coronavirus Disease 2019: A Randomized Double-Blind Controlled Trial. *Clin. Infect. Dis.* **2023**, *76*, 185–191. [CrossRef] [PubMed]
137. Razzaque, M.S. COVID-19 Pandemic: Can Maintaining Optimal Zinc Balance Enhance Host Resistance? *Tohoku J. Exp. Med.* **2020**, *251*, 175–181. [CrossRef] [PubMed]
138. Razzaque, M.S. COVID-19 pandemic: Can zinc supplementation provide an additional shield against the infection? *Comput. Struct. Biotechnol. J.* **2021**, *19*, 1371–1378. [CrossRef]
139. Zhang, Y.; Bi, J.; Wang, M.; Deng, H.; Yang, W. Correlation between helicobacter pylori infection and iron deficiency in children. *Pak. J. Med. Sci.* **2022**, *38*, 1188–1192. [CrossRef]
140. Yiannikourides, A.; Latunde-Dada, G.O. A Short Review of Iron Metabolism and Pathophysiology of Iron Disorders. *Medicines* **2019**, *6*, 85. [CrossRef]
141. Akimbekov, N.S.; Coban, S.O.; Atfi, A.; Razzaque, M.S. The role of magnesium in pancreatic beta-cell function and homeostasis. *Front. Nutr.* **2024**, *11*, 1458700. [CrossRef]
142. Mahdi, G.S. Chromium deficiency might contribute to insulin resistance, type 2 diabetes mellitus, dyslipidaemia, and atherosclerosis. *Diabet. Med.* **1996**, *13*, 389–390. [CrossRef]
143. Bai, J.; Xun, P.; Morris, S.; Jacobs, D.R., Jr.; Liu, K.; He, K. Chromium exposure and incidence of metabolic syndrome among American young adults over a 23-year follow-up: The CARDIA Trace Element Study. *Sci. Rep.* **2015**, *5*, 15606. [CrossRef] [PubMed]
144. Son, J.; Morris, J.S.; Park, K. Toenail Chromium Concentration and Metabolic Syndrome among Korean Adults. *Int. J. Environ. Res. Public. Health* **2018**, *15*, 682. [CrossRef] [PubMed]
145. Ahmad, R.; Shaju, R.; Atfi, A.; Razzaque, M.S. Zinc and Diabetes: A Connection between Micronutrient and Metabolism. *Cells* **2024**, *13*, 1359. [CrossRef] [PubMed]
146. Wang, H.; Hu, Y.F.; Hao, J.H.; Chen, Y.H.; Su, P.Y.; Wang, Y.; Yu, Z.; Fu, L.; Xu, Y.Y.; Zhang, C.; et al. Maternal zinc deficiency during pregnancy elevates the risks of fetal growth restriction: A population-based birth cohort study. *Sci. Rep.* **2015**, *5*, 11262. [CrossRef]
147. Tchounwou, P.B.; Yedjou, C.G.; Patlolla, A.K.; Sutton, D.J. Heavy metal toxicity and the environment. *Exp. Suppl.* **2012**, *101*, 133–164. [CrossRef]
148. Czarnek, K.; Terpiłowska, S.; Siwicki, A.K. Selected aspects of the action of cobalt ions in the human body. *Cent. Eur. J. Immunol.* **2015**, *40*, 236–242. [CrossRef]
149. Frimpong, E.; Ankapong, E.; Boakye, K.O.; Ansah, E.; Gyamfi, O.; Barnes, B.; Dodd, M.; Darko, G. Uptake and in-vitro bioaccessibility of toxic metals in cocoa beans: Human health risks. *Environ. Geochem. Health* **2024**, *47*, 33. [CrossRef]
150. Khayat, S.; Fanaei, H.; Ghanbarzehi, A. Minerals in Pregnancy and Lactation: A Review Article. *J. Clin. Diagn. Res.* **2017**, *11*, Qe01–Qe05. [CrossRef]
151. Grossklaus, R.; Liesenkötter, K.P.; Doubek, K.; Völzke, H.; Gaertner, R. Iodine Deficiency, Maternal Hypothyroxinemia and Endocrine Disrupters Affecting Fetal Brain Development: A Scoping Review. *Nutrients* **2023**, *15*, 2249. [CrossRef]
152. Makrides, M.; Crosby, D.D.; Bain, E.; Crowther, C.A. Magnesium supplementation in pregnancy. *Cochrane Database Syst. Rev.* **2014**, *2014*, CD000937. [CrossRef]
153. Farias, P.M.; Marcelino, G.; Santana, L.F.; de Almeida, E.B.; Guimarães, R.C.A.; Pott, A.; Hiane, P.A.; Freitas, K.C. Minerals in Pregnancy and Their Impact on Child Growth and Development. *Molecules* **2020**, *25*, 5630. [CrossRef] [PubMed]

154. Hansu, K.; Cikim, I.G. Vitamin and mineral levels during pregnancy. *Rev. Assoc. Med. Bras. (1992)* **2022**, *68*, 1705–1708. [CrossRef] [PubMed]
155. Wang, X.; Zhang, M.Y.; Miller, W.L.; Portale, A.A. Novel gene mutations in patients with 1alpha-hydroxylase deficiency that confer partial enzyme activity in vitro. *J. Clin. Endocrinol. Metab.* **2002**, *87*, 2424–2430. [CrossRef]
156. Papadopoulou, A.; Gole, E.; Melachroinou, K.; Meristoudis, C.; Siahanidou, T.; Papadimitriou, A. Identification and Functional Characterization of a Calcium-Sensing Receptor Mutation in an Infant with Familial Hypocalciuric Hypercalcemia. *J. Clin. Res. Pediatr. Endocrinol.* **2016**, *8*, 341–346. [CrossRef] [PubMed]
157. Iwaki, T.; Sandoval-Cooper, M.J.; Tenenhouse, H.S.; Castellino, F.J. A missense mutation in the sodium phosphate co-transporter Slc34a1 impairs phosphate homeostasis. *J. Am. Soc. Nephrol.* **2008**, *19*, 1753–1762. [CrossRef]
158. Yu, X.; White, K.E. FGF23 and disorders of phosphate homeostasis. *Cytokine Growth Factor. Rev.* **2005**, *16*, 221–232. [CrossRef]
159. Razzaque, M.S. The FGF23-Klotho axis: Endocrine regulation of phosphate homeostasis. *Nat. Rev. Endocrinol.* **2009**, *5*, 611–619. [CrossRef]
160. Nanba, K.; Usui, T.; Nakamura, M.; Toyota, Y.; Hirota, K.; Tamanaha, T.; Kawashima, S.T.; Nakao, K.; Yuno, A.; Tagami, T.; et al. A novel GATA3 nonsense mutation in a newly diagnosed adult patient of hypoparathyroidism, deafness, and renal dysplasia (HDR) syndrome. *Endocr. Pract.* **2013**, *19*, e17–e20. [CrossRef]
161. Marini, F.; Giusti, F.; Cioppi, F.; Maraghelli, D.; Cavalli, T.; Tonelli, F.; Brandi, M.L. Bone and Mineral Metabolism Phenotypes in MEN1-Related and Sporadic Primary Hyperparathyroidism, before and after Parathyroidectomy. *Cells* **2021**, *10*, 1895. [CrossRef]
162. Robinson-Cohen, C. Genetic variants of mineral metabolism in health and disease. *Curr. Opin. Nephrol. Hypertens.* **2020**, *29*, 387–393. [CrossRef]
163. Avnee; Sood, S.; Chaudhary, D.R.; Jhorar, P.; Rana, R.S. Biofortification: An approach to eradicate micronutrient deficiency. *Front. Nutr.* **2023**, *10*, 1233070. [CrossRef] [PubMed]
164. Wimalawansa, S. Rational food fortification programs to alleviate micronutrient deficiencies. *J. Food Process. Technol.* **2013**, *4*, 257–267. [CrossRef]
165. Yang, Z.; Laillou, A.; Smith, G.; Schofield, D.; Moench-Pfanner, R. A review of vitamin D fortification: Implications for nutrition programming in Southeast Asia. *Food Nutr. Bull.* **2013**, *34*, S81–S89. [CrossRef]
166. Smith, G.; Wimalawansa, S.J. Reconciling the irreconcilable: Micronutrients in clinical nutrition and public health. *Vitam. Miner.* **2015**, *4*, 1–4. [CrossRef]
167. Silva, P.; Araujo, R.; Lopes, F.; Ray, S. Nutrition and food literacy: Framing the challenges to health communication. *Nutrients* **2023**, *15*, 4708. [CrossRef]
168. Barker, M.; Dombrowski, S.U.; Colbourn, T.; Fall, C.H.D.; Kriznik, N.M.; Lawrence, W.T.; Norris, S.A.; Ngaiza, G.; Patel, D.; Skordis-Worrall, J.; et al. Intervention strategies to improve nutrition and health behaviours before conception. *Lancet* **2018**, *391*, 1853–1864. [CrossRef]
169. Wu, H.; Bai, M.; Li, X.; Xing, Y.; Sun, S. Diagnosis and treatment of brain injury complicated by hypernatremia. *Front. Neurol.* **2022**, *13*, 1026540. [CrossRef]
170. An, J.N.; Lee, J.P.; Jeon, H.J.; Kim, D.H.; Oh, Y.K.; Kim, Y.S.; Lim, C.S. Severe hyperkalemia requiring hospitalization: Predictors of mortality. *Crit. Care* **2012**, *16*, R225. [CrossRef]
171. Costa, D.; Patella, G.; Provenzano, M.; Ielapi, N.; Faga, T.; Zicarelli, M.; Arturi, F.; Coppolino, G.; Bolignano, D.; De Sarro, G.; et al. Hyperkalemia in CKD: An overview of available therapeutic strategies. *Front. Med.* **2023**, *10*, 1178140. [CrossRef]
172. Juurlink, D.N.; Tenenbein, M.; Koren, G.; Redelmeier, D.A. Iron poisoning in young children: Association with the birth of a sibling. *CMAJ* **2003**, *168*, 1539–1542.
173. Kozaki, K.; Egawa, H.; Garcia-Kennedy, R.; Cox, K.L.; Lindsay, J.; Esquivel, C.O. Hepatic failure due to massive iron ingestion successfully treated with liver transplantation. *Clin. Transpl.* **1995**, *9*, 85–87. [CrossRef]
174. Chandran, J.; Sanketh, R.; Vyasam, S.; Chrysolyte, A.; Ebenezer, K. Accidental iron poisoning in children—Experience from a teaching institution. *J. Fam. Med. Prim. Care* **2023**, *12*, 2520–2523. [CrossRef]
175. Sankar, J.; Shukla, A.; Khurana, R.; Dubey, N. Near fatal iron intoxication managed conservatively. *BMJ Case Rep.* **2013**, *2013*, bcr2012007670. [CrossRef] [PubMed]
176. Kandimalla, R.; Vallamkondu, J.; Corgiat, E.B.; Gill, K.D. Understanding aspects of aluminum exposure in Alzheimer's disease development. *Brain Pathol.* **2016**, *26*, 139–154. [CrossRef]
177. Charkiewicz, A.E.; Omeljaniuk, W.J.; Nowak, K.; Garley, M.; Niklinski, J. Cadmium toxicity and health effects-A brief summary. *Molecules* **2023**, *28*, 6620. [CrossRef]
178. Bandara, J.M.; Wijewardena, H.V.; Liyanege, J.; Upul, M.A.; Bandara, J.M. Chronic renal failure in Sri Lanka caused by elevated dietary cadmium: Trojan horse of the green revolution. *Toxicol. Lett.* **2010**, *198*, 33–39. [CrossRef]
179. Umair, M.; Alfadhel, M. Genetic disorders associated with metal metabolism. *Cells* **2019**, *8*, 1598. [CrossRef]
180. Brown, A.S.; Gwinn, M.; Cogswell, M.E.; Khoury, M.J. Hemochromatosis-associated morbidity in the United States: An analysis of the National Hospital Discharge Survey, 1979–1997. *Genet. Med.* **2001**, *3*, 109–111. [CrossRef]

181. Porter, J.L.; Rawla, P. Hemochromatosis. In *StatPearls*; StatPearls Publishing LLC.: Treasure Island, FL, USA, 2025.
182. de Bie, P.; van de Sluis, B.; Burstein, E.; van de Berghe, P.V.; Muller, P.; Berger, R.; Gitlin, J.D.; Wijmenga, C.; Klomp, L.W. Distinct Wilson's disease mutations in ATP7B are associated with enhanced binding to COMMD1 and reduced stability of ATP7B. *Gastroenterology* **2007**, *133*, 1316–1326. [CrossRef]
183. Lorincz, M.T. Wilson disease and related copper disorders. *Handb. Clin. Neurol.* **2018**, *147*, 279–292. [CrossRef]
184. Charkiewicz, A.E. Is Copper Still Safe for Us? What Do We Know and What Are the Latest Literature Statements? *Curr. Issues Mol. Biol.* **2024**, *46*, 8441–8463. [CrossRef] [PubMed]
185. Ojha, R.; Prasad, A.N. Menkes disease: What a multidisciplinary approach can do. *J. Multidiscip. Healthc.* **2016**, *9*, 371–385. [CrossRef] [PubMed]

Disclaimer/Publisher's Note: The statements, opinions and data contained in all publications are solely those of the individual author(s) and contributor(s) and not of MDPI and/or the editor(s). MDPI and/or the editor(s) disclaim responsibility for any injury to people or property resulting from any ideas, methods, instructions or products referred to in the content.

nutrients

MDPI

Review

The Role of Nutrition on Thyroid Function

Anna-Mariia Shulhai [1], Roberta Rotondo [1], Maddalena Petraroli [2], Viviana Patianna [2], Barbara Predieri [3], Lorenzo Iughetti [3], Susanna Esposito [1,2] and Maria Elisabeth Street [1,2,*]

[1] Department of Medicine and Surgery, University of Parma, 43121 Parma, Italy; annashulhai@gmail.com (A.-M.S.); robertarot.97@gmail.com (R.R.); susannamariaroberta.esposito@unipr.it (S.E.)

[2] Paediatric Clinic, University Hospital of Parma, 43121 Parma, Italy; m.petraroli@gmail.com (M.P.); viviana.patianna@gmail.com (V.P.)

[3] Pediatric Unit, Department of Medical and Surgical Sciences for Mothers, Children and Adults, University of Modena and Reggio Emilia, 41124 Modena, Italy; barbara.predieri@unimore.it (B.P.); lorenzo.iughetti@unimore.it (L.I.)

* Correspondence: mariaelisabeth.street@unipr.it; Tel.: +39-333-207-6294

Abstract: Thyroid function is closely linked to nutrition through the diet–gut–thyroid axis. This narrative review highlights the influence of nutritional components and micronutrients on thyroid development and function, as well as on the gut microbiota. Micronutrients such as iodine, selenium, iron, zinc, copper, magnesium, vitamin A, and vitamin B12 influence thyroid hormone synthesis and regulation throughout life. Dietary changes can alter the gut microbiota, leading not just to dysbiosis and micronutrient deficiency but also to changes in thyroid function through immunological regulation, nutrient absorption, and epigenetic changes. Nutritional imbalance can lead to thyroid dysfunction and/or disorders, such as hypothyroidism and hyperthyroidism, and possibly contribute to autoimmune thyroid diseases and thyroid cancer, yet controversial issues. Understanding these relationships is important to rationalize a balanced diet rich in essential micronutrients for maintaining thyroid health and preventing thyroid-related diseases. The synthetic comprehensive overview of current knowledge shows the importance of micronutrients and gut microbiota for thyroid function and uncovers potential gaps that require further investigation.

Keywords: thyroid; thyroid hormones; gut-thyroid axis; micronutrients; microbiome; trace elements; vitamins

1. Introduction

The thyroid gland is an essential endocrine organ that regulates physiological processes that are important for maintaining health and well-being. It is responsible for the synthesis and secretion of thyroid hormones, thyroxine (T4) and triiodothyronine (T3), which are involved in the function of all systems throughout the body [1]. The function of the thyroid gland is influenced by various factors, including diet and micronutrients [2,3]. Given its important role in maintaining homeostasis, any disruption in thyroid function can have far-reaching consequences, manifesting as a spectrum of disorders ranging from hypothyroidism to hyperthyroidism and thyroid cancer. In recent years, there has been a growing interest in the effects that nutrition can have on thyroid function. Research studies have highlighted, in particular, the influence of different dietary factors and micronutrients on thyroid hormone synthesis, metabolism, and function [2–5]. Understanding these relationships is important because diet and nutrition can be changeable factors that can potentially be improved and, as a result, improve thyroid function and reduce the incidence of thyroid disorders.

Nutrition is crucial for our well-being and the proper function of all organs and systems. Macronutrients and micronutrients are vital parts of nutrition. Lipids, carbohydrates, and proteins are macronutrients that produce energy and hormones, are required to synthesize molecules, and regulate metabolic processes [6]. Micronutrients are required in trace levels for biochemical processes such as gene transcription, enzymatic and hormonal reactions, cell protection, signal transduction, and primary and secondary metabolism [6]. Micronutrients are those essential vitamins and minerals required in small quantities that have been found to be key players in supporting optimal thyroid function [4,6], from iodine and selenium to iron and vitamin D.

To date, a thyroid–gut axis [2] is well recognized. The role of microbiota is also known to be influenced by micronutrient availability, drugs, and environmental contaminants, among other factors.

This paper synthesizes current research findings highlighting the mechanisms behind the interactions among nutrition, micronutrients, microbiota, and thyroid health.

2. Methods

The literature search was carried out using research strings in Pubmed, Scopus, and Mendeley via MeSH, with targeted keywords. Official sites of the WHO and European international organizations were also reviewed. The identified records were imported into Mendeley, and duplicates were removed. Reviewers independently screened the remaining papers by title and abstract, considering studies published from 1986 to March 2024. The following keywords were used for the search: thyroid, thyroid hormones, diet, micronutrients, nutrition, microbiome, gut microbiota, iodine, iron, selenium, vitamin D, vitamin A, vitamin B12, zinc, copper, magnesium, environmental pollution, combined in various ways to meet the research needs. A narrative synthesis was performed to organize and interpret the search findings.

3. Micronutrients and Thyroid Function

The overall effects of known nutrients and micronutrients on thyroid function are reported in Figure 1.

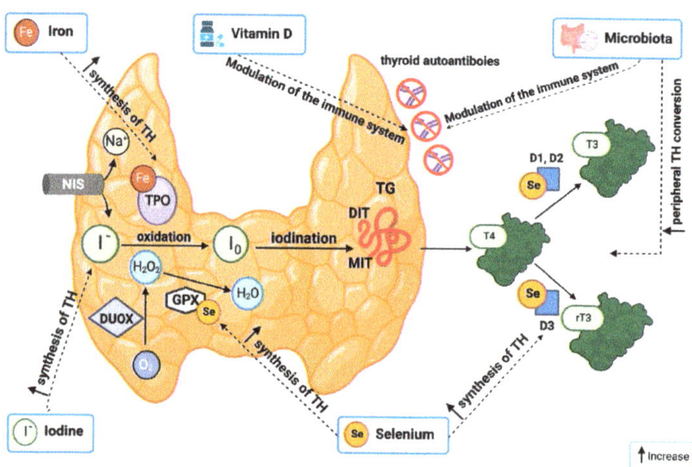

Figure 1. The overall effects of known nutrients and micronutrients on thyroid function. NIS: Na^+ –Iodid Symporter (iodide uptake); DUOX: dual oxidase (H_2O_2 generation); TPO: thyroperoxidase (hemoprotein); TG: thyroglobulin (synthesis and storage protein); GPX: glutathione peroxidase (antioxidant defense); thyroid hormones (TH); monoiodotyrosine (MIT) and diiodotyrosine (DIT); D1, 2, 3: deiodinase (TH in–/activation). Created with BioRender.com, accessed on 19 April 2024.

3.1. Iodine

Iodine plays an essential role in the synthesis of thyroid hormones (TH). It is absorbed in the small intestine, primarily in the form of iodide ions (I-), transported via the bloodstream to the thyroid gland, where it is actively taken up by thyroid follicular cells [7]. Within the thyroid follicular cells, iodine is enzymatically incorporated into tyrosine residues on thyroglobulin, a large glycoprotein synthesized by the thyroid gland. This process results in the formation of monoiodotyrosine (MIT) and diiodotyrosine (DIT), which then undergo further iodination to produce triiodothyronine and thyroxine, the active thyroid hormones [7]. These iodinated tyrosine residues are part of the thyroglobulin molecule, and when the hormone is required, thyroglobulin is broken down, releasing T3 and T4 into the bloodstream (Figure 1). Therefore, adequate iodine intake is necessary for maintaining normal thyroid function. The European Food Society Authority (EFSA) proposes age-specific intakes of iodine ranging from 70 µg/24 h for children to 150 µg/24 h at the end of adolescence [8,9] (Table 1). These are intakes calculated to ensure a urinary iodine concentration (UIC) $\geq$ 100 µg/L, which is the cut-off associated with the lowest prevalence of iodine-deficient goiter in school-aged children [10] (Table 1). It is calculated that in healthy children, the body contains around 15 to 20 mg of iodine, with the majority found in the thyroid gland.

Table 1. Iodine intake requirements according to age according to the regulatory authorities.

Age	Adequate Intake (ug/Day) REF * for EFSA	Recommended Iodine Intake (ug/Day) REF ** for WHO
0–6 months old	-	90
7–12 months old	70	90
1–6 years old	90	90
7–10 years old	90	120
11–14 years old	120	120–150
15–17 years old	130	150
$\geq$18 years old	150	150
During pregnancy	200	250
During lactation	200	250

REF, reference; * European Food Safety Authority Panel, EFSA J, 2014 [8]. ** World Health Organization and Food and Agriculture Organization of the United Nations. Vitamin and Mineral Requirements in Human Nutrition: Report of a Joint FAO/WHO Expert Consultation· 2nd ed. World Health Organization; Geneva, Switzerland: 2004. [(accessed on 11 February 2024)] [9].

Iodine deficiency in pregnancy can severely impact the cognitive development of offspring, as both maternal and fetal thyroid functions are affected. It is well known that in areas with severe and chronic iodine insufficiency, hypothyroidism can occur in both mothers and fetuses from the early stages of gestation, leading to irreversible brain damage, mental retardation, and neurological abnormalities [11,12]. The most severe form of in-utero deficiency is congenital hypothyroidism (formerly categorized under various "cretinism" labels) [13]. A meta-analysis evaluated iodine supplementation during pregnancy in areas with mild-to-moderate iodine deficiency; a lack of robust randomized controlled trial (RCT) evidence, with many observational studies being of poor quality, was described. It was concluded that maternal prepregnancy iodine status might influence thyroid adaptation during pregnancy, with mixed effects on thyroid volume and function. However, current evidence is insufficient to strongly support iodine supplementation recommendations in mildly-to-moderately iodine-deficient pregnant women (median UIC < 100 µg/L). Further research is needed to clarify the benefits and potential risks of iodine supplementation during pregnancy with consideration of maternal intrathyroidal iodine status [14].

Recent studies have investigated the impact of mild-to-moderate iodine deficiency during pregnancy on children's outcomes, and in addition, in countries such as Tasmania, maternal iodine deficiency was associated with lower literacy and numeracy scores in children at 9 years of age [15,16]. Even in postnatally iodine-rich environments, language and spelling outcomes remain lower if iodine deficiency during pregnancy has occurred. In a study from the UK, maternal iodine deficiency during the first trimester of pregnancy was associated with suboptimal verbal Intelligence Quotient (IQ) and reading scores in their children at 8 and 9 years of age, respectively [16,17].

Excess iodine consumption can also have a negative impact on thyroid function. Normally, large iodine dosages cause a transitory shutdown of thyroid hormone production (the acute Wolff-Chaikoff effect). Continuous exposure to high iodine levels causes down-regulation of the sodium iodide symporter (NIS), which transports iodine into the thyroid, enabling thyroid hormone production to continue; this is known as an escape from the acute Wolff-Chaikoff Effectf [18]. Failure of this effect can lead to iodine-induced hyperthyroidism or hypothyroidism, especially in individuals with pre-existing thyroid conditions. The fetus is particularly vulnerable to hypothyroidism, as the ability to fully escape this effect does not develop until about 36 weeks of gestation. Maximum safety limits for iodine intake in pregnant women, breastfeeding women, and infants, however, have not yet been comprehensively defined [16].

Universal salt iodization is considered the most effective method to ensure sufficient iodine intake in the population [12,19]. However, it has not been universally adopted, not implementing this strategy. Even in areas where salt iodization programs are in place, pregnant women who require higher iodine levels may still have insufficient intake. This has led various medical societies and government bodies, such as the American Thyroid Association, the Endocrine Society, the U.S. Teratology Society, the American Academy of Pediatrics, and the European Thyroid Association, to recommend iodine supplementation for pregnant, lactating, or women planning pregnancy [16,20]. The World Health Organization (WHO) provides iodine supplementation guidelines based on household iodized salt consumption [19]. In regions where less than 90% of households consume iodized salt and the median urinary iodine concentration is below 100 μg/L, iodine supplementation in the form of potassium iodide is advised, with an intake target of 250 μg/day. Alternatively, in severely iodine-deficient populations where daily supplementation may be impractical, an annual 400 mg dose of iodized oil supplement can be considered. Ideally, iodine supplementation should be started at least 3 months before conception to ensure that maternal thyroid iodine stores are sufficient for the entire pregnancy [12,16,20].

The role of iodine during early childhood is crucial for brain development and TH synthesis. Adequate iodine intake is particularly critical during the weaning period, as infants transition from exclusive breastfeeding to a mixed diet [21]. While breastfeeding is an excellent source of iodine for infants, the amount in breast milk largely depends on the mother's iodine status. In the United States, the iodine content of breast milk has been shown to vary geographically, with some regions experiencing mild-to-moderate deficiency. This can be attributed to factors such as decreased iodine intake in maternal diets, decreased iodine content in cow's milk if taken by the mothers during breastfeeding, or the absence of iodine supplementation during pregnancy [21]. A 2018 study conducted in a non-iodine-deficient region showed that although there is a correlation between Maternal Body Iodine Content (BMIC) and the iodine status of infants, the feeding type (exclusive breastfeeding vs. partial formula feeding) does not influence the UIC values of infants and mothers in analyses adjusted for BMIC [22]. The introduction of complementary nutrients during weaning is an opportunity to enhance iodine intake. However, studies have suggested that the iodine content of many infant foods is inadequate to meet the recommended daily intake. Commercially available baby food, including fruit, vegetables, and cereals, often lacks sufficient iodine levels. Moreover, organic baby foods, which may appeal to health-conscious parents, tend to have even lower iodine content. Low iodine intake during weaning can have long-term consequences in cognitive development and thyroid function in infants. Studies have shown that infants with insufficient iodine intake may exhibit lower IQ scores and an increased risk of thyroid-related disorders [13,22]. Strategies should be implemented at both the individual and public health levels to address the iodine nutrition gap in weaning infants. Healthcare providers can play a crucial role in educating parents about the importance of iodine-rich foods and the potential need for iodine supplementation. Manufacturers can also contribute by fortifying infant foods with iodine to ensure an adequate intake for weaning infants.

During adolescence, deficient iodine intake can lead to nodular goiter, while excess intake or iodine fortification can increase the risk of thyroid autoimmunity. The mechanisms underlying how iodine leads to thyroiditis remain unclear, and various immunological and non-immunological theories have been proposed. Among these, it is hypothesized that excess and accumulation of iodine in thyroid cells can generate high oxidative stress and cause cellular damage. This can lead to the production and secretion of cytokines and chemokines, recruiting lymphocytes to the thyroid, where they encounter thyroid auto-antigens, such as thyroglobulin. Excess iodine can alter the conformation of thyroglobulin, making it more easily recognized by immune system cells [23]. This can ultimately lead to the development of pathological hypersensitivity to thyroid auto-antigens, with subsequent development of thyroiditis [23,24]. In China, after 3 years of implementing salt iodization, the prevalence of autoimmune thyroid diseases varied based on iodine intake levels: 0.5% in mildly deficient areas, 1.7% in areas with more-than-adequate iodine, and 2.8% in regions with excessive-high consumption [25]. Meanwhile, in Denmark, which previously experienced mild-to moderate iodine deficiency, iodine status significantly improved after 5 years of mandatory iodine fortification of salt, transitioning from deficient to optimal iodine intake levels also leading to a decreased prevalence of autoimmune thyroid diseases [26]. Therefore, special attention must be paid to nutritional intake once adequate iodine uptake (aiming for ioduria >100 μg/L, a value associated with minimal goiter incidence in the growing population) has been ensured.

Nowadays, with global gastronomic influences significantly altering adolescent eating habits, this becomes even more crucial. In a study involving Italian children and adolescents [27], iodine requirements were met primarily through excessive salt intake (>10.2 g/day). This not only highlighted the issue of excessive salt consumption but also emphasized that iodized salt remains underutilized in industrial processes and catering services [27]. Furthermore, this study [27] provided estimated values for iodine intake from unsupplemented food in different age groups: 44 μg/1000 Kcal for children, 45 μg/1000 Kcal for teenage girls, and 35 μg/1000 Kcal for teenage boys. It is commonly assumed that certain foods, like those from the Brassicaceae family, cassava, and millet, contribute to iodine malabsorption [27]. However, this is only significant when consumed in large quantities under conditions of iodine deficiency. Of particular interest is the mechanism involving thiocyanates, which competitively inhibit iodine uptake by thyrocytes and facilitate the efflux of intrathyroidal iodine [28].

Notably, cigarette smoke contains substantial amounts of thiocyanates, explaining the association between smoking and goiter [28]. This association is even more pronounced in cases of obesity [28]. This underscores the importance of reinforcing prevention campaigns targeting smoking and obesity within this age group [10].

3.2. Selenium

Selenium (Se) is an essential trace element crucial for the thyroid system. The human body contains approximately 14 mg of Se [29], and it is an integral component of the enzymes glutathione peroxidase (GPx) [30] and iodothyronine deiodinases, where it is incorporated as selenocysteine. Selenium-dependent GPx catalyzes the breakdown of hydrogen peroxide (H_2O_2), shielding most tissues from oxidative stress and cell damage [31]. In the thyroid gland, H_2O_2 is produced in response to thyrotropin (TSH) stimulation and is required for thyroglobulin iodination by the enzyme thyroid peroxidase (TPO), which leads to T4 production. Excess H_2O_2 has been demonstrated to inactivate thyroid peroxidase in vitro, suggesting that its accumulation could also affect TH [32] (Figure 1).

Regarding the prenatal period, a significant RCT involving pregnant women positive for anti-thyroid peroxidase antibody (TPOAb), demonstrated that selenium supplementation led to a notable reduction in TPOAb levels during and after pregnancy [20,33]. This supplementation also reduced the incidence of postpartum thyroid disease and permanent hypothyroidism [33]. At variance with the study, another RCT examining the impact of selenium supplementation on autoimmune thyroid disease in pregnancy did not find any reduction in TPOAbs between the Se and placebo groups. However, this study had lower baseline TPOAb concentrations, mothers received a lower selenium dose (60 μg/d), and the study was overall underpowered [34]. There is a clear need for high-quality, well-powered RCT focusing on TPOAb-positive pregnant women to validate the findings of the Italian study.

In areas with combined selenium and iodine deficiencies, selenium deficiency may reduce fetal hypothyroidism by reducing maternal T4 metabolism, offering protection to fetal brain development. However, it is hypothesized that after birth, the iodine-deficient thyroid gland, which is hyperplastic and hyperstimulated, generates an excess of free oxygen radicals that, in addition to reduced GPx activity, could trigger the destruction of the thyroid gland. This intriguing hypothesis awaits formal validation and is not fully corroborated by epidemiological evidence [35].

Adequate iodine levels should be ensured before administering selenium compounds to prevent unintended loss of iodinated TH [13].

Selenium deficiency has been associated with various thyroid disorders, including hypothyroidism, subclinical hypothyroidism, thyroid cancer, and autoimmune thyroid diseases, including HT and GD [36]. In childhood and adolescence, there are few studies. For instance, some research has focused on children and adolescents who have phenylketonuria (PKU). Van Bakel et al. have found significantly higher fT4 and fT3 concentrations in children with PKU compared to control subjects [37]. Furthermore, Calomme et al. studied the effect of selenium supplementation on TH concentrations in ten severely deficient subjects: Se supplementation led to a significant increase in plasma selenium concentrations and GPx activity, and this was associated with a notable decrease in plasma T4 (−16%) and rT3 (−29%) compared to initial values. Discontinuing Se treatment resulted in a significant increase in T4 and rT3 levels, which returned to their initial values. Plasma TSH and T3 concentrations were unaffected by selenium status [38]. Both studies demonstrated a negative connection between plasma selenium and plasma T4 and rT3 levels [37,38]. However, these investigations found no evidence that selenium impacted TH feedback at the pituitary level or produced clinically detectable hypothyroidism. These findings are consistent with those from animal research [39,40]. Prolonged and isolated selenium deprivation in rodents results in a complete loss of liver 5′-deiodinase (5′DI) activity and an increase in circulating T4, with little to no change in T3 or TSH concentrations. Interestingly, selenium insufficiency is linked to higher blood rT3 concentrations in young rats but not in adult males or non-pregnant females. This might represent a gradual increase in 5′DI expression early in life. The anticipated significant increases in cerebrocortical and pituitary 5′DII activities in response to hypothyroidism, as well as thyroid GPx an′ 5′DI and pituitar′ 5′DI activities, are mostly unaffected by selenium deficiency in rats. This supports the absence of biochemical or clinical signs of hypothyroidism in selenium-deficient PKU patients [35].

In humans, the impact of selenium supplementation (without iodine supplementation) on thyroid function was evaluated in 23 school-age children residing in an area severely deficient in both iodine and selenium. Selenium supplementation elevated plasma selenium levels and erythrocyte GPx activity but led to a substantial 30–50% reduction in serum T4 and free T4 concentrations, transitioning from normal to hypothyroid levels. Plasma TSH and T3 concentrations stayed within the euthyroid range at baseline and remained unaffected by selenium supplementation. This suggests that in iodine-deficient regions, addressing concurrent selenium deficiency without iodine supplementation heightens the conversion of T4 to T3 in the periphery [41].

A recent observational cohort study on the adult population documented an increased incidence of Hashimoto's thyroiditis (HTs) after a 6-year follow-up in 1254 individuals with low selenium status. This study provided epidemiological support for a strategy to prevent autoimmune thyroiditis by ensuring adequate Se nutritional intake, especially in women [42].

Various systematic reviews/meta-analyses have examined the effects of selenium treatment in autoimmune thyroid diseases and/or Hashimoto's disease. One meta-analysis of 16 trials concluded that selenium supplementation reduced TPOAb levels, especially in patients already on L-T4 treatment [43]. The systematic review from 2017 identified only three studies showing some improvement in the quality of life and thyroid echogenicity on ultrasound in patients treated with levothyroxine replacement therapy with selenium supplementation. However, it was not possible to synthesize them in a meta-analysis due to the small sample sizes [44]. A recent meta-analysis revealed a substantial decrease in TSH levels and anti-TPO antibodies in euthyroid and subclinical hypothyroid individuals with HTs without hormone replacement therapy (THRT) after selenium supplementation [45,46]. These findings suggest a potential benefit of selenium supplementation for autoimmune thyroid conditions. However, larger, more comprehensive studies are required to conclusively confirm these effects. To reduce the risk of selenium poisoning, physicians should follow recommended selenium doses (80–400 µg/day) and avoid extended high-dose supplementation. Potential adverse effects should be considered, especially in non-selenium-deficient people [45].

3.3. *Iron*

Iron (Fe) is the third essential trace element important for the normal biosynthesis and function of THs. An adult human body contains about 4 g of iron (0.005% of body weight). Most of this iron is found in hemoglobin and myoglobin, but it is also present in various cytochromes and other heme-containing proteins. Fe is the central atom in the active sites of these proteins, including TPO, which is uniquely expressed by thyroid cells and is related to myeloperoxidase and lactoperoxidase [13] (Figure 1). TPO is a crucial multifunctional enzyme in the synthesis of thyroid hormones [47].

During pregnancy, there is an increased demand for iron to support the expansion of maternal blood volume and meet the needs of the developing fetus. Previous studies have shown that iron deficiency can be a risk factor for thyroid disorders during pregnancy [48]. The effect of this insufficiency on TH may be correlated with changes in the activity of TPO or the reduction in the conversion of T4 to T3 by influencing the activity of thyroxine deiodinase [49]. Several studies have observed lower levels of fT3 and fT4 in pregnant women with iron deficiency [50–52]. Iron and hemoglobin have been found to be positively correlated with fT3 and fT4 but negatively correlated with TSH, suggesting that the fT4 level might change with iron status. Regular monitoring of iron status and thyroid function, along with appropriate supplementation and interventions, is essential to mitigate potential risks associated with deficiencies or dysregulation during pregnancy [53].

Iron deficiency and anemia are associated with hypothyroidism, likely due to the reduced biosynthesis of the hemoprotein TPO [54]. However, the interplay between iron and TH is bidirectional, as TH directly stimulates erythropoiesis, predominantly through the TR-alpha receptor [55,56]. Iron supplements and improving iron status in areas with endemic deficiency increased TH biosynthesis, as demonstrated in children [57] and anemic patients (especially women) [5].

The association between iron deficiency and thyroid autoimmunity has not been firmly established. Many patients with HTs have iron deficiency due to comorbidities such as autoimmune gastritis, which leads to reduced iron absorption, or celiac disease, which results in iron loss [58]. In a recent retrospective study, TSH, TPOAb, and anti-thyroglobulin antibody (TgAb) levels were significantly higher, whereas hematocrit, mean corpuscular volume, hemoglobin, ferritin, and iron were significantly lower in 180 female patients with positive thyroid autoantibodies compared to 81 healthy controls [59]. Combined treatment with T4 and sufficiently bioavailable iron supplements may be necessary, and clinical studies have shown better effects than those obtained with each factor taken individually [13,60], further investigations of iron status on the efficacy and absorption of thyroid hormone replacement therapy.

3.4. Vitamin D

Vitamin D receptors are present in the thyroid gland, indicating a potential role in thyroid function [61]. Vitamin D modulates the immune system by boosting the innate immune response and inhibiting the adaptive immune system [62] (Figure 1). Vitamin D's capacity to suppress the adaptive immune system improves immune tolerance and appears to help with a variety of autoimmune disorders [62]. Pleiotropic activities have been identified through preclinical and observational research, indicating a positive role in thyroid disease treatment. However, few equivocal causal correlations and a few intervention studies have been documented thus far, particularly in the juvenile population. Therefore, the preventative and therapeutic potential of vitamin D or its analogs in thyroid illnesses is still being debated.

Animal studies have provided evidence for the role of vitamin D in autoimmune thyroid diseases.

Active vitamin D (25-hydroxyvitamin D—25(OH)D) crosses the placenta, and the mother's adequate vitamin D status is important for the health of both the mother and the child [63]. There are significant regional and ethnic differences in the prevalence of vitamin D insufficiency in pregnancy, ranging from 18 to 84% [63–65]. There is limited data regarding the relationship between vitamin D status and thyroid autoimmunity during pregnancy. In some studies, the prevalence of TPOAb was higher in the first trimester [66,67].

In childhood/adolescence, clinical studies have reported low vitamin D levels in autoimmune thyroid disease (AITD), indicating an association between vitamin D deficiency and thyroid autoimmunity [68]. Otherwise, the association between vitamin D and HTs remains controversial. Some studies, including observations from the Croatian Biobank of HTs patients and other comparative studies, have not detected an association between vitamin D levels and the prevalence of HTs [69–71]. Other studies have observed higher rates of vitamin D deficiency in children with HTs (inverse correlation between 25(OH)D and TPO- Ab levels) [68]. This difference was more evident in patients with overt hypothyroidism compared to those with subclinical hypothyroidism. Furthermore, significant negative correlations were found between serum levels of 25(OH)D and patient age, disease duration, BMI, and TPOAb, TgAb, and TSH levels. The authors proposed low serum vitamin D to be significantly associated with autoimmune thyroid disease, but this was not an independent risk factor for the progression of the condition [72]. It is worth noting that recently, studies evaluating whether vitamin D supplementation is beneficial for autoimmune thyroid disease have been published. After 4 months of oral vitamin D3 supplementation (1200–4000 IU/day) in 186 deficient patients, there was a significant decrease (20.3%) in serum TPOAb levels [73,74]. In a study by Taheriniya et al., which examined 42 studies involving patients with AITD, it was found that vitamin D deficiency correlates with the onset of AITD, including HTs and hypothyroidism [75]. A relationship between vitamin D and Graves' disease has only been reported in older people [75]. The Vitamin D Receptor (VDR) and 1α-hydroxylase are present in various immune cells including T and B lymphocytes, dendritic cells, neutrophils, and monocytes [73,74]. This allows these cells to produce calcitriol, the active form of vitamin D3. It inhibits the production of proinflammatory cytokines such as IL-6, IL-8, IL-9, IL-12, IFN-γ, and TNF-α, while enhancing the production of anti-inflammatory cytokines such as IL-10, IL-5, and IL-4. Overall, vitamin D is considered to have an anti-inflammatory effect [76,77].

Low vitamin D status may be a consequence rather than a cause of thyroid diseases, and it can be influenced by various factors such as low intake of vitamin D, malabsorption, lack of sun exposure, or reduced outdoor activity [78]. The last meta-analyses showed an increased risk of thyroid cancer in subjects with vitamin D deficiency [79]. Study design limitations, inter-assay and inter-laboratory variability in vitamin D measurements, and heterogeneity of the study populations also contribute to the conflicting results [80]. The preventive and therapeutic potential of vitamin D or its analogs in thyroid diseases remains a topic of debate, and ongoing research will provide insight into their efficacy and safety as therapeutic tools.

A recent meta-analysis showed that vitamin D supplementation significantly decreases TPOAb and TgAb titers in patients undergoing thyroid therapy, leading to improved thyroid function with reduced TSH levels and increased FT3 and FT4 levels [81]. Vitamin D supplementation at a dosage of 1500 to 2000 IU/d for more than 12 weeks resulted in more significant decreases in TPOAb levels and higher increases in FT3 and FT4 compared to shorter treatments. Vitamin D administration may reduce TPOAb and TgAb titers by improving thyroid function through immunomodulation. Therefore, vitamin D supplementation might be beneficial for patients with HTs [81]. However, further research is needed to understand the specific mechanisms behind these effects and long-term benefits.

3.5. Zinc

Zinc (Zn) is a trace mineral that contributes to gene expression and cell growth and serves as a cofactor for numerous enzymes involved in various physiological processes, including thyroid hormone synthesis and metabolism [82,83]. The most common nutrition rich in zinc are oysters and fish, legumes, nuts, red meat, whole grains, and dairy [6]. Within the thyroid gland, zinc plays a critical role in the activity of TPO, an enzyme essential for the synthesis of thyroid hormones. TPO catalyzes the iodination of thyroglobulin, a protein precursor of thyroid hormones, as well as the pairing of iodotyrosine residues to form T4 and T3 [84]. Experimental studies in vitro and in vivo have provided evidence that zinc behaves as an antioxidant, reducing the oxidation of macromolecules like DNA/RNA and proteins. It also mitigates the inflammatory response, leading to a decrease in the formation of reactive oxygen species (ROS) [85]. Autoimmune disorders are linked to abnormal zinc levels and can lead to disrupted signal transduction, affecting the immune response, cell function, and differentiation. Zinc deficiency can weaken both the innate and adaptive immune responses. Additionally, T cell activation and differentiation into specific subpopulations (Th1, Th2, Th17, and Treg) significantly influence zinc balance [86–88].

Zinc deficiency can impair thyroid hormone receptor function, affecting their ability to bind thyroid hormones effectively and regulate gene expression, leading to disruptions in thyroid hormone signaling pathways [89], may influence decreased thyroid hormone synthesis, and potentially contribute to hypothyroidism development [90,91]. Many studies and systematic reviews report on the association between Zn deficiency and hypothyroidism [90,92–94]. Acquired zinc deficiency can cause another common symptom in patients with hypothyroidism, which is hair loss [90]. The loss of pigment, dryness, brittleness, and hair loss appear due to Zn deficiency as a cofactor of metalloenzyme. Some researchers suggest using a diet that is rich in zinc to achieve euthyroidism, which can be beneficial for alopecia [90,95–97]. However, further randomized clinical trials are required for understanding the role of zinc supplementation on hair loss in hypothyroid patients. Nevertheless, there is limited data about the zinc–hyperthyroidism relation, which is controversial and requires further investigation. Some studies have demonstrated zinc's influence on papillary thyroid cancer and follicular carcinoma [98,99]. However, a recent Mendelian randomized study found no evidence of this influence [100], indicating the need for further understanding.

Zinc deficiency has been linked with thyroid enlargement in children and adults. The study of 68 school-age children showed a positive inverse correlation between zinc concentration and thyroid size [101]. Researchers hypothesized the main cause of this is Zn loss and impaired T3 action and T3 binding to its nuclear receptor [101]. The same results were found in adults, where Zn concentration was negatively correlated with thyroid volume [91], and adults with nodular goiter, especially in iodine deficiency regions [102]. Subjects with low zinc levels have shown a positive correlation with autoimmune thyroid disease [91]. However, this finding is controversial, as an investigation of Zn level differences between women with Hashimoto thyroiditis and healthy controls did not show any differences [103], highlighting further need for investigation of the association of HTs and zinc relations.

It is worth noting that recently, the RCT of daily zinc supplementation for 12 weeks on serum thyroid auto-antibody levels in children and adolescents with autoimmune thyroiditis has found no difference in thyroid levels, TPOAb, TgAb, and oxidative stress markers [104]. Interestingly, despite the absence of thyroid hormones and auto-antibodies changes, levothyroxine dose requirements significantly increased in the control (placebo) group compared to the group with zinc supplementation [104], but further long-term trials are required. So, zinc supplementation might be useful in hypothyroid patients [104,105].

4. Other Micronutrients: Copper, Magnesium, Vitamin A, and Vitamin B12

4.1. Copper

Copper (Cu) is a trace mineral that can be found in organ meats, fish and shellfish, seeds and whole grains, chocolate, and leafy greens [6]. It is involved in several key processes related to thyroid hormone synthesis and regulation. The recommended dietary allowance (RDA) is approximately 1 mg/d for both adult men and women [6]. Copper is a cofactor for tyrosinase and is involved in the conversion of inactive T4 to the biologically active T3, necessary for the synthesis of TPO, a precursor to thyroid hormones, and the subsequent coupling of iodotyrosine residues to form T4 and T3 [6,89]. Moreover, Cu has pro-oxidant and antioxidant characteristics, and its imbalance leads to oxidative stress and potentially may lead to thyroid dysfunction [106]. As a pro-oxidant, Cu can participate in Fenton-like reactions, producing highly reactive hydroxyl radicals from hydrogen peroxide. These hydroxyl radicals are potent oxidizing agents that can damage cellular components, including lipids, proteins, and DNA. Antioxidant actions occur through Cu's role in superoxide dismutase for oxidative stress mitigation and copper-dependent enzymes (cytochrome c oxidase and ceruloplasmin) [106]. Cu is associated with the regulation of body calcium levels, which in turn prevents the over-absorption of T4 in blood cells and is essential for supporting optimal thyroid function and hormone regulation [89]. A deficiency of Cu is related to subclinical hypothyroidism [107] or hypothyroidism [108]. However, a large USA survey showed that Cu elevation is related to higher levels of fT4 and TT4 in males and TT3 and TT4 in females [109], which requires further studies. There are limited studies about Cu and HTs. Szczepanik et al. (2021) have not found significant differences between Cu levels in patients with HTs and controls [103]. In contrast, Rasic-Milutinovic et al. (2017) defined higher Cu levels in patients with HTs [109]. In this single study, researchers hypothesized that elevated serum selenium levels and reduced copper levels in patients with HTs and overt hypothyroidism might result in lower L-thyroxine dosages or reaching a euthyroid state with decreased free T4 due to the direct impact of the copper-selenium ratio on redox balance and as a result of thyroid health [109]. Moreover, excessive copper intake can be detrimental to thyroid function and indicate proliferative process progression in the thyroid gland. Thus, the meta-analysis showed that higher Cu was found in patients with thyroid cancer in China and Turkey. As copper is a cofactor in tumor angiogenesis, and its high levels influence tumor growth [110], Cu agents potentially can be used for BRAFV600E mutation-positive malignancies treatment and tumors resistant to BRAFV600E and MEK1/2 inhibitors [111].

4.2. Magnesium

Magnesium (Mg) is a mineral that can be found in leafy green vegetables, legumes, nuts, seeds, milk, whole grains, etc. [6]. It is involved in various aspects of thyroid function, as it is required for the activation of adenosine triphosphate (ATP) and DNA replication and transcription. Magnesium acts as a cofactor for several enzymes and enzymatic reactions and is also involved in thyroid hormone metabolism. It can indirectly influence deiodination, which catalyzes the conversion of T4 to the more active T3 form [6,112]. Additionally, it is involved in the regulation of thyroid hormone receptor sensitivity as a second messenger, which affects the receptivity of target tissues to thyroid hormones and balances oxidative phosphorylation [113]. Magnesium shortage impacts the bioavailability and tissue distribution of selenium, resulting in reduced levels [114]. Experimental studies showed an increase in follicular-colloidal index, amount of thyrocytes in the follicle and resorption vacuoles in the colloid, and activation of thyroid gland synthetic activity by ingestion of magnesium chloride [115]. Mg deficiency is common in the general population and has been associated with impaired thyroid function, metabolic disorders, and carcinogenesis. Studies involving 1257 subjects demonstrated elevated risks of TgAb positivity and higher HTs and hypothyroidism prevalence in case of severe Mg deficiency (0.55 mmol/L), suggesting that blood magnesium levels should be investigated in individuals with autoimmune thyroiditis and hypothyroidism [116]. Hypothesized that magnesium deficiency, in combination with coenzyme Q10 and selenium and inefficient oxidative phosphorylation, leads to mitochondrial dysfunction and potentiates hyperthyroidism development [114]. Moreover, magnesium deficiency has an impact on cancerogenesis by linking with inflammation and/or free radicals, which can cause DNA oxidative damage and cancer formation. The cross-sectional study analyzing 5709 who underwent thyroidectomy found that patients with papillary thyroid cancer had lower serum magnesium levels within the common range compared to benign nodules [117]. Notably, magnesium level > 935 μmol/L was found to be a potential independent protective factor against papillary thyroid cancer [117]. Based on this, more studies should be done to find Mg's possible potential protective role against thyroid cancer and establish guidelines to prevent cancerogenesis. Additionally, comparing premenopausal and postmenopausal women, researchers have defined a significant decrease in magnesium in postmenopausal women [112]. There is an indirect relation between Mg and thyroid hormone levels, particularly with TSH levels. Hypomagnesemia development can be one of the factors contributing to the development of thyroid disorders in the postmenopausal period [112]. However, a high dose of magnesium can increase thyroid activity [110]. Animal studies on albino Wistar strain rats showed that chronic high magnesium intake leads to thyroid disruption as it elevates thyroid peroxidase and Na(+)-K(+)-ATPase, changes iodothyronine 5′-deiodinase type I activities, increases serum T4, tT3, and TSH, resulting in lowering fT3 and T3 and enlarging of the thyroid gland [118,119]. Therefore, a balanced level of Mg, by diet or supplementation intake, may be beneficial for supporting thyroid health and overall well-being.

4.3. *Vitamin A*

The relationship between vitamin A and thyroid function is a topic of considerable scientific interest due to the intricate interplay between these two essential elements for human health. Vitamin A, a fat-soluble micronutrient, plays critical roles in various physiological processes, including vision, immune function, reproduction, and cellular growth and differentiation [120]. Within the context of thyroid function, emerging evidence suggests that vitamin A status can profoundly influence thyroid gland homeostasis through multiple mechanisms. Vitamin A deficiency (VAD) has been implicated in thyroid gland impairment, particularly when coupled with iodine deficiency [83,121,122]. Animal studies utilizing mouse models have provided valuable insights into the effects of VAD on thyroid function. These studies have demonstrated that VAD can lead to reduced uptake of iodine by the thyroid gland, impaired synthesis of thyroid hormones, and decreased production of thyroglobulin [123]. Consequently, these disruptions in thyroid hormone synthesis and secretion can contribute to thyroid hypertrophy, goiter formation, and alterations in intrathyroidal hormone levels [83,122]. Moreover, studies have highlighted the role of vitamin A in modulating the sensitivity of thyroid cells to TSH, a key regulator of thyroid function. Specifically, vitamin A has been shown to influence the expression of genes involved in TSH receptor signaling pathways, thereby affecting thyroid hormone production and secretion [83,122,124]. Additionally, vitamin A status appears to modulate the peripheral metabolism of thyroid hormones, including the conversion of T4 to the biologically active T3 hormone [119,125]. Notably, the interaction between vitamin A and thyroid function extends beyond hormonal regulation to encompass the role of transport proteins such as transthyretin (TTR). Vitamin A binds to retinol-binding protein (RBP) to form a complex that interacts with TTR, a carrier protein for both retinol and thyroid hormones in the bloodstream. Studies have suggested that changes in vitamin A status can influence the binding capacity and affinity of TTR for thyroid hormones, potentially impacting their transport and distribution in the body [126]. In regions with endemic malnutrition and micronutrient deficiencies, investigations into the effects of vitamin A supplementation on thyroid function have yielded diverse findings [127,128]. While some studies have reported improvements in thyroid volume and hormone levels following vitamin A supplementation, others have shown mixed or inconclusive results [127,128]. Factors such as the severity of VAD, the presence of concomitant ID, and individual variations in response to supplementation may contribute to the variability observed in these studies.

Moreover, vitamin A plays a crucial role in modulating immune functions, particularly through the modulation of regulatory T cells (Treg) and various cytokines [129]. Dysregulation of immune function and alterations in gut microbiota composition have been implicated in the pathogenesis of AITD. Polymorphisms in genes like CYP26B1 and RAR are associated with HTs severity and Th17 polarization, indicating a potential role for vitamin A in modulating immune responses in this condition [129]. Dysbiosis in HTs affects nutrient absorption and contributes to autoimmunity, suggesting a regulatory role for vitamin A in maintaining gut barrier integrity and potentially influencing HTs pathogenesis [129–131]. In summary, the relationship between vitamin A and thyroid function is complex and multifaceted, involving intricate molecular interactions and physiological pathways. While considerable progress has been made in elucidating the role of vitamin A in thyroid homeostasis, many questions remain unanswered. Future research efforts aimed to find an association between vitamin A, RBP, and TTR, and their impact on the transport and distribution of thyroid hormones in the body and the potential of vitamin A supplementation in patients with thyroid disorders.

4.4. Vitamin B12

Vitamin B12, also known as cobalamin, plays a crucial role in thyroid function and overall metabolic health. It is essential for the synthesis of thyroid hormones and metabolism. Dietary sources rich in vitamin B12 include animal-based foods such as meat, fish, eggs, and dairy products [132]. Vitamin B12 deficiency also contributes to AITD, such as Hashimoto's thyroiditis, as it plays a role in immune function and regulation. A recent study demonstrates that the level of vitamin B12 shows a significant correlation with AITD [133]. Thus, assessing the concentration of vitamin B12 in patients with AITD is essential, as it serves as a diagnostic test with high sensitivity and good specificity [133]. Vitamin B12 also plays a crucial role in addressing specific challenges faced by patients with HTs, like the risk of anemia that may be heightened by coexisting autoimmune conditions like pernicious anemia or atrophic gastritis [134,135]. Research involving 130 patients diagnosed with autoimmune hypothyroidism revealed that 46% of them had vitamin B12 deficiency, which correlated with the presence of the disease [136]. Furthermore, patients with low vitamin B12 levels exhibited significantly higher levels of TPOAb, indicating a negative correlation between vitamin B12 and TPOAb antibodies [136]. Studies involving patients with AIT found that a considerable percentage had serum vitamin B12 levels below the reference range, although there was no significant correlation between B12 and TPOAb [137]. Moreover, a study comparing TgAb- or TPOAb-positive patients with healthy controls revealed a higher frequency of hemoglobin, iron, or vitamin B12 deficiency among patients [137], highlighting the need for further investigation into the risk of vitamin B12 deficiency in HTs patients. However, a recent systematic review and meta-analysis showed no difference between B12 levels in patients with HTs, hyperthyroidism, or subclinical hypothyroidism, but individuals with hypothyroidism had lower B12 levels compared to healthy adults [138]. This suggests that screening for anemia and B12 deficiency in patients with hypothyroidism is of importance. Further studies should focus on the association between vitamin B12 deficiency and AITD and other comorbid conditions.

5. Nutrition, Gut Microbiota, and Thyroid Function

The intestinal microbiota plays a crucial role in regulating metabolism and energy balance by extracting nutrients from food and producing metabolites that can influence host metabolic processes. Several studies have illustrated changes in microbiota composition following dietary modifications. The intestinal bacterial flora plays a crucial role in maintaining metabolic, nutritional, and even immunological homeostasis [139,140]. The hypothesis of a thyroid–gut axis is becoming increasingly concrete, suggesting that the gut microbiota, consisting of trillions of microorganisms inhabiting the gastrointestinal tract, may influence thyroid function through various mechanisms. In addition to influencing the absorption of minerals important for thyroid function (i.e., iodine, selenium, zinc, and iron), the microbiota is involved in the endogenous and exogenous metabolism of thyroid hormones. The presence of iodothyronine deiodinase (i.e., an enzyme that converts thyroxine into its active form, triiodothyronine, or into reverse T3 [rT3]) has been found in the intestinal wall, thus influencing total body T3 levels [2].

Exposure to endocrine disruptors (EDCs) can modify the composition of the gut microbiota, leading to changes in energy metabolism and weight gain. During the first 1000 days, breast milk represents one of the primary sources of EDCs exposure for newborns, as these compounds, metabolized and accumulated in the human body, have been detected in maternal milk, serum, hair, and placenta [3,141,142]. The entero-mammary circle is believed to facilitate the transfer of these compounds from the maternal bloodstream to infants via breast milk. Phthalates, including Di-2-ethylhexyl phthalate (DEHP), Di-isononyl phthalate (DINP), and Bisphenol A (BPA), are among the most common EDCs detected in breast milk [141,143–145]. Maternal exposure to EDCs can be influenced by various factors such as lifestyle, place of residence, occupation, and types of personal products used [141,142,146]. Studies have hypothesized that higher EDCs exposure can stimulate the synthesis of pathogenic microbial species [146,147], which may impact the development of thyroid disorders. A review by our group reported the role of the microbiota in metabolizing EDCs through a bidirectional interaction that leads to dysbiosis and alteration of pathways involved in the development of various metabolic diseases [141]. Dysbiosis and the consequent negative influence on the immune system and intestinal permeability can promote the development of inflammatory and autoimmune diseases, including thyroid diseases.

Diet plays a multifaceted role in the connection between the epigenome, thyroid function, and gut microbiota, exerting both direct and indirect effects on each of these systems [148,149]. Epigenetic modifications can influence the expression of genes involved in thyroid hormone synthesis, metabolism, and signaling pathways, thereby shaping thyroid function and hormone responsiveness [149,150]. The gut microbiota is known to produce metabolites that can influence epigenetic modifications in host cells, thereby exerting indirect effects on thyroid function. For example, short-chain fatty acids produced by gut bacteria have been shown to modulate histone acetylation and DNA methylation, which may impact thyroid hormone metabolism and signaling [151]. The methyl-donor nutrients folate and vitamin B12, which are rich in dark leafy greens, beans, peas, lentils, lemons, bananas, melons, and cereals, e.g., are involved in DNA methylation and chromatin remodeling, a key epigenetic mechanism [152]. It could be hypothesized that potentially consuming a diet rich in methyl-donor nutrients may promote optimal thyroid health. Similarly, dietary polyphenols found in fruits, vegetables, and tea have been shown to modulate histone modifications and can be used as a preventive recommendation for thyroid disorders [153]. Conversely, thyroid hormones can also influence the composition and function of the gut microbiota. Thyroid hormone receptors are expressed in the gut epithelium and immune cells, suggesting a direct role for thyroid hormones in gut physiology and immune function [154].

Supplementation with probiotics has exhibited positive impacts on thyroid hormone levels and thyroid function, as this approach may restore intestinal balance and promote the growth of beneficial microorganisms [155]. However, recent studies indicate the need for further research in the pediatric field regarding the interaction between microbiota and AITD [155]. Educating patients about the associations between gut and thyroid health can promote strategies to improve gut microbiota, especially in pregnant women and infants. Moreover, there is a strong need for further experimental studies to determine the potential benefits of nutritional interventions in HTs.

6. Conclusions

The importance of an adequate iodine intake with the diet is confirmed at all ages, while selenium supplementation must be carried out carefully after having corrected previous iodine insufficiency, as selenium has relationships with the development of autoimmune thyroiditis. Iron, vitamin D, zinc, and other micronutrients also play a role in thyroid function, but the studies currently available are insufficient for definite conclusions and recommendations. The main key points of current studies suggest that:

- iodine is essential for thyroid hormone synthesis and beneficial for the prevention of thyroid disorders;
- selenium may reduce thyroid autoantibodies and improve thyroid function, even though careful monitoring is required;
- hypothyroid individuals should be assessed for anemia, iron, and vitamin B12 deficiency;
- vitamin D supplementation might be beneficial for patients with Hashimoto thyroiditis;
- the evaluation of zinc levels is recommended for subjects with thyroid enlargement; and
- copper and magnesium should be studied for their potential protective role against thyroid cancer, and vitamin A for autoimmune thyroid disease.

Nutrition and gut microbiota can influence thyroid function by modulating immune responses, producing microbial metabolites, affecting nutrient absorption and epigenetic changes regulating thyroid hormone synthesis and metabolism, leading to the development of thyroid disease or protecting against it. Balanced nutrition is an essential source for maintaining healthy, optimal thyroid function and health. The diet–gut–thyroid axis should be considered for the possibility of modulating gut microbiota and thyroid health.

Author Contributions: Conceptualization, M.E.S.; methodology, A.-M.S. and R.R.; validation, M.P., V.P. and B.P.; investigation, A.-M.S., R.R., M.P. and V.P.; resources, A.-M.S. and R.R. data curation, M.P. and V.P.; writing—original draft preparation, A.-M.S., R.R., M.P. and M.E.S.; writing—review and editing, M.E.S., L.I., B.P. and S.E.; visualization, A.-M.S. and R.R.; supervision, M.E.S. and L.I. All authors have read and agreed to the published version of the manuscript.

Funding: This research received no external funding.

Acknowledgments: We sincerely extend our gratitude to the Università degli Studi di Parma and LIFE18 ENV/IT/000460 "Life MILCH mother and infant dyads: Lowering the impact of endocrine disrupting chemicals in milk for a healthy life" project for providing support for the research fellowship to A-M Shulhai.

Conflicts of Interest: The authors declare no conflicts of interest.

References

1. Armstrong, M.; Asuka, E.; Fingeret, A. Physiology, Thyroid Function. In *StatPearls*; [Updated 13 March 2023]; StatPearls Publishing: Treasure Island, FL, USA, 2024. Available online: https://www.ncbi.nlm.nih.gov/books/NBK537039/ (accessed on 12 February 2024).
2. Knezevic, J.; Starchl, C.; Tmava Berisha, A.; Amrein, K. Thyroid-Gut-Axis: How Does the Microbiota Influence Thyroid Function? *Nutrients* **2020**, *12*, 1769. [CrossRef] [PubMed]
3. Street, M.E.; Shulhai, A.-M.; Petraroli, M.; Patianna, V.; Donini, V.; Giudice, A.; Gnocchi, M.; Masetti, M.; Montani, A.G.; Rotondo, R.; et al. The impact of environmental factors and contaminants on thyroid function and disease from fetal to adult life: Current evidence and future directions. *Front. Endocrinol.* **2024**, *15*, 1429884. [CrossRef]

4. Krishnamurthy, H.K.; Reddy, S.; Jayaraman, V.; Krishna, K.; Song, Q.; Rajasekaran, K.E.; Wang, T.; Bei, K.; Rajasekaran, J.J. Effect of Micronutrients on Thyroid Parameters. *J. Thyroid. Res.* **2021**, *2021*, 1865483. [CrossRef]
5. Rayman, M.P. Multiple nutritional factors and thyroid disease, with particular reference to autoimmune thyroid disease. *Proc. Nutr. Soc.* **2019**, *78*, 34–44. [CrossRef] [PubMed]
6. Espinosa-Salas, S.; Gonzalez-Arias, M. Nutrition: Micronutrient Intake, Imbalances, and Interventions. In *StatPearls*; [Updated 21 September 2023]; StatPearls Publishing: Treasure Island, FL, USA, 2024. Available online: https://www.ncbi.nlm.nih.gov/books/NBK597352/ (accessed on 12 February 2024).
7. Sorrenti, S.; Baldini, E.; Pironi, D.; Lauro, A.; D'Orazi, V.; Tartaglia, F.; Tripodi, D.; Lori, E.; Gagliardi, F.; Praticò, M.; et al. Iodine: Its Role in Thyroid Hormone Biosynthesis and Beyond. *Nutrients* **2021**, *13*, 4469. [CrossRef] [PubMed] [PubMed Central]
8. European Food Safety Authority Panel 2014. Scientific Opinion on Dietary Reference Values for Iodine. *EFSA J.* **2014**, *12*, 3660.
9. World Health Organization and Food and Agriculture Organization of the United Nations. *Vitamin and Mineral Requirements in Human Nutrition: Report of a Joint FAO/WHO Expert Consultation*, 2nd ed.; World Health Organization: Geneva, Switzerland, 2004.
10. Iannuzzo, G.; Campanozzi, A.; Trevisani, V.; Rutigliano, I.; Abate, V.; Rendina, D.; De Filippo, G. Iodine Requirements in Pediatrics: From Fetal Life to Adolescence. *Front. Endocrinol.* **2022**, *13*, 929176. [CrossRef]
11. Morreale de Escobar, G.; Obregon, M.J.; Escobar del Rey, F. Role of thyroid hormone during early brain development. *Eur. J. Endocrinol.* **2004**, *151* (Suppl. S3), U25–U37. [CrossRef] [PubMed]
12. Bath, S.C. Thyroid function and iodine intake: Global recommendations and relevant dietary trends. *Nat. Rev. Endocrinol.* **2024**, *20*, 474–486. [CrossRef]
13. Köhrle, J. Selenium, Iodine and Iron-Essential Trace Elements for Thyroid Hormone Synthesis and Metabolism. *Int. J. Mol. Sci.* **2023**, *24*, 3393. [CrossRef]
14. Dineva, M.; Fishpool, H.; Rayman, M.P.; Mendis, J.; Bath, S.C. Systematic review and meta-analysis of the effects of iodine supplementation on thyroid function and child neurodevelopment in mildly-to-moderately iodine-deficient pregnant women. *Am. J. Clin. Nutr.* **2020**, *112*, 389–412. [CrossRef] [PubMed]
15. Hynes, K.L.; Otahal, P.; Hay, I.; Burgess, J.R. Mild iodine deficiency during pregnancy is associated with reduced educational outcomes in the offspring: 9-year follow-up of the gestational iodine cohort. *J. Clin. Endocrinol. Metab.* **2013**, *98*, 1954–1962. [CrossRef] [PubMed]
16. Rodriguez-Diaz, E.; Pearce, E.N. Iodine status and supplementation before, during, and after pregnancy. *Best Pract. Res. Clin. Endocrinol. Metab.* **2020**, *34*, 101430. [CrossRef] [PubMed]
17. Bath, S.C.; Steer, C.D.; Golding, J.; Emmett, P.; Rayman, M.P. Effect of inadequate iodine status in UK pregnant women on cognitive outcomes in their children: Results from the Avon Longitudinal Study of Parents and Children (ALSPAC). *Lancet* **2013**, *382*, 331–337. [CrossRef] [PubMed]
18. Leung, A.; Braverman, L. Consequences of excess iodine. *Nat. Rev. Endocrinol.* **2014**, *10*, 136–142. [CrossRef] [PubMed]
19. World Health Organization. *Guideline: Fortification of Food-Grade Salt with Iodine for the Prevention and Control of Iodine Deficiency Disorders*; World Health Organization: Geneva, Switzerland, 2014. Available online: https://www.ncbi.nlm.nih.gov/books/NBK254243/ (accessed on 12 February 2024).
20. Alexander, E.K.; Pearce, E.N.; Brent, G.A.; Brown, R.S.; Chen, H.; Dosiou, C.; Grobman, W.A.; Laurberg, P.; Lazarus, J.H.; Mandel, S.J.; et al. 2017 Guidelines of the American Thyroid Association for the Diagnosis and Management of Thyroid Disease During Pregnancy and the Postpartum. *Thyroid* **2017**, *27*, 315–389, Erratum in *Thyroid* **2017**, *27*, 1212. [CrossRef] [PubMed]
21. Fallah, R.; Du, L.; Braverman, L.E.; He, X.; Segura-Harrison, M.; Yeh, M.W.; Pearce, E.N.; Chiu, H.K.; Mittelman, S.D.; Leung, A.M. Iodine Nutrition in Weaning Infants in the United States. *Thyroid* **2019**, *29*, 573–576. [CrossRef] [PubMed]
22. Nazeri, P.; Dalili, H.; Mehrabi, Y.; Hedayati, M.; Mirmiran, P.; Azizi, F. Is there any difference between the iodine statuses of breast-fed and formula-fed infants and their mothers in an area with iodine sufficiency? *Br. J. Nutr.* **2018**, *119*, 1012–1018. [CrossRef] [PubMed]
23. Luo, Y.; Kawashima, A.; Ishido, Y.; Yoshihara, A.; Oda, K.; Hiroi, N.; Ito, T.; Ishii, N.; Suzuki, K. Iodine excess as an environmental risk factor for autoimmune thyroid disease. *Int. J. Mol. Sci.* **2014**, *15*, 12895–12912. [CrossRef]
24. Duntas, L.H. The Role of Iodine and Selenium in Autoimmune Thyroiditis. *Horm. Metab. Res.* **2015**, *47*, 721–726. [CrossRef]
25. Teng, W.; Shan, Z.; Teng, X.; Guan, H.; Li, Y.; Teng, D.; Jin, Y.; Yu, X.; Fan, C.; Chong, W.; et al. Effect of iodine intake on thyroid diseases in China. *N. Engl. J. Med.* **2006**, *354*, 2783–2793. [CrossRef] [PubMed]
26. Pedersen, I.B.; Knudsen, N.; Carlé, A.; Vejbjerg, P.; Jørgensen, T.; Perrild, H.; Ovesen, L.; Rasmussen, L.B.; Laurberg, P. A cautious iodization programme bringing iodine intake to a low recommended level is associated with an increase in the prevalence of thyroid autoantibodies in the population. *Clin. Endocrinol.* **2011**, *75*, 120–169. [CrossRef]
27. Iacone, R.; Iaccarino Idelson, P.; Campanozzi, A.; Rutigliano, I.; Russo, O.; Formisano, P.; Galeone, D.; Macchia, P.E.; Strazzullo, P.; MINISAL-GIRCSI Study Group. Relationship Between Salt Consumption and Iodine Intake in a Pediatric Population. *Eur. J. Nutr.* **2021**, *60*, 2193–2202. [CrossRef]
28. Rendina, D.; De Palma, D.; De Filippo, G.; De Pascale, F.; Muscariello, R.; Ippolito, R.; Fazio, V.; Fiengo, A.; Benvenuto, D.; Strazzullo, P.; et al. Prevalence of Simple Nodular Goiter and Hashimoto's Thyroiditis in Current, Previous, and Never Smokers in a Geographical Area with Mild Iodine Deficiency. *Horm. Metab. Res.* **2015**, *47*, 214–219. [CrossRef] [PubMed]
29. Santi, C.; Bagnoli, L. Celebrating Two Centuries of Research in Selenium Chemistry: State of the Art and New Prospective. *Molecules* **2017**, *22*, 2124. [CrossRef]

30. Chada, S.; Whitney, C.; Newburger, P.E. Post-transcriptional regulation of GPx gene expression by selenium in the HL-60 human myeloid cell line. *Blood* **1989**, *74*, 2535–2541. [CrossRef]
31. Flohe, L.; Günzler, W.A.; Schock, H.H. Glutathione peroxidase: A selenoenzyme. *FEBS Lett.* **1973**, *32*, 132–134. [CrossRef] [PubMed]
32. Wildberger, E.; Kohler, H.; Jenzer, H.; Kämpf, J.; Studer, H. Inactivation of peroxidase and glucose oxidase by H_2O_2 and iodide during in vitro thyroglobulin iodination. *Mol. Cell Endocrinol.* **1986**, *46*, 149–154. [CrossRef]
33. Negro, R.; Greco, G.; Mangieri, T.; Pezzarossa, A.; Dazzi, D.; Hassan, H. The influence of selenium supplementation on postpartum thyroid status in pregnant women with thyroid peroxidase autoantibodies. *J. Clin. Endocrinol. Metab.* **2007**, *92*, 1263–1268. [CrossRef]
34. Mao, J.; Pop, V.J.; Bath, S.C.; Vader, H.L.; Redman, C.W.; Rayman, M.P. Effect of low-dose selenium on thyroid autoimmunity and thyroid function in UK pregnant women with mild-to-moderate iodine deficiency. *Eur. J. Nutr.* **2016**, *55*, 55–61. [CrossRef]
35. Chanoine, J.P. Selenium and thyroid function in infants, children and adolescents. *Biofactors* **2003**, *19*, 137–143. [CrossRef] [PubMed]
36. Wang, F.; Li, C.; Li, S.; Cui, L.; Zhao, J.; Liao, L. Selenium and thyroid diseases. *Front. Endocrinol.* **2023**, *14*, 1133000. [CrossRef] [PubMed]
37. van Bakel, M.M.; Printzen, G.; Wermuth, B.; Wiesmann, U.N. Antioxidant and thyroid hormone status in selenium-deficient phenylketonuric and hyperphenylalaninemic patients. *Am. J. Clin. Nutr.* **2000**, *72*, 976–981. [CrossRef] [PubMed]
38. Calomme, M.R.; Vanderpas, J.B.; François, B.; Van Caillie-Bertrand, M.; Herchuelz, A.; Vanovervelt, N.; Van Hoorebeke, C.; Vanden Berghe, D.A. Thyroid function parameters during a selenium repletion/depletion study in phenylketonuric subjects. *Experientia* **1995**, *51*, 1208–1215. [CrossRef] [PubMed]
39. Chanoine, J.P.; Veronikis, I.; Alex, S.; Stone, S.; Fang, S.L.; Leonard, J.L.; Braverman, L.E. The postnatal serum 3,5,3′-triiodothyronine (T3) surge in the rat is largely independent of extrathyroidal 5′-deiodination of thyroxine to T3. *Endocrinology* **1993**, *133*, 2604–2609. [CrossRef] [PubMed]
40. Chanoine, J.P.; Alex, S.; Stone, S.; Fang, S.L.; Veronikis, I.; Leonard, J.L.; Braverman, L.E. Placental 5-deiodinase activity and fetal thyroid hormone economy are unaffected by selenium deficiency in the rat. *Pediatr. Res.* **1993**, *34*, 288–292. [CrossRef] [PubMed]
41. Contempré, B.; Duale, N.L.; Dumont, J.E.; Ngo, B.; Diplock, A.T.; Vanderpas, J. Effect of selenium supplementation on thyroid hormone metabolism in an iodine and selenium deficient population. *Clin. Endocrinol.* **1992**, *36*, 579–583. [CrossRef] [PubMed]
42. Wu, Q.; Wang, Y.; Chen, P.; Wei, J.; Lv, H.; Wang, S.; Wu, Y.; Zhao, X.; Peng, X.; Rijntjes, E.; et al. Increased Incidence of Hashimoto Thyroiditis in Selenium Deficiency: A Prospective 6-Year Cohort Study. *J. Clin. Endocrinol. Metab.* **2022**, *107*, e3603–e3611. [CrossRef]
43. Wichman, J.; Winther, K.H.; Bonnema, S.J.; Hegedüs, L. Selenium Supplementation Significantly Reduces Thyroid Autoantibody Levels in Patients with Chronic Autoimmune Thyroiditis: A Systematic Review and Meta-Analysis. *Thyroid* **2016**, *26*, 1681–1692. [CrossRef]
44. Winther, K.H.; Wichman, J.E.; Bonnema, S.J.; Hegedüs, L. Insufficient documentation for clinical efficacy of selenium supplementation in chronic autoimmune thyroiditis, based on a systematic review and meta-analysis. *Endocrine* **2017**, *55*, 376–385. [CrossRef]
45. Huwiler, V.V.; Maissen-Abgottspon, S.; Stanga, Z.; Mühlebach, S.; Trepp, R.; Bally, L.; Bano, A. Selenium Supplementation in Patients with Hashimoto Thyroiditis: A Systematic Review and Meta-Analysis of Randomized Clinical Trials. *Thyroid* **2024**, *34*, 295–313. [CrossRef] [PubMed] [PubMed Central]
46. Kong, X.Q.; Qiu, G.Y.; Yang, Z.B.; Tan, Z.X.; Quan, X.Q. Clinical efficacy of selenium supplementation in patients with Hashimoto thyroiditis: A systematic review and meta-analysis. *Medicine* **2023**, *102*, e33791. [CrossRef] [PubMed] [PubMed Central]
47. Godlewska, M.; Banga, P.J. Thyroid peroxidase as a dual active site enzyme: Focus on biosynthesis, hormonogenesis and thyroid disorders of autoimmunity and cancer. *Biochimie* **2019**, *160*, 34–45. [CrossRef]
48. Moreno-Reyes, R.; Corvilain, B.; Daelemans, C.; Wolff, F.; Fuentes Peña, C.; Vandevijvere, S. Iron Deficiency Is a Risk Factor for Thyroid Dysfunction During Pregnancy: A Population-Based Study in Belgium. *Thyroid* **2021**, *31*, 1868–1877. [CrossRef] [PubMed]
49. Brigham, D.E.; Beard, J.L. Effect of thyroid hormone replacement in iron-deficient rats. *Am. J. Physiol.* **1995**, *269 Pt 2*, R1140–R1147. [CrossRef] [PubMed]
50. Maldonado-Araque, C.; Valdés, S.; Lago-Sampedro, A.; Lillo-Muñoz, J.A.; Garcia-Fuentes, E.; Perez-Valero, V.; Gutierrez-Repiso, C.; Goday, A.; Urrutia, I.; Peláez, L.; et al. Iron deficiency is associated with Hypothyroxinemia and Hypotriiodothyroninemia in the Spanish general adult population: Di@bet.es study. *Sci. Rep.* **2018**, *8*, 6571. [CrossRef] [PubMed]
51. Zimmermann, M.B.; Burgi, H.; Hurrell, R.F. Iron deficiency predicts poor maternal thyroid status during pregnancy. *J. Clin. Endocrinol. Metab.* **2007**, *92*, 3436–3440. [CrossRef] [PubMed]
52. Li, S.; Gao, X.; Wei, Y.; Zhu, G.; Yang, C. The Relationship between Iron Deficiency and Thyroid Function in Chinese Women during Early Pregnancy. *J. Nutr. Sci. Vitaminol.* **2016**, *62*, 397–401. [CrossRef] [PubMed]
53. Cao, C.; O'Brien, K.O. Pregnancy and iron homeostasis: An update. *Nutr. Rev.* **2013**, *71*, 35–51. [CrossRef] [PubMed]
54. Luo, J.; Wang, X.; Yuan, L.; Guo, L. Iron Deficiency, a Risk Factor of Thyroid Disorders in Reproductive-Age and Pregnant Women: A Systematic Review and Meta-Analysis. *Front. Endocrinol.* **2021**, *12*, 629831. [CrossRef]

55. van Gucht, A.L.M.; Meima, M.E.; Moran, C.; Agostini, M.; Tylki-Szymanska, A.; Krajewska, M.W.; Chrzanowska, K.; Efthymiadou, A.; Chrysis, D.; Demir, K.; et al. Anemia in Patients with Resistance to Thyroid Hormone α: A Role for Thyroid Hormone Receptor α in Human Erythropoiesis. *J. Clin. Endocrinol. Metab.* **2017**, *102*, 3517–3525. [CrossRef] [PubMed]
56. Pastori, V.; Pozzi, S.; Labedz, A.; Ahmed, S.; Ronchi, A.E. Role of Nuclear Receptors in Controlling Erythropoiesis. *Int. J. Mol. Sci.* **2022**, *23*, 2800. [CrossRef] [PubMed]
57. Zimmermann, M.B.; Köhrle, J. The impact of iron and selenium deficiencies on iodine and thyroid metabolism: Biochemistry and relevance to public health. *Thyroid* **2002**, *12*, 867–878. [CrossRef] [PubMed]
58. Fujisawa, H.; Korwutthikulrangsri, M.; Fu, J.; Liao, X.H.; Dumitrescu, A.M. Role of the Thyroid Gland in Expression of the Thyroid Phenotype of Sbp2-Deficient Mice. *Endocrinology* **2020**, *161*, bqz032. [CrossRef]
59. Koç, Ş.; Güngör, K.; Güngör, N.D.; Uzunlulu, M. Iron Deficiency in Women with Thyroid-Specific Autoantibodies: A Case Control Study. *J. Exp. Clin. Med.* **2022**, *39*, 194–198. [CrossRef]
60. Ravanbod, M.; Asadipooya, K.; Kalantarhormozi, M.; Nabipour, I.; Omrani, G.R. Treatment of iron-deficiency anemia in patients with subclinical hypothyroidism. *Am. J. Med.* **2013**, *126*, 420–424. [CrossRef] [PubMed]
61. Czarnywojtek, A.; Florek, E.; Pietrończyk, K.; Sawicka-Gutaj, N.; Ruchała, M.; Ronen, O.; Nixon, I.J.; Shaha, A.R.; Rodrigo, J.P.; Tufano, R.P.; et al. The Role of Vitamin D in Autoimmune Thyroid Diseases: A Narrative Review. *J. Clin. Med.* **2023**, *12*, 1452. [CrossRef] [PubMed]
62. Bikle, D. Nonclassic actions of vitamin D. *J. Clin. Endocrinol. Metab.* **2009**, *94*, 26–34. [CrossRef] [PubMed]
63. Ginde, A.A.; Sullivan, A.F.; Mansbach, J.M.; Camargo, C.A., Jr. Vitamin D insufficiency in pregnant and nonpregnant women of childbearing age in the United States. *Am. J. Obstet. Gynecol.* **2010**, *202*, 436.e1–436.e8. [CrossRef]
64. van der Meer, I.M.; Karamali, N.S.; Boeke, A.J.; Lips, P.; Middelkoop, B.J.; Verhoeven, I.; Wuister, J.D. High prevalence of vitamin D deficiency in pregnant non-Western women in The Hague, Netherlands. *Am. J. Clin. Nutr.* **2006**, *84*, 350–353, quiz 468–469. [CrossRef]
65. Mulligan, M.L.; Felton, S.K.; Riek, A.E.; Bernal-Mizrachi, C. Implications of vitamin D deficiency in pregnancy and lactation. *Am. J. Obstet. Gynecol.* **2010**, *202*, 429.e1–429.e9. [CrossRef] [PubMed]
66. Smyth, P.P.; Wijeyaratne, C.N.; Kaluarachi, W.N.; Smith, D.F.; Premawardhana, L.D.; Parkes, A.B.; Jayasinghe, A.; de Silva, D.G.; Lazarus, J.H. Sequential studies on thyroid antibodies during pregnancy. *Thyroid* **2005**, *15*, 474–477. [CrossRef]
67. Panesar, N.S.; Chan, K.W.; Li, C.Y.; Rogers, M.S. Status of anti-thyroid peroxidase during normal pregnancy and in patients with hyperemesis gravidarum. *Thyroid* **2006**, *16*, 481–484. [CrossRef] [PubMed]
68. Evliyaoğlu, O.; Acar, M.; Özcabı, B.; Erginöz, E.; Bucak, F.; Ercan, O.; Kucur, M. Vitamin D Deficiency and Hashimoto's Thyroiditis in Children and Adolescents: A Critical Vitamin D Level for This Association? *J. Clin. Res. Pediatr. Endocrinol.* **2015**, *7*, 128–133. [CrossRef] [PubMed]
69. Cvek, M.; Kaličanin, D.; Barić, A.; Vuletić, M.; Gunjača, I.; Torlak Lovrić, V.; Škrabić, V.; Punda, A.; Boraska Perica, V. Vitamin D and Hashimoto's Thyroiditis: Observations from CROHT Biobank. *Nutrients* **2021**, *13*, 2793. [CrossRef]
70. Hanna, H.W.Z.; Rizzo, C.; Abdel Halim, R.M.; El Haddad, H.E.; Salam, R.; El-Sayed Abou-Youssef, H. Vitamin D status in Hashimoto's thyroiditis and its association with vitamin D receptor genetic variants. *J. Steroid Biochem. Mol. Biol.* **2021**, *212*, 105922. [CrossRef]
71. Filipova, L.; Lazurova, Z.; Fulop, P.; Lazurova, I. Vitamin D insufficiency is not associated with thyroid autoimmunity in Slovak women with Hashimoto's disease. *Bratisl. Lek. Listy* **2023**, *124*, 182–186. [CrossRef] [PubMed]
72. Metwalley, K.A.; Farghaly, H.S.; Sherief, T.; Hussein, A. Vitamin D status in children and adolescents with autoimmune thyroiditis. *J. Endocrinol. Investig.* **2016**, *39*, 793–797. [CrossRef]
73. Mazokopakis, E.E.; Papadomanolaki, M.G.; Tsekouras, K.C.; Evangelopoulos, A.D.; Kotsiris, D.A.; Tzortzinis, A.A. Is vitamin D related to pathogenesis and treatment of Hashimoto's thyroiditis? *Hell. J. Nucl. Med.* **2015**, *18*, 222–227.
74. Chaudhary, S.; Dutta, D.; Kumar, M.; Saha, S.; Mondal, S.A.; Kumar, A.; Mukhopadhyay, S. Vitamin D supplementation reduces thyroid peroxidase antibody levels in patients with autoimmune thyroid disease: An open-labeled randomized controlled trial. *Indian. J. Endocrinol. Metab.* **2016**, *20*, 391–398.
75. Taheriniya, S.; Arab, A.; Hadi, A.; Fadel, A.; Askari, G. Vitamin D and thyroid disorders: A systematic review and Meta-analysis of observational studies. *BMC Endocr. Disord.* **2021**, *21*, 171. [CrossRef] [PubMed] [PubMed Central]
76. Bizzaro, G.; Shoenfeld, Y. Vitamin D and thyroid autoimmune diseases: The known and the obscure. *Immunol. Res.* **2015**, *61*, 107–109. [CrossRef] [PubMed]
77. Bui, L.; Zhu, Z.; Hawkins, S.; Cortez-Resendiz, A.; Bellon, A. Vitamin D regulation of the immune system and its implications for COVID-19: A mini review. *SAGE Open Med.* **2021**, *9*, 20503121211014073. [CrossRef] [PubMed]
78. Palanca, A.; Ampudia-Blasco, F.J.; Real, J.T. The Controversial Role of Vitamin D in Thyroid Cancer Prevention. *Nutrients* **2022**, *14*, 2593. [CrossRef] [PubMed]
79. Hu, Y.; Xue, C.; Ren, S.; Dong, L.; Gao, J.; Li, X. Association between vitamin D status and thyroid cancer: A meta-analysis. *Front. Nutr.* **2024**, *11*, 1423305. [CrossRef] [PubMed] [PubMed Central]
80. Lebiedziński, F.; Lisowska, K.A. Impact of Vitamin D on Immunopathology of Hashimoto's Thyroiditis: From Theory to Practice. *Nutrients* **2023**, *15*, 3174. [CrossRef]
81. Tang, J.; Shan, S.; Li, F.; Yun, P. Effects of vitamin D supplementation on autoantibodies and thyroid function in patients with Hashimoto's thyroiditis: A systematic review and meta-analysis. *Medicine* **2023**, *102*, e36759. [CrossRef] [PubMed] [PubMed Central]

82. Costa, M.I.; Sarmento-Ribeiro, A.B.; Gonçalves, A.C. Zinc: From Biological Functions to Therapeutic Potential. *Int. J. Mol. Sci.* **2023**, *24*, 4822. [CrossRef] [PubMed]
83. O'Kane, S.M.; Mulhern, M.S.; Pourshahidi, L.K.; Strain, J.J.; Yeates, A.J. Micronutrients, iodine status and concentrations of thyroid hormones: A systematic review. *Nutr. Rev.* **2018**, *76*, 418–431. [CrossRef] [PubMed]
84. Severo, J.S.; Morais, J.B.S.; de Freitas, T.E.C.; Andrade, A.L.P.; Feitosa, M.M.; Fontenelle, L.C.; de Oliveira, A.R.S.; Cruz, K.J.C.; do Nascimento Marreiro, D. The Role of Zinc in Thyroid Hormones Metabolism. *Int. J. Vitam. Nutr. Res.* **2019**, *89*, 80–88. [CrossRef]
85. Prasad, A.S.; Bao, B. Molecular mechanisms of zinc as a pro-antioxidant mediator: Clinical therapeutic implications. *Antioxidants* **2019**, *8*, 164. [CrossRef] [PubMed]
86. Maywald, M.; Rink, L. Zinc homeostasis and immunosenescence. *J. Trace Elem. Med. Biol.* **2015**, *29*, 24–30. [CrossRef] [PubMed]
87. Maywald, M.; Wang, F.; Rink, L. The intracellular free zinc level is vital for treg function and a feasible tool to discriminate between treg and activated th cells. *Int. J. Mol. Sci.* **2018**, *19*, 3575. [CrossRef] [PubMed]
88. Kravchenko, V.; Zakharchenko, T. Thyroid hormones and minerals in immunocorrection of disorders in autoimmune thyroid diseases. *Front. Endocrinol.* **2023**, *14*, 1225494. [CrossRef] [PubMed]
89. Zhou, Q.; Xue, S.; Zhang, L.; Chen, G. Trace elements and the thyroid. *Front. Endocrinol.* **2022**, *13*, 904889. [CrossRef] [PubMed]
90. Betsy, A.; Binitha, M.; Sarita, S. Zinc deficiency associated with hypothyroidism: An overlooked cause of severe alopecia. *Int. J. Trichology* **2013**, *5*, 40–42. [CrossRef] [PubMed]
91. Ertek, S.; Cicero, A.F.; Caglar, O.; Erdogan, G. Relationship between serum zinc levels, thyroid hormones and thyroid volume following successful iodine supplementation. *Hormones* **2010**, *9*, 263–268. [CrossRef]
92. Rezaei, M.; Javadmoosavi, S.Y.; Mansouri, B.; Azadi, N.A.; Mehrpour, O.; Nakhaee, S. Thyroid dysfunction: How concentration of toxic and essential elements contribute to risk of hypothyroidism, hyperthyroidism, and thyroid cancer. *Environ. Sci. Pollut. Res.* **2019**, *26*, 35787–35796. [CrossRef] [PubMed]
93. Regmi, D.; Gautam, N.; Shahi, A.; Bohara, S.; Subedi, S.; Jayan, A. Status of Serum Zinc Level in Hypothyroid Patients with Normal Serum Albumin Level: A Case Control Study. *J. Univers. Coll. Med. Sci.* **2019**, *7*, 34–38. [CrossRef]
94. Talebi, S.; Ghaedi, E.; Sadeghi, E.; Mohammadi, H.; Hadi, A.; Clark, C.C.T.; Askari, G. Trace Element Status and Hypothyroidism: A Systematic Review and Meta-analysis. *Biol. Trace Elem. Res.* **2020**, *197*, 1–14. [CrossRef]
95. Chang, H.C.; Chang, Y.S. Association between serum zinc levels and androgenetic alopecia: A systematic review and meta-analysis. *J. Cosmet. Dermatol.* **2022**, *21*, 1774–1777. [CrossRef] [PubMed]
96. Jin, W.; Zheng, H.; Shan, B.; Wu, Y. Changes of serum trace elements level in patients with alopecia areata: A meta-analysis. *J. Dermatol.* **2017**, *44*, 588–591. [CrossRef] [PubMed]
97. Manisha, A.; Roshan, K.M.; Sudeep, K.; Imran, M.; Sumesh, P.S. Study of Trace Elements in Patients of Hypothyroidism with Special Reference to Zinc and Copper. *Biomed. J. Sci. Tech. Res.* **2018**, *6*, 5190–5194. [CrossRef]
98. Zhang, C.; Wu, H.B.; Cheng, M.X.; Wang, L.; Gao, C.B.; Huang, F. Association of exposure to multiple metals with papillary thyroid cancer risk in China. *Environ. Sci. Pollut. Res. Int.* **2019**, *26*, 20560–20572. [CrossRef] [PubMed]
99. Baltaci, A.K.; Dundar, T.K.; Aksoy, F.; Mogulkoc, R. Changes in the serum levels of trace elements before and after the operation in thyroid cancer patients. *Biol. Trace Elem. Res.* **2017**, *175*, 57–64. [CrossRef] [PubMed]
100. Shen, J.; Zhang, H.; Jiang, H.; Lin, H.; He, J.; Fan, S.; Yu, D.; Yang, L.; Tang, H.; Lin, E.; et al. The effect of micronutrient on thyroid cancer risk: A Mendelian randomization study. *Front. Nutr.* **2024**, *11*, 1331172. [CrossRef]
101. El-Fadeli, S.; Bouhouch, S.; Skalny, A.V.; Barkouch, Y.; Pineau, A.; Cherkaoui, M.; Sedki, A. Effects of Imbalance in Trace Element on Thyroid Gland from Moroccan Children. *Biol. Trace Elem. Res.* **2016**, *170*, 288–293. [CrossRef]
102. Turan, E.; Turksoy, V.A. Selenium, Zinc, and Copper Status in Euthyroid Nodular Goiter: A Cross-Sectional Study. *Int. J. Prev. Med.* **2021**, *12*, 46. [CrossRef]
103. Szczepanik, J.; Podgórski, T.; Domaszewska, K. The Level of Zinc, Copper and Antioxidant Status in the Blood Serum of Women with Hashimoto's Thyroiditis. *Int. J. Environ. Res. Public Health* **2021**, *18*, 7805. [CrossRef]
104. Sivakumar, R.R.; Chinnaiah Govindareddy, D.; Sahoo, J.; Bobby, Z.; Chinnakali, P. Effect of daily zinc supplementation for 12 weeks on serum thyroid auto-antibody levels in children and adolescents with autoimmune thyroiditis—A randomized controlled trial. *J. Pediatr. Endocrinol. Metab.* **2024**, *37*, 137–143. [CrossRef]
105. Lu, L.; Huang, Z.; Wang, X.; Chen, J. Interaction Between Dietary Selenium and Zinc Intakes on Hypothyroidism. *Biol. Trace Elem. Res.* **2023**, *201*, 4667–4676. [CrossRef] [PubMed]
106. Wróblewski, M.; Wróblewska, J.; Nuszkiewicz, J.; Pawłowska, M.; Wesołowski, R.; Woźniak, A. The Role of Selected Trace Elements in Oxidoreductive Homeostasis in Patients with Thyroid Diseases. *Int. J. Mol. Sci.* **2023**, *24*, 4840. [CrossRef] [PubMed]
107. Georgieva Bacelova, M.; Dimitrova Gatseva, P.; Ivanova Deneva, T.; Miteva Davcheva, D.; Veselinova Bivolarska, A. Are the elements zinc, copper, magnesium, and rubidium related to nutrition and iodine deficiency in pregnant Bulgarian women from iodine deficient region? *Cent. Eur. J. Public Health* **2024**, *32*, 31–38. [CrossRef]
108. Jain, R.B. Thyroid Function and Serum Copper, Selenium, and Zinc in General U.S. Population. *Biol. Trace Elem. Res.* **2014**, *159*, 87–98. [CrossRef] [PubMed]
109. Rasic-Milutinovic, Z.; Jovanovic, D.; Bogdanovic, G.; Trifunovic, J.; Mutic, J. Potential Influence of Selenium, Copper, Zinc and Cadmium on L-Thyroxine Substitution in Patients with Hashimoto Thyroiditis and Hypothyroidism. *Exp. Clin. Endocrinol. Diabetes* **2017**, *125*, 79–85. [CrossRef] [PubMed]

110. Shen, F.; Cai, W.S.; Li, J.L.; Feng, Z.; Cao, J.; Xu, B. The Association Between Serum Levels of Selenium, Copper, and Magnesium with Thyroid Cancer: A Meta-analysis. *Biol. Trace Elem. Res.* **2015**, *167*, 225–235. [CrossRef] [PubMed]
111. Brady, D.C.; Crowe, M.S.; Greenberg, D.N.; Counter, C.M. Copper Chelation Inhibits BRAFV600E-Driven Melanomagenesis and Counters Resistance to BRAFV600E and MEK1/2 Inhibitors. *Cancer Res.* **2017**, *77*, 6240–6252. [CrossRef]
112. Kolanu, B.R.; Vadakedath, S.; Boddula, V.; Kandi, V. Activities of Serum Magnesium and Thyroid Hormones in Pre-, Peri-, and Post-menopausal Women. *Cureus* **2020**, *12*, e6554. [CrossRef] [PubMed]
113. Moncayo, R.; Moncayo, H. The WOMED model of benign thyroid disease: Acquired magnesium deficiency due to physical and psychological stressors relates to dysfunction of oxidative phosphorylation. *BBA Clin.* **2014**, *3*, 44–64. [CrossRef]
114. Ige, A.O.; Chidi, R.N.; Egbeluya, E.E.; Jubreel, R.O.; Adele, B.O.; Adewoye, E.O. Amelioration of thyroid dysfunction by magnesium in experimental diabetes may also prevent diabetes-induced renal impairment. *Heliyon* **2019**, *5*, e01660. [CrossRef]
115. Yanko, R.V.; Chaka, E.G.; Levashov, M.I. Histomorphological changes in the thyroid gland ofrats after magnesium chloride ingestion. *Clin. Exp. Morphology.* **2019**, *8*, 41–47. [CrossRef]
116. Wang, K.; Wei, H.; Zhang, W.; Li, Z.; Ding, L.; Yu, T.; Tan, L.; Liu, Y.; Liu, T.; Wang, H.; et al. Severely low serum magnesium is associated with increased risks of positive anti-thyroglobulin antibody and hypothyroidism: A cross-sectional study. *Sci. Rep.* **2018**, *8*, 9904. [CrossRef] [PubMed]
117. Xu, H.; Hu, X.; Li, J.; Nie, Z.; Kang, S.; Liu, H.; Wang, Y.; Jia, X.; Lyu, Z. The Inverse Association of Serum Magnesium with Papillary Thyroid Cancer in Thyroid Nodules: A Cross-Sectional Survey Based on Thyroidectomy Population. *Biol. Trace Elem. Res.* **2023**, *201*, 3279–3289. [CrossRef]
118. Chandra, A.K.; Goswami, H.; Sengupta, P. Effects of magnesium on cytomorphology and enzyme activities in thyroid of rats. *Indian. J. Exp. Biol.* **2014**, *52*, 787–792. [PubMed]
119. Carazo, A.; Macáková, K.; Matoušová, K.; Krčmová, L.K.; Protti, M.; Mladěnka, P. Vitamin A Update: Forms, Sources, Kinetics, Detection, Function, Deficiency, Therapeutic Use and Toxicity. *Nutrients* **2021**, *13*, 1703. [CrossRef]
120. Brossaud, J.; Pallet, V.; Corcuff, J.B. Vitamin A, endocrine tissues and hormones: Interplay and interactions. *Endocr. Connect.* **2017**, *6*, R121–R130. [CrossRef]
121. Grignard, E.; Håkansson, H.; Munn, S. Regulatory needs and activities to address the retinoid system in the context of endocrine disruption: The European viewpoint. *Reprod. Toxicol.* **2020**, *93*, 250–258. [CrossRef] [PubMed]
122. Biebinger, R.; Arnold, M.; Koss, M.; Kloeckener-Gruissem, B.; Langhans, W.; Hurrell, R.F.; Zimmermann, M.B. Effect of concurrent vitamin A and iodine deficiencies on the thyroid-pituitary axis in rats. *Thyroid* **2006**, *16*, 961–965. [CrossRef]
123. Wolf, G. The regulation of the thyroid-stimulating hormone of the anterior pituitary gland by thyroid hormone and by 9-cis-retinoic acid. *Nutr. Rev.* **2002**, *60*, 374–377. [CrossRef]
124. Zimmermann, M.B. Interactions of vitamin A and iodine deficiencies: Effects on the pituitary-thyroid axis. *Int. J. Vitam. Nutr. Res.* **2007**, *77*, 236–240. [CrossRef]
125. van Bennekum, A.M.; Wei, S.; Gamble, M.V.; Vogel, S.; Piantedosi, R.; Gottesman, M.; Episkopou, V.; Blaner, W.S. Biochemical basis for depressed serum retinol levels in transthyretin-deficient mice. *J. Biol. Chem.* **2001**, *276*, 1107–1113. [CrossRef] [PubMed]
126. Ravaglia, G.; Forti, P.; Maioli, F.; Nesi, B.; Pratelli, L.; Savarino, L.; Cucinotta, D.; Cavalli, G. Blood micronutrient and thyroid hormone concentrations in the oldest-old. *J. Clin. Endocrinol. Metab.* **2000**, *85*, 2260–2265. [CrossRef] [PubMed]
127. Cooper, D.S.; Biondi, B. Subclinical thyroid disease. *Lancet* **2012**, *379*, 1142–1154. [CrossRef] [PubMed]
128. Homma, H.; Watanabe, M.; Inoue, N.; Isono, M.; Hidaka, Y.; Iwatani, Y. Polymorphisms in Vitamin A-Related Genes and Their Functions in Autoimmune Thyroid Disease. *Thyroid* **2021**, *31*, 1749–1756. [CrossRef] [PubMed]
129. Virili, C.; Fallahi, P.; Antonelli, A.; Benvenga, S.; Centanni, M. Gut microbiota and Hashimoto's thyroiditis. *Rev. Endocr. Metab. Disord.* **2018**, *19*, 293–300. [CrossRef] [PubMed]
130. Brown, E.M.; Kenny, D.J.; Xavier, R.J. Gut Microbiota Regulation of T Cells During Inflammation and Autoimmunity. *Annu. Rev. Immunol.* **2019**, *37*, 599–624. [CrossRef] [PubMed]
131. Obeid, R.; Heil, S.G.; Verhoeven, M.M.A.; van den Heuvel, E.G.H.M.; de Groot, L.C.P.G.M.; Eussen, S.J.P.M. Vitamin B12 Intake From Animal Foods, Biomarkers, and Health Aspects. *Front. Nutr.* **2019**, *6*, 93. [CrossRef] [PubMed]
132. Kacharava, T.; Giorgadze, E.; Janjgava, S.; Lomtadze, N.; Taboridze, I. Correlation Between Vitamin B12 Deficiency and Autoimmune Thyroid Diseases. *Endocr. Metab. Immune Disord. Drug Targets* **2023**, *23*, 86–94. [CrossRef] [PubMed]
133. Szczepanek-Parulska, E.; Hernik, A.; Ruchała, M. Anemia in Thyroid Diseases. *Pol. Arch. Intern. Med.* **2017**, *127*, 352–360. [CrossRef]
134. Aktaş, H.Ş. Vitamin B12 and Vitamin D Levels in Patients with Autoimmune Hypothyroidism and Their Correlation with Anti-Thyroid Peroxidase Antibodies. *Med. Princ. Pract.* **2020**, *29*, 364–370. [CrossRef]
135. Jaya Kumari, S.; Bantwal, G.; Devanath, A.; Aiyyar, V.; Patil, M. Evaluation of serum vitamin B12 levels and its correlation with anti-thyroperoxidase antibody in patients with autoimmune thyroid disorders. *Indian. J. Clin. Biochem.* **2015**, *30*, 217–220. [CrossRef] [PubMed]
136. Wang, Y.P.; Lin, H.P.; Chen, H.M.; Kuo, Y.S.; Lang, M.J.; Sun, A. Hemoglobin, iron, and vitamin B12 deficiencies and high blood homocysteine levels in patients with anti-thyroid autoantibodies. *J. Formos. Med. Assoc.* **2014**, *113*, 155–160. [CrossRef] [PubMed]
137. Mikulska, A.A.; Karaźniewicz-Łada, M.; Filipowicz, D.; Ruchała, M.; Główka, F.K. Metabolic Characteristics of Hashimoto's Thyroiditis Patients and the Role of Microelements and Diet in the Disease Management-An Overview. *Int. J. Mol. Sci.* **2022**, *23*, 6580. [CrossRef] [PubMed]

138. Benites-Zapata, V.A.; Ignacio-Cconchoy, F.L.; Ulloque-Badaracco, J.R.; Hernandez-Bustamante, E.A.; Alarcón-Braga, E.A.; Al-Kassab-Córdova, A.; Herrera-Añazco, P. Vitamin B12 levels in thyroid disorders: A systematic review and meta-analysis. *Front. Endocrinol.* **2023**, *14*, 1070592. [CrossRef] [PubMed] [PubMed Central]
139. Petraroli, M.; Castellone, E.; Patianna, V.; Esposito, S. Gut Microbiota and Obesity in Adults and Children: The State of the Art. *Front. Pediatr.* **2021**, *9*, 657020. [CrossRef] [PubMed]
140. Raspini, B.; Porri, D.; De Giuseppe, R.; Chieppa, M.; Liso, M.; Cerbo, R.M.; Civardi, E.; Garofoli, F.; Monti, M.C.; Vacca, M.; et al. Prenatal and postnatal determinants in shaping offspring's microbiome in the first 1000 days: Study protocol and preliminary results at one month of life. *Ital. J. Pediatr.* **2020**, *46*, 45. [CrossRef] [PubMed]
141. Street, M.E.; Shulhai, A.M.; Rotondo, R.; Giannì, G.; Caffarelli, C. Current knowledge on the effects of environmental contaminants in early life nutrition. *Front. Nutr.* **2023**, *10*, 1120293. [CrossRef]
142. Biagi, E.; Quercia, S.; Aceti, A.; Beghetti, I.; Rampelli, S.; Turroni, S.; Faldella, G.; Candela, M.; Brigidi, P.; Corvaglia, L. The Bacterial Ecosystem of Mother's Milk and Infant's Mouth and Gut. *Front. Microbiol.* **2017**, *8*, 1214. [CrossRef] [PubMed]
143. Ercan, O.; Tarcin, G. Overview on Endocrine disruptors in food and their effects on infant's health. *Glob. Pediatrics* **2022**, *2* 100019. [CrossRef]
144. Mendonca, K.; Hauser, R.; Calafat, A.M.; Arbuckle, T.E.; Duty, S.M. Bisphenol A concentrations in maternal breast milk and infant urine. *Int. Arch. Occup. Environ. Health* **2014**, *87*, 13–20. [CrossRef]
145. Witczak, A.; Pohoryło, A.; Abdel-Gawad, H. Endocrine-Disrupting Organochlorine Pesticides in Human Breast Milk: Changes during Lactation. *Nutrients* **2021**, *13*, 229. [CrossRef] [PubMed]
146. Di Napoli, I.; Tagliaferri, S.; Sommella, E.; Salviati, E.; Porri, D.; Raspini, B.; Cena, H.; Campiglia, P.; La Rocca, C.; Cerbo, R.M.; et al. Lifestyle Habits and Exposure to BPA and Phthalates in Women of Childbearing Age from Northern Italy: A Pilot Study. *Int. J. Environ. Res. Public Health* **2021**, *18*, 9710. [CrossRef] [PubMed]
147. Vacca, M.; Calabrese, F.M.; Loperfido, F.; Maccarini, B.; Cerbo, R.M.; Sommella, E.; Salviati, E.; Voto, L.; De Angelis, M.; Ceccarelli, G.; et al. Maternal Exposure to Endocrine-Disrupting Chemicals: Analysis of Their Impact on Infant Gut Microbiota Composition. *Biomedicines* **2024**, *12*, 234. [CrossRef] [PubMed]
148. Virili, C.; Stramazzo, I.; Bagaglini, M.F.; Carretti, A.L.; Capriello, S.; Romanelli, F.; Trimboli, P.; Centanni, M. The relationship between thyroid and human-associated microbiota: A systematic review of reviews. *Rev. Endocr. Metab. Disord.* **2024**, *25*, 215–237. [CrossRef] [PubMed]
149. Zhou, F.; Wang, X.; Wang, L.; Sun, X.; Tan, G.; Wei, W.; Zheng, G.; Ma, X.; Tian, D.; Yu, H. Genetics, Epigenetics, Cellular Immunology, and Gut Microbiota: Emerging Links with Graves' Disease. *Front. Cell Dev. Biol.* **2022**, *9*, 794912. [CrossRef] [PubMed]
150. Sibuh, B.Z.; Quazi, S.; Panday, H.; Parashar, R.; Jha, N.K.; Mathur, R.; Jha, S.K.; Taneja, P.; Jha, A.K. The Emerging Role of Epigenetics in Metabolism and Endocrinology. *Biology* **2023**, *12*, 256. [CrossRef] [PubMed]
151. Mendoza-León, M.J.; Mangalam, A.K.; Regaldiz, A.; González-Madrid, E.; Rangel-Ramírez, M.A.; Álvarez-Mardonez, O.; Vallejos, O.P.; Méndez, C.; Bueno, S.M.; Melo-González, F.; et al. Gut microbiota short-chain fatty acids and their impact on the host thyroid function and diseases. *Front. Endocrinol.* **2023**, *14*, 1192216. [CrossRef] [PubMed]
152. Poursalehi, D.; Lotfi, K.; Mirzaei, S.; Asadi, A.; Akhlaghi, M.; Saneei, P. Association between methyl donor nutrients and metabolic health status in overweight and obese adolescents. *Sci. Rep.* **2022**, *12*, 17045. [CrossRef]
153. Habza-Kowalska, E.; Kaczor, A.A.; Bartuzi, D.; Piłat, J.; Gawlik-Dziki, U. Some Dietary Phenolic Compounds Can Activate Thyroid Peroxidase and Inhibit Lipoxygenase-Preliminary Study in the Model Systems. *Int. J. Mol. Sci.* **2021**, *22*, 5108. [CrossRef]
154. Jiang, W.; Lu, G.; Gao, D.; Lv, Z.; Li, D. The relationships between the gut microbiota and its metabolites with thyroid diseases. *Front. Endocrinol.* **2022**, *13*, 943408. [CrossRef]
155. Calcaterra, V.; Mameli, C.; Rossi, V.; Magenes, V.C.; Massini, G.; Perazzi, C.; Verduci, E.; Zuccotti, G. What we know about the relationship between autoimmune thyroid diseases and gut microbiota: A perspective on the role of probiotics on pediatric endocrinology. *Minerva Pediatr.* **2022**, *74*, 650–671. [CrossRef] [PubMed]

Disclaimer/Publisher's Note: The statements, opinions and data contained in all publications are solely those of the individual author(s) and contributor(s) and not of MDPI and/or the editor(s). MDPI and/or the editor(s) disclaim responsibility for any injury to people or property resulting from any ideas, methods, instructions or products referred to in the content.

MDPI AG
Grosspeteranlage 5
4052 Basel
Switzerland
Tel.: +41 61 683 77 34

Nutrients Editorial Office
E-mail: nutrients@mdpi.com
www.mdpi.com/journal/nutrients

Disclaimer/Publisher's Note: The title and front matter of this reprint are at the discretion of the Guest Editors. The publisher is not responsible for their content or any associated concerns. The statements, opinions and data contained in all individual articles are solely those of the individual Editors and contributors and not of MDPI. MDPI disclaims responsibility for any injury to people or property resulting from any ideas, methods, instructions or products referred to in the content.

www.ingramcontent.com/pod-product-compliance
Lightning Source LLC
Chambersburg PA
CBHW061623210926
51850CB00009B/2420